W0107142

Wheat Rusts
AN ATLAS OF RESISTANCE GENES

RA McIntosh, CR Wellings and RF Park
Plant Breeding Institute, The University of Sydney

CSIRO
AUSTRALIA

Springer-Science+Business Media, B.V.

Adult plant symptoms of A. leaf rust, B. stripe rust and C. stem rust. D. Symptoms on paired seedling leaves. Upper and lower leaf surface of (L–R): leaf rust, stem rust and stripe rust.

FOREWORD

World food security will depend on increased production of two major cereal crops — wheat and rice. Of these, wheat is of greater importance, in terms of tonnage if not in financial value. One significant constraint to increased wheat production is the variety of rust diseases attacking this crop — leaf rust, stem rust and stripe rust.

Sources of resistance to these diseases are known, and have been utilised by wheat breeders for a long time. However, achieving durable resistance can be difficult, and the rust diseases continue to evolve and circumvent the breeders' achievements.

This book provides an important and powerful tool for breeders in their continuing fight to outwit the rusts. In a valuable introductory section, *Wheat Rusts: An Atlas of Resistance Genes* gives a concise summary of the rusts, their interactions with the host plant and the nature and genetic bases of host resistance; then the main body of the work provides a comprehensive and descriptive catalogue of most known sources of rust resistance in wheat. Catalogues of the resistance genes in wheat have been published and updated regularly by RA McIntosh, but in this book they are presented together for the first time with full descriptions and colour photographs of their phenotypes. *Wheat Rusts: An Atlas of Resistance Genes* will, therefore, be a unique and invaluable aid to all involved in developing wheats for resistance to the rust pathogens.

Such a compilation has only been possible because of the breadth and depth of experience of the authors. RA McIntosh is an international leader and authority on the genes for resistance to rust fungi in wheat. He is Director of Rust Research in the Plant Breeding Institute of The University of Sydney. He has recently been awarded a personal chair in the University as Professor in Cereal Genetics and Cytogenetics, and has been elected a Fellow of the Australian Academy of Sciences. Dr Wellings and Dr Park are experts on resistance genes to the stripe (yellow) and leaf rust fungi in wheat. The rust program at the Plant Breeding Institute plays a major role in monitoring the evolution of virulence in the rust fungi and in assessing resistance to them in Australia and New Zealand. The Plant Breeding Institute is world renowned and provides valuable advice and leadership for similar endeavours in other countries.

The Australian Centre for International Agricultural Research (ACIAR) has been pleased to be associated with this team in a wheat rust project between Australia, India and Pakistan. In the course of this project Professor McIntosh and his colleagues were untiring and generous in sharing their experience with their overseas colleagues, and ACIAR is now very happy to be able to make the authors' knowledge and expertise more widely available through support for the present publication. We trust that *Wheat Rusts: An Atlas of Resistance Genes* will be of value in the continuing battle to feed a growing world.

Paul Ferrar

Paul Ferrar

Research Program Coordinator (Crop Sciences)
Australian Centre for International Agricultural Research (ACIAR)
Canberra, Australia

Wheat rusts : an atlas of resistance genes / edited by R.A. McIntosh
    and C.R. Wellings and R.F. Park.
            p. cm.
        Includes index.

    1. Wheat--Disease and pest resistance--Genetic aspects--Atlases.
2. Wheat rusts--Atlases. I. McIntosh, R.A. II. Wellings, C R.
III. Park, R. F.
SB608.W5W45 1995
633.1 ' 1423--dc20   95-7869

No part of the material protected by this copyright may be reproduced or utilized in any form or by any means, electronic or mechanical, including photocopying, recording or by any information storage and retrieval system, without written permission from the copyright owner.

ISBN 978-94-010-4041-9   ISBN 978-94-011-0083-0 (eBook)
DOI 10.1007/978-94-011-0083-0

© Springer Science+Business Media Dordrecht 1995
Originally published by Kluwer Academic Publishers in 1995

SPONSORS

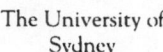

The University of
Sydney

PREFACE

The production of this book that describes the individual genes for resistance to wheat rusts and illustrates them with various aspects of host:pathogen interactions has required many years of research and development. Therefore, in bringing this book to fruition, we wish to recognise our colleagues, both past and present especially the late EP Baker, NH Luig, TT The and the late IA Watson, who made significant contributions to the direction and development of cereal rust research at the Plant Breeding Institute (PBI), The University of Sydney.

The catalogue and illustrations in *Wheat Rusts: An Atlas of Resistance Genes* provide a basis for the identification of the genes for resistance to the wheat rusts. The actual low seedling and adult plant responses to the wheat rust pathogens conferred by resistance genes provide a preliminary indication of the identity of a particular resistance gene. The pattern of response to an array of pathogen cultures, the effects of environment on response, and pedigree of the host line or cultivar provide further insight to support or contradict those early indications. These considerations usually reduce the number of genes for postulation to three or four, or even less — being able to exclude candidate genes is an important aspect of resistance gene identification. The reduction of choices to three or four represents a substantial saving in time and resources and can be achieved before a single cross is made for genetic analysis.

Whereas the major aim of the book was to illustrate infection types, we have also attempted to provide detailed genetic and physiological information, and indicate past and potential use of each gene. Potential resistance sources are often those that are effective now, but not yet genetically analysed, and the breeder can ill-afford to wait until genetic analyses are completed before attempting to utilise them as breeding donors. Indeed, the breeder may not have the luxury of genetic advice and must work in the absence of such information or depend on information from other places. In Chapter 1 we provide some basic information on principles, methods and problems, and outline some of the questions that must be addressed by breeders in order to achieve effective and hopefully lasting resistance. Our aim was to complement rather than duplicate some of the excellent and detailed publications available on the cereal rusts and breeding for rust resistance. In this regard, we direct readers to the following:

1984   The Cereal Rusts, Volume 1 (Bushnell and Roelfs, 1984)

1985   The Cereal Rusts, Volume 2 (Roelfs and Bushnell, 1985)

1988   Breeding Strategies for Resistance to the Rusts of Wheat (Simmonds and Rajaram, 1988)

1989   The Wheat Rusts: Breeding for Resistance (Knott, 1989)

1992   Rust Diseases of Wheat: Concepts and Methods of Disease Management (Roelfs *et al.*, 1992)

In choosing particular photographs we have attempted to emphasise different aspects of resistance gene:avirulence gene behaviour, while attempting to maintain an interesting variation in approach. The production of the more than 500 photographs from which we selected examples was interrupted by the move of the Institute from Castle Hill to Cobbitty, New South Wales, in 1991 and the retirement and resignation of assistants. Nevertheless, the pictures are a credit to the three persons involved, namely DJS Gow, DL Milne and P Abell, and we sincerely acknowledge their contributions. The collection of photographs was expanded by the generous donation of illustrations from R Johnson, AP Roelfs and our Plant Breeding Institute colleagues. In addition, CR Wellings was able to obtain further stripe rust photographs while on study leave at the Research Institute for Plant Protection, Wageningen, The Netherlands. The technical work of planning, planting and inoculation conducted by L Ferrari, NM Maseyk, DB McDonald, DL Milne and RM Pfeiffer are especially acknowledged. We thank PL Dyck, R Johnson, GHJ Kema, DR Knott, RG Rees, AP Roelfs and the late RW Stubbs for reviewing various sections of the text, and for offering many suggestions as well as published and unpublished genotype information. Assistance in preparing the many drafts of the manuscript was provided by Mrs Betty Gibson and Ms Hun Sun Hwang and we thank them for their durable patience. The support of our PBI colleagues during the course of this project, particularly JD Oates, is greatly appreciated.

Finally, the project would not have been possible without financial support. The Australian Centre for International Agricultural Research (ACIAR) is acknowledged for underwriting the project and permitting provision of a limited number of copies free of charge to scientists in developing countries. The Australian Grains Research and Development Corporation (GRDC) support the positions of RA McIntosh and RF Park and finance much of our research. CR Wellings is an employee of New South Wales Agriculture on secondment to the Plant Breeding Institute, Cobbitty.

RA McIntosh
CR Wellings
RF Park

Plant Breeding Institute, The University of Sydney,
Cobbitty, NSW, Australia.

CONTENTS

CHAPTER 1

Wheat Rusts and the Genetic Bases of Disease Resistance

INTRODUCTION

Economic Importance

The economic and social upheavals resulting from crop losses following epidemics have been the dominant influences on research and advisory activities directed at the wheat rusts. Although cereal rust diseases were clearly of major significance historically, estimates of the yield of actual losses incurred received attention only in the 20th century due to a better understanding of disease biology and an increasing need to justify economically the financial investment in control programs (Plate 1-1).

Roelfs (1978) compiled an overview of losses due to the cereal rusts in the United States of America (USA) from 1918 to 1976, noting statewide yield reductions of 50% or more in epidemic years due to stem (black) rust and leaf (brown) rust. Although stripe (yellow) rust was more restricted in distribution, losses of up to 70% in commercial fields were recorded. Green and Campbell (1979) estimated that resistant wheat cultivars grown in the stem rust-liable areas of Canada provided crop protection valued at $C217 million annually. During the 1960s, the rusts were conservatively estimated to have reduced North American wheat yields by over one million tonnes (2%) annually (Wiese, 1977).

Yield losses due to cereal rusts have also been reported from the Indian subcontinent and the Middle East. Severe epidemics have been recorded since the early 1800s in India (Joshi, 1976). A severe leaf rust epidemic in 1978 resulted in an estimated national loss of $US86 million in Pakistan (Hussain et al., 1980). Studies in Egypt estimated crop losses of up to 50% due to leaf rust infection (Abdel Hak et al., 1980). In 1993, stripe rust epidemics in selections of the International Maize and Wheat Improvement Centre (Centro Internacional de Mejoramiento de Maiz y Trigo, CIMMYT)-generated wheat line, Veery, caused significant yield losses in Yemen, Ethiopia and Iran (OF Mamluk, pers. comm. 1993).

Cereal rust losses in Europe are primarily associated with stripe rust and leaf rust. The effective barberry eradication campaign in the early 20th century and the infrequent occurrence of favourable temperatures has resulted in a decline in importance of stem rust. Economic assessments in the United Kingdom (UK) by Priestley and Bayles (1988) provided estimates of losses in susceptible winter wheats due to stripe rust and leaf rust of £83 million with the value of disease resistance estimated at £79.8 million. However, the value of resistance may be rapidly lost as illustrated by cultivar (cv.) Slejpner, which became susceptible in the UK and northern Europe in 1988 following a sudden change in the pathogen population (Bayles et al., 1989).

Winter wheat production in China is affected by recurring epidemics of stripe rust (Stubbs, 1985). Epidemics in 1950, 1964 and 1990 in China were estimated to have caused losses of 6, 3 and 2.5 million tonnes, respectively (S Xie and W Chen, unpublished 1993).

In Australia, stem rust and leaf rust epidemics have caused serious damage since European settlement. A sequence of severe epidemics in the 1880s resulted in sufficient public concern and subsequent political pressure to establish State Departments of Agriculture in New South Wales (NSW) and Victoria. Several attempts have been made to assess yield losses due to rust epidemics in Australia. Estimates of crop losses varied from 30% in leaf rust susceptible cultivars (Rees and Platz, 1975) to 55% in wheats susceptible to both stem and leaf rust (Keed and White, 1971). Field plots of commercial cultivars with relatively low levels of natural leaf rust infection were shown to sustain up to 15% yield loss in northern NSW (Wellings et al., 1985). A widespread leaf rust epidemic in Western Australia in 1992 caused yield losses of up to 37% in susceptible cultivars (Plate 1-2C) with average losses of 15% across many fields (R Loughman, pers. comm. 1993). Economic appraisals of national losses have also ranged from $A100–200 million due to the 1973 stem rust epidemic (Australian Wheat Board, 1975) to an estimated $A8 million cost of chemical application for disease control in NSW during an epidemic of stripe rust in 1983 (Wellings and Luig, 1984). Brennan and Murray (1988) attempted to develop an economic analysis of losses due to a range of wheat diseases in Australia and estimated that the annual value of control strategies for stem, stripe and leaf rust was $A124, 139 and 26 million, respectively. These estimates represent the annual national benefits derived primarily from resistance breeding activities directed towards the control of wheat rusts.

An overview of global crop losses caused by the three wheat rusts indicated varying regional significances (Saari and Prescott, 1985). Stripe rust assumed more importance in west Asia, southern Africa, the Far East (China), South America and northern Europe. Leaf rust caused more serious losses in south Asia, north Africa, southeast Asia and South America. Stem rust has traditionally been important in North America, Australasia, northern Africa, South Africa and, to some extent, Europe. Crop losses inevitably reflect the interplay between pathogen, host and environment at local, regional and global levels. Cereal rust diseases will continue to demand the attention of research and advisory personnel because of the dynamic nature of this relationship.

A

B

D

C

PLATE 1-1

Although major rust epidemics are now less common due to the use of resistance genes, their occurrence often results in considerable public attention.

A. United States of America barberry eradication poster. Courtesy JJ Burdon.

B. Stem rust in Australia, 1973.

C. Stripe rust in eastern Australia, 1983.

D. Leaf rust in Western Australia, 1992. Courtesy R Loughman.

A

B

C

PLATE 1-2

Effects of rust infection on grain size and quality.

A. Stem rust in triticale, northern New South Wales, 1982. This crop of Coorong (right) included a low frequency of off-type resistant plants (left).

B. Stripe rust in wheat, Victoria, 1979. Seed from a field of moderately resistant Condor (upper) is compared with seed from an adjacent crop of susceptible Zenith (lower). The Zenith crop yielded 30% of its estimated potential.

C. Leaf rust in wheat, Western Australia, 1992. The disease was so severe that this grain crop was abandoned. Note the spore cloud rising above the hay cutting equipment. Courtesy R Loughman.

Historical Perspectives

The occurrence of rust diseases in cultivated cereals has significantly influenced the development of human civilisation (Large, 1940; Carefoot and Sprott, 1967; Roelfs *et al.*, 1992). Urediospores of *Puccinia graminis* taken from excavations in Israel were dated at about 1300 BC (Kislev, 1982); biblical accounts, at about 1870 BC, indicate that rust epidemics forced the family of patriarch Jacob to seek refuge in Egypt; and the scourge of rust is recorded in the early Greek and Roman literatures where, in about 500 BC, ceremonial details indicate liturgies to appease Robigus, the Corn God, in an attempt to prevent crop failure.

Although the biology was not understood, an observed relationship between rusted cereals and the proximity of diseased barberry was appreciated from early times. In 1660, a law enacted in Rouen, France, required the removal of barberries from the vicinity of grain production fields. Measures that followed to control barberry in other countries such as England and the USA, culminated in the expensive and often controversial barberry eradication program in the USA in the early years of this century (Roelfs, 1982; Christensen, 1984) (see Plate 1-1A).

According to Chester (1946), the first person to recognise that rust was caused by a fungal parasite was Felice Fontana in 1767. However, it was not until well into the 19th century that a distinction among the rust diseases was made. This was noted by de Candolle, who described the leaf rust pathogen as *Uredo rubigo-vera* and thereby distinguished it from Persoon's *Puccinia graminis* (Chester, 1946). De Bary in the 1860s (1866; cited in Walker, 1976) provided proof of heteroecism of *P. graminis* on cereals and barberry. Eriksson (1894; cited in Walker, 1976) in Sweden defined *formae speciales* to describe "special forms" (f. sp.) of the wheat stem rust and stripe rust pathogens that showed specialisation on different host species.

Thus, by the beginning of the 20th century, it was generally accepted that the different rusts, also described as 'mildew' or 'blast' in some publications, were caused by distinctive fungal species with contrasting host ranges. The massive effort directed at cereal rusts and their control since the 1880s, both in terms of basic science and in practical efforts to curtail losses, led Large (1940) to observe that "the greatest single undertaking in the history of plant pathology was to be the attack on rust in cereals". This effort continues.

Twentieth Century Landmarks in Cereal Rust Biology

Stem rust studies by EC Stakman and colleagues at the University of Minnesota, USA, had a significant influence, not only on cereal rust research, but on plant pathology generally. They demonstrated that stable variants, originally termed physiologic races, and later known as races, strains, biotypes or pathotypes, occurred within *Puccinia graminis* f. sp. *tritici* (Stakman and Piemeisel, 1917; Stakman *et al.*, 1918). This discovery permitted previously observed variation in the responses of single host lines to different pathogen isolates and/or at different geographic sites to be explained in terms of pathogenic variation. A key was subsequently developed, based on 12 differential indicators, to describe the pathogen response phenotypes (Stakman and Levine, 1922; Stakman *et al.*, 1962). Surveys based on this differential set were initiated wherever stem rust was a problem, and differential sets were soon proposed on a similar basis for the study of the other wheat rust pathogens (Mains and Jackson, 1926 for leaf rust; Gassner and Straib, 1932 for stripe rust). It was later realised that these widely adopted differential sets only partially described the genetic variation of the pathogen and were not particularly relevant to practical breeding programs. Moreover, the responses of some individual differentials were so significantly affected by environment that they were either abandoned or replaced. Despite the shortcomings, the information from early pathotype surveys was useful in demonstrating the significance, role and epidemiology of different pathogen clones or closely related groups of clones.

The discovery of the role of pycnia in sexual reproduction of rust fungi (Craigie, 1927) provided an initial explanation for variability in rust fungi. However, in situations where sexual reproduction was absent, other mechanisms such as somatic hybridisation (Watson, 1957; Watson and Luig, 1958) and mutation had to be sought. It is now known that forces of migration, mutation, recombination (both sexual and asexual), selection and chance play significant roles in determining gene frequencies and evolutionary pathways in cereal rust fungi (McIntosh, 1992*b*).

The gene-for-gene relationship, established by Flor (1942, 1956) for flax and its rust pathogen *Melampsora lini* L. (Ehrenb) Lev., has been shown to have general application to the genetics of many disease and pest situations (Day, 1974). This hypothesis forms the basic model for work on both conventional and molecular genetic analyses in the cereal rusts.

BIOLOGY OF PATHOGEN AND HOST

Rust Pathogen Nomenclature

The currently accepted names for the three basidiomycete rust fungi found predominantly on wheat are:
Puccinia graminis Pers. f. sp. *tritici*, the pathogen of stem rust, also known as black rust;
Puccinia recondita Rob. ex Desm. f. sp. *tritici*, the pathogen of leaf rust, also known as brown rust; and
Puccinia striiformis Westend. f. sp. *tritici*, the pathogen of stripe rust, also known as yellow rust.

Puccinia graminis and *P. recondita* are macrocyclic and heteroecious. The alternate hosts of *P. graminis* include *Berberis* spp. and *Mahonia* spp., and those of *P. recondita* include species of *Thalictrum*, *Anchusa*, *Isopyrum* and *Clematis*. *P. striiformis* has a microcyclic life cycle with no known alternate host. Where present, alternate hosts are important in disease epidemiology in providing local inoculum to initiate rust development in adjacent wheat crops and in acting as a source of new pathotypes by hosting the sexual stage of the fungal life cycle.

Rust Pathogen Host Range

The relationships of the wheat-attacking form of each species to the forms occurring on other cereals and grasses differ considerably. *P. graminis* f. sp. *tritici* is closely related to *P. graminis* f. sp. *secalis*, and they both readily hybridise with each other both sexually and asexually. On the other hand, these forms rarely, if at all, hybridise with *P. graminis* f. sp. *avenae*. Similarly, the cereal and grass host ranges of *P. graminis* f. sp. *tritici* and f. sp. *secalis* are more alike than the hosts of *P. graminis* f. sp. *avenae*. Thus *P. graminis* f. sp. *tritici* readily infects cereal rye (*Secale cereale* L.)

B

PLATE 1-3

Stripe rust in barley grass.

A. Close view of barley grass infected with stripe rust.

B. Heavy infestations of barley grass by stripe rust in headlands, pasture areas and crops are common in much of the wheatbelt of southern Australia. Barley grass has become increasingly affected by stripe rust.

A

and may produce significant disease levels on some genotypes. Similarly, *P. graminis* f. sp. *secalis* infects and may produce high levels of disease on certain wheat genotypes. In addition, both *formae speciales* and their hybrids may be significant pathogens of barley (*Hordeum vulgare* L.) and various species of *Aegilops* and *Agropyron*. However *P. graminis* f. sp. *avenae* is not found on any of these hosts under field conditions.

The relationships among the cereal-infecting *formae speciales* of *Puccinia recondita* appear to be more distant. Our experience is that *P. recondita* f. sp. *tritici* does not attack cereal rye and *P. recondita* f. sp. *secalis* is avirulent on all wheats. Of perhaps greater interest is the likelihood of distinctive groups of *P. recondita* f. sp. *tritici* showing greater specialisation to tetraploid wheats, and possibly species of *Aegilops*, than to common wheats (Huerta-Espino, 1992; Roelfs *et al.*, 1992). The relationships of these groups to each other and to their sexual hosts appear largely unknown.

The sources of pathogenic variability in *Puccinia striiformis* are restricted to mutation and asexual recombination because there is no known alternate host. As with the other rust pathogens of wheat, this species is highly variable both within and between geographical areas. *P. striiformis* f. sp. *tritici* is probably most closely related to *formae speciales hordei* although they can be readily distinguished in field nurseries involving variable germplasm of both host species (Zadoks, 1961). Relatively few barley genotypes become rusted with *P. striiformis* f. sp. *tritici* and relatively few wheats become rusted with *P. striiformis* f. sp. *hordei*. However, the relationships between *formae speciales* are not always clear; for example, several cultures of f. sp. *hordei*, which are highly avirulent on wheat, produce an intermediate to high response on the wheat differential Chinese 166 (Stubbs, 1985). Such anomalies may have led Gassner and Straib (e.g. Gassner and Straib, 1932) to remain with race descriptions and avoid the use of *formae speciales*.

A number of accessory (alternative) hosts for both *P. striiformis* f. sp. *hordei* and f. sp. *tritici* are known (Zadoks, 1961; Holmes and Dennis, 1985). There appears to have been an increasing incidence of *P. striiformis* f. sp. *tritici* on the widely established barley grass species (Plate 1-3), *Hordeum leporinum* Link and *H. glaucum* Steud., following its introduction into Australia in 1979 (O'Brien *et al.*, 1980) and there is evidence for pathogenic specialisation on certain genotypes of these species (CR Wellings, JJ Burdon and DJ Kull, unpublished 1993). Zadoks (1961) mentioned infection of *H. murinum* L. with "highly specialised" races of *P. striiformis* f. sp. *hordei*, but gave no indication of rusting of these grasses with *P. striiformis* f. sp. *tritici*.

Wheat Taxonomy

The names and synonyms with authorities, genome designations and chromosome numbers of common wheat and relatives mentioned in this book are given in Appendix I. Wheat (*Triticum aestivum*) is a complex allohexaploid species (genomes AABBDD, 2n = 42) derived from a hybrid between cultivated emmer (*T. turgidum* AABB, 2n = 28) and the grass species, *T. tauschii* (DD,

$2n = 14$). Such hybridisation events were presumably limited in number, but various natural and artificial hybrids and subsequent genetic introgressions have enlarged the range of variability.

Common wheat shares its AABB genomes with durum and the cultivated and wild emmers (*T. turgidum*, $2n = 28$). These tetraploids have a relatively close relationship to cultivated *T. timopheevii* and its wild relative, *T. araraticum* (AAGG). Each of these species or subspecies has contributed, or is contributing, genetic variation to present day cultivated hexaploids. In addition, diploid wheats (*T. monococcum*) and various diploid and polyploid relatives have acted as germplasm sources in wheat breeding.

HOST:PATHOGEN GENETICS

Definitions

In order to study and communicate aspects of host:pathogen genetics, terms and concepts are required and, as Loegering (1966, 1972) pointed out, these must be directed separately at the levels of the disease (i.e. the interaction), the pathogen and the host.

Basic Compatibility
Plant pathology addresses disease problems. In crop production, a disease problem must occur on a significant scale before it receives attention. The concept of basic compatibility represents a starting point for problem-solving and recognises that a particular host species (or genotype group) is susceptible to a particular pathogen species (or genotype group). Although most host species are not affected by most pathogen species, the interesting topic of 'non-host' resistance is of little direct concern to the practicing agriculturalist. With basic compatibility, the host is said to be susceptible and the pathogen is defined as virulent. At the level of interaction between particular host genotypes and pathogen genotypes, the disease response may be described as incompatible, or low response, in contrast to compatible, or high response (Loegering, 1972).

Reaction
Host phenotype is described as reaction, where the contrasting states are termed resistant, or low reaction, and susceptible, or high reaction.

Pathogenicity
Pathogen phenotype is described as avirulent, or low pathogenicity, and virulent, or high pathogenicity. Pathogenicity is defined more broadly in the plant pathology literature, but here we follow the examples of Loegering (1966), Moseman (1970) and Browder (1971). The use of virulence in both generic and specific contexts is not acceptable, and no alternate term other than pathogenicity has been proposed.

The components of disease response (i.e. host reaction and pathogenicity) have been defined above in contrasting qualitative, or binary, terms. However, the reality is that each character is continuous and a decision as to whether the disease response is low or high may be difficult or arbitrary. Moreover, that decision will vary with the viewpoint of the person making the decision. For example, a geneticist may determine if a particular resistance (gene) source has any affect on the growth of the pathogen relative to appropriate controls, whereas a plant pathologist/breeder may be more interested in the effectiveness or value of the particular resistance (gene) source in reducing crop losses.

Principles

Flor (1942) was first to realise that incompatibility between a host and a pathogen involved corresponding genes in each organism. Consideration of the gene-for-gene relationship that he proposed leads to two fundamental rules which parallel the basic rules of genetics formulated by Mendel in the 19th century (Mendel, 1865). The first rule relates to the interaction of products of single genes in hosts and pathogens, whereas the second considers the second order interactions. Formally stated:

1. Incompatibility between a host and pathogen is the consequence of interaction between the products of at least one host resistance gene and at least one corresponding pathogen avirulence gene, that is: $LIT = LP{:}LR$.
   where: LIT is low infection type; LP is low pathogenicity; and LR is low reaction.

2. When more than one interacting gene pair are involved, the level of incompatibility is as low as, or lower than, the level produced by the most incompatible interacting gene pair acting alone, that is: $LIT_{1,2} \leq LIT1$ where: $LIT_1 < LIT_2$.

Experimental Designs

The above rules lead directly to four basic experimental designs that are applied to studies in host:pathogen genetics (Browder, 1971). These experimental designs are based on assumptions that flow from the gene-for-gene relationship. They can be summarised as follows.

1. Unknown host–unknown pathogen combinations. In this design host:pathogen response matrices can be generated without prior knowledge or assumption of variability in either host or pathogen species. Similar row or column patterns indicate genetic similarities between isolates of the relevant organism.

2. Known host–unknown pathogen combinations. This design involves studies of genetically characterised or fixed sets of host lines tested with unknown arrays of pathogen samples. These experiments include the common pathotype surveys undertaken by most research groups working on the application of host:pathogen genetics in breeding.

3. Unknown host–known pathogen combinations. In this design genetically uncharacterised sets of host lines can be tested with a genetically known or selected array of pathogen isolates. This is the method usually adopted to postulate genes for resistance, thereby identifying genes or gene combinations that represent novel and potentially useful sources of resistance.

4. Known host–known pathogen combinations. This experimental design is applied to investigations aimed at the detailed effects of individual corresponding gene pairs, such as their response to environmental variability or their effectiveness in reducing pathogen inoculum and yield loss.

METHODOLOGIES IN WHEAT RUST DISEASES

As Roelfs *et al.* (1992) pointed out, many methods have been developed for the study of rust diseases and any review of them must be restricted to principles, rather than detail which is often dictated by local conditions and facilities.

Accessions, Inoculum Increase and Storage

Isolates of rust pathogens to be stored as reference cultures for either research or breeding are usually accessioned in a collection on a similar basis to a plant germplasm collection. Phenotypically uniform cultures can be generated from single urediospores, single uredia or small

A

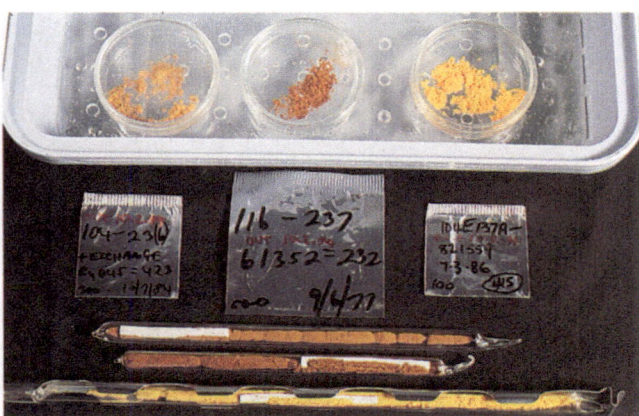

B

C

PLATE 1-4

Urediospore collection and storage.

A. Urediospores can be collected by shaking them from infected leaves (as shown) or by the use of vacuum apparatus.

B. Urediospore samples of *P. recondita* f. sp. *tritici* (left), *P. graminis* f. sp. *tritici* (centre) and *P. striiformis* f. sp. *tritici* collected for drying prior to long-term storage in aluminium foil pouches (centre) or evacuated glass tubes (lower).

C. View of holding compartment for foil pouches removed from a liquid nitrogen refrigerator (-196°C).

bulks through asexual reproduction. Cultures established from single urediospores or uredia should be genetically uniform, whereas those established from bulks may be genetically heterogeneous. Each culture should carry a unique accession number that permits its identification. Where there are multiple accessions of the same phenotype (pathotype, strain or race), such distinction is essential for future reference because the phenotype is specified by a limited set of pathogenicities. Whereas many collections tend to be represented only by the first isolate of each pathotype, our experience has indicated a need to ensure a series of accessions over time and space for pathotypes that prove particularly important. Because 'importance' is not easily defined or predicted, accessioning is best undertaken by means of an established protocol.

Inoculum increase is undertaken in isolation on susceptible genotypes (Roelfs *et al.*, 1992). After collection, urediospores must be dried prior to subpackaging in quantities relevant to their anticipated future use. The means of storage in paper, plastic or aluminium pouches, or in glass vials, is dictated by the anticipated period of storage and available facilities. Major rust research laboratories maintain working collections almost indefinitely in vacuum, ultra-low refrigeration or liquid nitrogen (Plate 1-4). At The University of Sydney, inoculum is stored in aluminium pouches in a liquid nitrogen facility, with each pouch containing sufficient urediospores to infect 100–200 9-cm-diameter pots of seedlings, or in larger amounts suitable for establishing field epidemics. Once removed from liquid nitrogen and temperature-shocked (four minutes at 42°C), pouches can be safely stored in a domestic refrigerator for at least one (stripe rust) to three months (leaf rust and stem rust).

Inoculation

Inoculation undertaken in the greenhouse or field will be successful only if moisture is available for spore germination and infection. The main methods of inoculation include: dusting or brushing with dry spores, with or without a carrier such as talc or spores of the club moss *Lycopodium*; spraying with water or isoparaffinic oil-based suspensions; or plant tissue injection using water-based suspensions (Plate 1-5). The last method is the most labour intensive, although it is very efficient with minimal contamination during infection and exogenous moisture is not necessary for successful infection.

PLATE 1-5

Infection can be achieved by several methods of inoculation.

A. Seedlings removed from a misting chamber following inoculation with urediospores suspended in mineral oil or mixed with talc.

B. Hypodermic injections of elongating stems with aqueous urediospore suspensions provide reliable contamination-free infections without need for dew formation. The technique is suitable for the greenhouse or field.

C. Inoculation of field-grown plants with water or mineral oil suspension of urediospores in late afternoon in anticipation of overnight dew.

Field epidemics are commonly initiated by injection of susceptible plants, placement of infected potted plants at intervals throughout the target nursery, or by dusting or spraying on days prior to overnight dew development. Pots of healthy susceptible plants positioned in field plots prior to inoculation, and returned to a greenhouse after the anticipated dew treatment, are a convenient and rapid means of determining the outcome of the inoculation.

The facilities for successful inoculation and infection can be rather crude provided there is adequate spore viability, dew development and suitable temperature and light conditions. In some laboratories, sophisticated facilities allow for quantitative control of inoculum application, while dew chambers and fixed light regimes are directed at reliability and repeatibility of experiments.

Post-infection Rust Development

Once infection has occurred, plants can be placed on benches in growth rooms or greenhouses where conditions are favourable for rust development. In the temperature controlled greenhouse rooms currently used at The University of Sydney Plant Breeding Institute, Cobbitty (PBI), stripe rust develops successfully within the range 15–20°C, leaf rust in the range 15–25°C, and stem rust in the range 18–30°C. Periods of exposure beyond these ranges are usually not detrimental to rust development. Under Australian conditions, supplementary lighting of greenhouses is usually unnecessary, but extra lighting is common in many Centres. Plants are normally assessed 10–18 days after inoculation, depending on the disease and temperature.

Disease Assessment

Disease responses for the cereal rust diseases can be assessed by either qualitative or quantitative means, or a combination of both. The rationale for this book is based on the knowledge that each resistance gene produces characteristic responses in terms of infection type, infection site variability on a single leaf and environmental variation. This is only one of the four methods proposed by Roelfs (1985) to assess disease, the others being numbers of uredia per unit of inoculum (i.e. host receptivity), length of the latent period (i.e. time for pustule development) and duration of sporulation.

Seedling Scales

The infection type descriptions still in use are based on the original scales proposed by Stakman *et al.* (1962) for stem rust and Gassner and Straib (1932) for stripe rust. Our interpretations of these scales are listed in Tables 1-1 and 1-2. Browder and Young (1975) proposed a three-digit 0–7 description for leaf rust, but this scale was not widely adopted. In stripe rust research the 0–9 scale of McNeal *et al.* (1971) has been more commonly used, especially in North America.

At the PBI, where ongoing research involves all three wheat rusts, it has been more convenient to use a common scale for each disease (Tables 1-1 and 1-2; Plate 1-6). The host response descriptions that are listed with seedling infection types can be considered 'average' correlations. Very high levels of seedling resistance are almost invariably associated with high or acceptable levels of adult plant resistance under field conditions. However, intermediate seedling responses may give variable responses in the field and thus each gene must be assessed separately.

TABLE 1-1	Major infection type classes for stem rust and leaf rust	
Infection type[a]	**Host Response**	**Symptoms**
0	Immune	No visible uredia
;	Very resistant	Hypersensitive flecks
1	Resistant	Small uredia with necrosis
2	Resistant to moderately resistant	Small to medium sized uredia with green islands and surrounded by necrosis or chlorosis
3	Moderately resistant/ moderately susceptible	Medium sized uredia with or without chlorosis
4	Susceptible	Large uredia without chlorosis
X	Resistant	Heterogeneous, similarly distributed over the leaves
Y	?	Variable size with larger uredia towards the tip
Z	?	Variable size with larger uredia towards the leaf base

[a] Variations are indicated by the use of ⁻ (less than average for the class) and ⁺ (more), as well as C and N to indicate more than usual degrees of chlorosis and necrosis. Heterogeneity on a leaf not adequately described with X, Y or Z may be written as a sequence, for example 12⁻. A comma is used to indicate heterogeneity between plants in a single test, for example 1⁺,X.

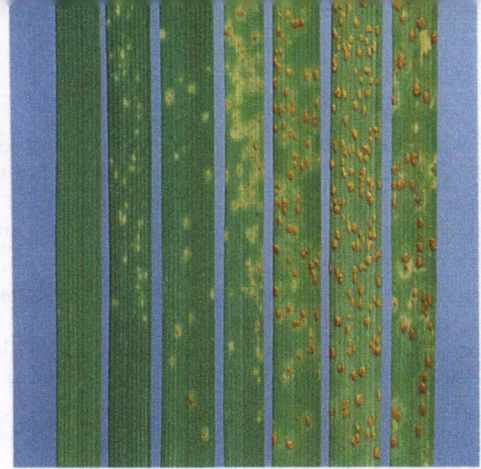

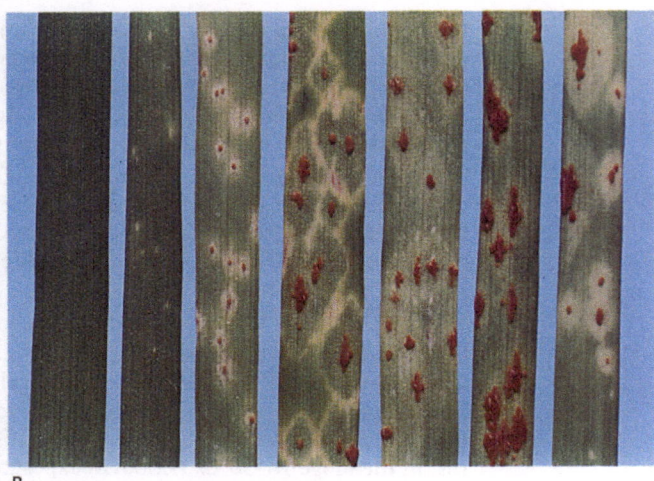

PLATE 1-6

Range of seedling infection types (IT) for the three rust diseases (see Tables 1-1 and 1-2).

A. Leaf rust (L to R): IT 0, ; , 1, 2, 33+, 4, X.

B. Stem rust (L to R): IT 0, ; , 1+, 2+, 3, 4, X.

C. Stripe rust (L to R): IT 0, ; , ;N, 1+, 2C, 3, 4.

TABLE 1-2   Major infection type classes for stripe rust

Infection type[a]			Host response	Symptoms
Gassner and Straib (1932)	McNeal et al. (1971)	PBI[a]		
i	0	0	Immune	No visible uredia
00	1	;	Very resistant	Necrotic flecks
0	2	;N	Resistant	Necrotic areas without sporulation
I	3–4	1	Resistant	Necrotic and chlorotic areas with restricted sporulation
II	5–6	2	Moderately resistant	Moderate sporulation with necrosis and chlorosis
III	7–8	3	Moderately susceptible	Sporulation with chlorosis
IV	9	4	Susceptible	Abundant sporulation without chlorosis

[a] −, +, C and N are used in the same way as for leaf rust and stem rust.

The most critical decision in host:pathogen genetics is the distinction of incompatibility and compatibility, resistance and susceptibility, or avirulence and virulence. Interactions considered significantly and repeatably different from compatibility as defined by a selected control will be judged as incompatible. Once a decision for incompatibility has been made, one concludes the presence of corresponding gene pairs for avirulence and resistance. However, some resistance genes readily detected in seedling tests may confer levels of resistance in adult plants that are considered inadequate for commercial use. In other situations, significant changes in the avirulence phenotype of the pathogen may be observed. This was described by Watson and Luig (1968) as progressive increase in virulence, that is, pathogen cultures may exhibit different infection types

PLATE 1-7

A. and B. Stem rust: upper and lower sections of peduncles (L to R): 0R, TR, 5R, 10R-MR, 25MR, 60MS, 90S.

C. Leaf rust: flag leaves (L to R): TR, 25R, 40MR, 60MR-MS, 40MS, 90S.

D. and E. Stripe rust: flag leaves (L to R): D. 0R, 30R, 20R-MR, 40MR, 30MR-MS; E. 60MR-MS, 50MR-MS, 70MS, 90MS-S, 100S.

within the range of incompatibility for a particular host gene. Under field conditions such progressive increases in virulence may result in either adequate or inadequate disease control. Samborski (1963) showed that *P. recondita* isolates heterozygous for the locus corresponding to host gene *Lr9* gave intermediate infection types. On the other hand, Watson and Luig (1968) noted up to five or six distinct levels of interaction involving lines with *Sr15* and different cultures of the pathogen. In genetic experiments the same host gene *Sr15* was shown to be involved.

Adult Plant Assessment

The McNeal *et al.* (1971) scale for stripe rust has been used to score adult plant response, whereas the Gassner and Straib (1932) scale is usually considered unsuitable for scoring adult plants. The Stakman *et al.* (1962) descriptions for leaf and stem rust can be used for adult plants, but a more common approach under field conditions is to use the modified Cobb scale (Peterson *et al.*, 1948) as a quantitative measure of disease. Alternatively, a coefficient of infection (CI) which weights the modified Cobb scale rating by the disease response (R, MR, MS, S) has been used in several laboratories (Stubbs *et al.*, 1986; Roelfs *et al.*, 1992). A range of adult plant reactions to each of the three rusts is illustrated in Plate 1-7. The CI is probably more closely correlated with crop loss than is either scale from which it is calculated, but it does have a problem in that the two variables from which it is estimated are not independent.

In adult plant assessments, stem rust scores are usually based on stem infection, whereas leaf rust and stripe rust are based on the upper leaves, even though other tissues may be affected. Under

Australian conditions the total leaf area affected (modified Cobb scale or McNeal scale score) with stripe rust appears to be more closely correlated with yield loss than the CI, mainly because the response can be significantly and rapidly affected by temperature.

A more labour intensive method of gaining a quantitative measure of disease is to use the area under the disease progress curve (AUDPC), but for cereal rusts, the major contribution made to the area value is the last one or two observations. Consequently, appropriately timed observations result in a very high correlation between a single score and AUDPC (Broers, 1989). The main benefit of multiple scores in our experience is the value of replication, and the likelihood of detecting early and potentially severe rusting genotypes. Also with differing maturities of host materials, terminal disease ratings for leaf rust and stem rust will be obtained at different scoring dates.

Selection of Pathotypes for Disease Nurseries

Opinion varies widely as to the rationale for selecting pathotypes for use in disease nurseries. The primary reason for this relates to the objective of particular tests and the level of genetic knowledge of the material. There is no doubt that a breeder's aim is to achieve maximum resistance with assured durability. On the other hand, a genetic study is ideally satisfied by use of single pathotypes. Because breeding studies and genetic studies often compete for limited field areas, compromises must be reached.

Many nurseries in the Indian subcontinent are set up with the objective of testing with mixtures of all available pathotypes. While this strategy potentially offers assurance that lines having single resistance genes with a limited range of effectiveness will not be selected, it provides danger of escape of rare pathotypes to neighbouring farming areas. Moreover, a basic assumption of multiple pathotype nurseries is that all component pathotypes will develop to adequate levels for effective screening. In order to ensure that this constraint is overcome, strategies including detailed monitoring and selective inoculation of carefully chosen susceptible host genotypes are necessary.

In Australia where there have been long-term nation-wide pathogenicity surveys, rust nurseries are grown at one major centre which is well removed from wheat-growing areas, and either one, or a mixture of two, pathotypes are released at each of two sites. Pathotypes selected for field nursery inoculation are chosen to represent either a pathotype predominant in wheat-growing areas, or a pathotype considered capable of gaining predominance in the near future. Such latter pathotypes usually possess virulence genes matching resistance genes that are represented in high frequency in leading cultivars and breeding populations. The use of these pathotypes allows selection for resistance in anticipation of projected future changes in the pathogen population. Less frequent, unpredictable events, such as the occurrence of exotic pathogen phenotypes or the preferential survival of pathotypes for reasons other than a selective advantage on deployed resistances, may limit the usefulness of this approach.

A CO-ORDINATED STRATEGY FOR RUST CONTROL

The two main methods of wheat rust control are resistance breeding and chemical control. Some very effective chemicals are now available for the control of rusts, especially stripe rust and leaf rust, and with timely application effective stem rust control can also be achieved. These fungicides include flutriafol, propiconazole, tebuconazole and triadimefon. In high yielding situations with opportunities to vary inputs, chemicals can be applied prophylactically and repeatedly with the expectation of returns on investment. However, in low yielding/low priced situations chemical use becomes difficult to justify. For example, the use of chemicals to control stripe rust in Australia can be justified only if the yield expectation of a crop exceeds 3 t/ha, which is relatively uncommon in a country where average yields are less than 2 t/ha.

Effective control of rust diseases is best achieved by a co-ordinated resistance breeding effort across the epidemiological area of the pathogen. Control by individual breeding organisations or countries within major geographic areas will be much more difficult in the absence of similar control strategies by neighbouring organisations or countries. A strategy for resistance breeding encompasses the following:

1. monitoring pathogen variability by means of pathotype surveys;
2. searching for effective sources of resistance and understanding how to manipulate those sources in breeding programs;
3. breeding and release of cultivars; and
4. post release monitoring of resistant cultivars.

Pathotype Surveys

Most wheat-producing countries undertake some kind of pathogen monitoring in crops whereby rust samples are collected and taken to a central laboratory to analyse pathogenicity phenotype. However, it is relevant to ask why pathogenicity surveys are conducted in the first instance. The reasons include:

1. peer pressure, that is, other laboratories undertake surveys;
2. to determine the degree and range of pathogenic variation in particular regions;
3. to understand the mechanisms and monitor the origin and spread of new pathotypes in a particular region;
4. to monitor the resistances used in commercial cultivars and to confirm suspected pathogenicity changes from field observations; and
5. to obtain new or relevant pathotypes for use in breeding programs and in research studies of relevance to national breeding programs.

Although the detail and scope of the analysis varies with laboratory or country because of the above reasons, the prime objective is to monitor the pathogen population and to recognise pathogenic changes if and when they occur. Attempts are made to correlate changes in the pathogen population with changes in the response of cultivars, or to correlate changes in crop cultivar responses with changes in response of one or more of the tester genotypes. If a changed response in the field involves virulence for a source of adult plant resistance, it is unlikely to be genetically correlated with a change in seedling response, although phenotypic correlations could be involved. If the genes involved in resistance in the predominant cultivars are known, then those genes should be included in the host lines used as testers in the survey. If the genes are unknown, then the resistant cultivar can be used as a tester. Where disease control is based on adult plant resistance, adult plant tests should be used in the survey.

Surveys in most countries involve an arbitrary set of testers chosen from published accounts in the international literature, or selected on various criteria such as the ability to differentiate the local pathogen population. In these circumstances, the survey may become a monitoring project *per se* with little attempt to relate the results to variation in current commercial cultivars. Thus the results may become a series of phenotypic, rather than genotypic, correlations.

Survey Procedures

COLLECTION OF SAMPLES: Procedures used to collect rust samples are similar in most parts of the world. Samples are collected by researchers, farmers or other interested persons from commercial wheat crops, experimental nurseries and other areas such as stubble regrowth, self-sown plants along roadsides and protected areas and certain grasses. These sample sources are usually supplemented with collections made on survey tours involving random crop inspections at regular intervals along predetermined routes. In looking for rust in a crop, particular attention may be given to potentially susceptible off-type wheat plants, and grasses known to harbour wheat-attacking rust pathogens. Plots of selected genotypes are deployed in some countries as trap plots to improve detection of specific pathotypes and to provide indication of viable inoculum movement. This has been the procedure in the USA and Canada, where data from the plots are processed separately from the data obtained from commercial crops (Roelfs, 1984).

INOCULATION PROCEDURES: Two approaches are used in determining the pathotypes present in collections of rusted plant material. The first has been used in the USA, Mexico and South Africa for leaf rust and stem rust, and involves inoculation of a susceptible genotype with spores from the original collection and establishing three random single pustule isolates from the inoculated plants. The single pustule isolates are used to inoculate seedlings of differential sets. Residues of the original collections from a particular region may be bulked and applied to a series of wheats possessing resistance genes for which virulence has not been detected. This technique simplifies the identification procedure by providing pure cultures of single pathotypes, but reduces the probability of detecting variants which may be present at low frequencies in samples comprising more than one pathotype. The second approach, used in Australia, is to apply inoculum, either directly from bulk samples or following increase, to differential sets. Where pathotype mixtures are suspected, inoculum is subsampled from those differentials most likely to yield new or potentially important pathotypes.

Differential Sets and Pathotype Nomenclatures

Systems of pathotype nomenclature which gained early acceptance were proposed for stem rust by Stakman and Levine (1922), for leaf rust by Mains and Jackson (1926) and for stripe rust by Gassner and Straib (1932). A problem with these systems was that the genetic bases for

TABLE 1-3 Differential genotypes used to identify pathotypes of *Puccinia recondita* f. sp. *tritici* in different geographic regions of the world

Australasia [a]		North America [b]		South Africa [c]		India [d]	
Differential	Resistance gene	Differential	Resistance gene	Differential	Resistance gene	Differential	Resistance gene
International Series		*Set 1*		Tc + Lr1	Lr1	*Set A*	
Tarsa	Lr1	Tc + Lr1	Lr1	Tc + Lr2a	Lr2a	Tc + Lr14a	Lr14a
Webster	Lr2a	Tc + Lr2a	Lr2a	Tc + Lr2b	Lr2b	Prelude*6/Agent	Lr24
Mediterranean	Lr2a, Lr3a	Tc + Lr2c	Lr2c	Tc + Lr2c	Lr2c	Timvera	Lr18
Democrat	Lr3a	Tc + Lr3a	Lr3a	Tc + Lr3a	Lr3a	Egret	Lr13
				Tc + Lr3bg	Lr3bg	Tc + Lr17	Lr17
Australian Series		*Set 2*		Tc + Lr3ka	Lr3ka	Kenya 1483	Lr15
1. Thew	Lr20	Tc + Lr3ka	Lr3ka	Tc + Lr11	Lr10	Tc + Lr10	Lr10
2. Gaza	Lr23	Tc + Lr9	Lr9	Tc + Lr11	Lr11	Agatha	Lr19
3. Spica	Lr14a	Tc + Lr11	Lr11	Tc + Lr14a	Lr14a		
4. Kenya 1483	Lr15	Tc + Lr16	Lr16	Tc + Lr15	Lr15	*Set B*	
5. Klein Titan	Lr3ka, Lr13			Tc + Lr16	Lr16	Loros	Lr2c
6. Gatcher [e]	Lr27 + Lr31[f]	*Set 3*		Tc + Lr17	Lr17	Tc + Lr2a	Lr2a
7. Songlen	Lr17	Tc + Lr17	Lr17	Thew	Lr20	Democrat	Lr3a
8. CS 2A/2M	Lr28	Tc + Lr24	Lr24	Agent	Lr24	Thew	Lr20
9. Mildress	Lr26	Tc + Lr26	Lr26	Gamtoos	Lr26	Benno	Lr26
10. Egret	Lr13	Tc + Lr30	Lr30	Tc + Lr 30	Lr30		
11. Exchange [e]	Lr16						
		Mexico					
		Set 4					
		Tc + Lr3bg	Lr3bg				
		Manitou	Lr13				
		Tc + Lr15	Lr15				
		Tc + Lr18	Lr18				
		Set 5					
		Tc + Lr10	Lr10				
		Tc + Lr19	Lr19				
		Tc + Lr23	Lr23				
		Gatcher	Lr27 + Lr31[f]				

a Park and Wellings (1992).
b Long and Kolmer (1989); Singh (1991).
c Pretorius and Le Roux (1988).
d Nagarajan et al. (1986b).
e Also carry *Lr10* which is generally not effective against Australian pathotypes.
f Complementary genes.

resistance were unknown and new differentials could not be added. It soon became apparent that they did not account for all of the variability that was present, and thus supplementary sets of testers were added in different geographic regions.

Leaf Rust

In Australasia, pathotypes of *P. recondita* f. sp. *tritici* are routinely identified using a selection of four host genotypes which permit a standard race designation (Johnston and Browder, 1966), and an additional 11 wheats chosen as supplementary differentials (Table 1-3). Pathotype 104-2,3,6,(7) therefore complies closely with the description for standard race 104, and is virulent for genes in the second (*Lr23*), third (*Lr14a*) and sixth (*Lr27* + *Lr31*) differentials. Partial virulence on the seventh (*Lr17*) differential is indicated by inclusion of the differential number in parentheses. This procedure still fails to describe the total variability that may be present where there are more than three interaction phenotypes.

In the USA and Canada, a series of 12 near-isogenic lines in a Thatcher background is used (Table 1-3) (Long and Kolmer, 1989). The differentials are grouped into sets of four, with the avirulence/virulence combinations for each group being coded by a consonant ranging from B (avirulent on all four differentials) to T (virulent on all four differentials). Provision is made for supplementary differential sets, as is currently used in Mexico (Singh, 1991).

The pathogenicity of South African isolates of *P. recondita* f. sp. *tritici* is determined from infection of 13 near-isogenic lines of Thatcher and additional testers for *Lr20*, *Lr24*, *Lr26* and *Lr30* (Table 1-3). The isolates are characterised by avirulence/virulence formulae, and each is referred to by a race number preceded by 3SA, the 3 indicating *P. recondita* f. sp. *tritici* and SA indicating South Africa (e.g. 3SA129) (Pretorius *et al.*, 1987). In India, three differential sets are used in annual pathogenicity surveys, the first of which (the O set, not listed in Table 1-3) represents current cultivars and is not used in pathotype classification. The A and B sets comprise eight and five genotypes, respectively (Table 1-3), with race numbers determined for each set by adding decanery (2^0, 2^1, 2^2, 2^3 ...) values corresponding to host lines rendered ineffective by the isolates being tested. The final formula comprises two race numbers separated by the letter R to indicate *P. recondita*, for example 132R7 (Nagarajan *et al.*, 1986*b*).

Stem Rust

The nomenclatural system used for *P. graminis* f. sp. *tritici* in the respective countries have essentially the same bases as those for *P. recondita* f. sp. *tritici*. The Australasian system uses a selection of six genotypes to assign a standard race designation based on Stakman *et al.* (1962), and an additional 11 wheats and 2 triticales comprising an Australian supplementary differential set (Table 1-4). The addition of triticale became necessary when a significant stem rust problem developed in triticale, and the pathogen responsible was shown to be *P. graminis* f. sp. *tritici* (McIntosh *et al.*, 1983). Twelve near-isogenic or single gene lines in various genetic backgrounds are used in the USA and Canada (North America, Table 1-4) (Roelfs and Martens, 1988). In South Africa, 22 differentials are used (Table 1-4) with the prefix 2 to indicate the wheat stem rust pathogen, followed by a further standard race number *sensu* Stakman *et al.* (1962); for example 2SA6 (Le Roux and Rijkenberg, 1987*a*). The Indian system is based on the responses of eight wheats in an A set and six wheats in a B set of differentials (Table 1-4), with the set numbers separated by G to indicate the wheat stem rust pathogen (Nagarajan *et al.*, 1986*b*).

Stripe Rust

The original differential series and nomenclatural system proposed for *P. striiformis* f. sp. *tritici* by Gassner and Straib (1932) was abandoned due to difficulties in reproducing the reported responses of standard races, the need to subdivide and/or combine race descriptions, and the need to use supplementary differentials (Stubbs, 1985). The most widely used system is based on the proposal of Johnson *et al.* (1972). This system has gained acceptance throughout Europe and, with certain additional differentials, in Australia (Wellings *et al.*, 1988). An 'International' series and a 'European' series are employed (Table 1-5) with a number assigned to each series by the addition of decanery values corresponding to each differential rendered susceptible. The second number is preceded by the letter E to indicate the European series. In Australasia, the pathotype formula is followed by A– or A+ to indicate avirulence or virulence, respectively, for a distinctive resistance present in a selection of the Australian cultivar, Avocet (Wellings *et al.*, 1988) (e.g. 104 E137 A+). A similar system based on different host lines is used in India (Table 1-5) with two numbers in the pathotype formula separated by the letter S (Nagarajan *et al.*, 1986*b*). In the UK, stripe rust pathotypes with virulences for adult plant resistances have been detected, necessitating the adoption of a modification to the standard nomenclature. Thus, Johnson and Taylor (1972)

TABLE 1-4 Differential genotypes used to identify pathotypes of *Puccinia graminis* f. sp. *tritici* in different geographic regions of the world

Australasia [a] Differential	Resistance gene	North America [b] Differential	Resistance gene	South Africa [c] Differential	Resistance gene	India [d] Differential	Resistance gene
International Series		*Set 1*		*Set A*		*Set A*	
Marquis	Sr7b	ISr5-Ra	Sr5	Little Club	SrLC	Not specified	Sr13
Reliance	Sr5	T. monococcum derivative	Sr21	Marquis	Sr7b	"	Sr9b
Acme [e]	Sr9g	Vernstein	Sr9e	Reliance	Sr5	"	Sr11
Einkorn [f]	Sr21	ISr7b-Ra	Sr7b	Kota	Sr7b + Sr28	"	Sr28
Vernal [e]	Sr9e	*Set 2*		Arnautka [e]	Sr9d	"	Sr8
Australian Series		ISr11-Ra	Sr11	Mindum [e]	Sr9d	"	Sr9e
1. McMurachy	Sr6	ISr6-Ra	Sr6	Spelmar [e]	Sr9d	"	Sr30
2. Yalta	Sr11	ISr8-Ra	Sr8a	Kubanka [e]	Sr9g	"	Sr37
3. W2402	Sr7b + Sr9b	CnSSr9g	Sr9g	Acme [e]	Sr9g	*Set B*	
4. W1656	Sr7b + Sr36	*Set 3*		*Set B*		Marquis	Sr7b
5. Renown	Sr7b + Sr17	W2691 SrTt-1	Sr36	Einkorn [f]	Sr21	Einkorn [f]	Sr21
6. Mentana	Sr8a	W2691 Sr9b	Sr9b	Vernal [e]	Sr9e	Kota	Sr7b + Sr28
Charter		Bt Sr30 Wst	Sr30	Khapli [e]	Sr13 + Sr14	Reliance	Sr5
7. Norka	Sr15			I Sr5-Ra	Sr5	Charter	Sr11 + Sr
8. Festiguay	Sr30	Combination VII	Sr17	I Sr6-Ra	Sr6	Khapli [e]	Sr13 + Sr14
9. Taf 2	SrAgi			ISr7b-Ra	Sr7b		
10. Entrelargo de Montijo	SrEm			Mentana	Sr8a		
11. Barleta Benvenuto	Sr8b			W2691 + Sr9b	Sr9b		
12. Coorong triticale	Sr27			Yalta	Sr11		
13. Satu triticale	SrSatu			Renown	Sr7b + Sr17		
				Bt Sr24Ag	Sr24		
				Festiguay	Sr30		
				W2691 + SrTt1	Sr36		

[a] Park and Wellings (1992). [b] Roelfs and Martens (1988); Singh (1991). [c] Le Roux and Rijkenberg (1987). [d] Nagarajan *et al.* (1986b). [e] 2n = 28. [f] 2n = 14.

TABLE 1-5 Differential genotypes used to identify pathotypes of *Puccinia striiformis* f. sp. *tritici* in different geographic regions of the world

Europe/Australasia [a] Differential	Resistance gene	North America [b] Differential	Resistance gene	India [c] Differential	Resistance gene	China [d] Differential	Resistance gene
International Series		*International Series*		*Set A*			
Chinese 166	Yr1	1. Lemhi	Yr1	Chinese 166	Yr1	Trigo Eureka	Yr6
Lee	Yr7	2. Chinese 166	Yr1	Lee	Yr7	Fulhard	Yr1
Heines Kolben	Yr2 Yr6	3. Heines VII	Yr2	Heines Kolben	Yr2 Yr6	Bima 1	
Vilmorin 23	Yr3	4. Moro	Yr10	Vilmorin	Yr3	Lutescens 128	
Moro	Yr10	5. Paha		Moro	Yr10	Xibei Fenshou	
Strubes Dickkopf		6. Druchamp	Yr6	Strubes Dickkopf		Xibei 54	Yr1
Suwon 92/Omar	Yr2 Yr9	7. Riebesel 47-51	Yr9	Suwon 92/Omar		Quality	
Clement		8. Produra	Yr6	*Set B*		Mentana	
Triticum spelta album	Yr5[e]	9. Yamhill	Yr2	Hybrid 46	Yr4	Kansu 96	
		10. Stephens		Heines VII	Yr2	Virgilio	
European Series		11. Lee	Yr7	Compair	Yr8	Abbondanza	
Hybrid 46	Yr4	12. Fielder	Yr6	*Triticum spelta album*		Beijing 8	Yr1
Reichersberg 42	Yr7	13. Tyee	Yr9	Riebesel	Yr5	Early Premium	
Heines Peko	Yr2 Yr6			Sonalika	Yr2 YrA	Funo	YrAf
Nord Desprez				Kalyansona	Yr2	Danish 1	Yr3
Compair	Yr8					Jubilejna 2	
Carstens V						Feng Chan 3	Yr1
Spaldings Prolific						Strubes Dickkopf	
Heines VII	Yr2					Lovrin 13	Yr9
						Kangyin 655	
Australian Series						Taishan I	
Avocet R	YrAf					Shuiyun 11	
						Zhong 4	

a Johnson *et al.* (1972).
b Line and Qayoun (1991).
c Nagarajan *et al.* (1986*b*).
d Stubbs (1985).
e Wellings and McIntosh (1990).
f Wellings *et al.* (1988).

described 104 E137 Type 1 as the original pathotype, and 104 E137 Type 2 as the pathotype with virulence for the adult plant resistance in Joss Cambier. Nine variants of standard UK pathotypes were reported in a review by Wellings (1986).

The North American and Chinese systems for identifying pathogenic variation in *P. striiformis* f. sp. *tritici* are unique to those regions. In North America, 13 differentials are used (Table 1-5) and isolates are characterised by an avirulence/virulence formula comprised of differential numbers rather than resistance genes; for example 2,4,5,6,7/1,3. A Cereal Disease Laboratory (CDL) number is applied to an isolate with a distinctive formula, for example CDL-39 (Line and Qayoum, 1991). Pathogenicity surveys of *P. striiformis* f. sp. *tritici* have been conducted in China since the 1940s (Fang, 1944; Hu and Roelfs, 1985). Currently, 23 differentials are used, although the resistance genes present in many of these differentials have not been determined (Table 1-5). Pathotypes have been sequentially numbered since surveys began, with Hu and Roelfs (1985) proposing that pathotype numbers be preceded by the letters CS.

Factors Limiting International Agreement on Differential Sets

Historical and biological reasons have prevented the use of common sets of differentials and nomenclatures worldwide. Some of these reasons are discussed in the following sections.

Geographical Differences in Virulence Frequencies

Whereas most reported differential series for any particular host:pathogen system include common genes, if not common tester lines, there are other sources of resistance that are of more value as differentials in local areas. For example, lines with *Lr20* and *Lr15* might be considered essential for the study of variation in *P. recondita* f. sp. *tritici* in Australasia, but they are of little use in North America where the population is almost entirely virulent. On the other hand, lines with *Lr10* and *Lr30* that may be of value in North America are of little use in Australasia where the pathogen would be considered virulent on seedlings with these genes.

Relevance of Differential Genotypes to Deployed Resistances

Pathogenicity surveys are often completed with limited resources and only a selection of potential differentials can be used on a routine basis. Because most laboratories impose limits of 15–20 host genotypes, individual decisions will be made to give priority to different genotypes. Surveys are usually justified by their value to local breeding programs, and therefore, must be kept relevant to those programs.

Genetic Background of Differential Testers

Lines selected as efficient differentials in one area or time may not be equally useful in other locations or times. For example, certain lines traditionally used as differentials for *P. graminis* f. sp. *tritici* in Australia carry *Sr7b* as well as the originally targeted genes (e.g. W2402, the differential for *Sr9b*, possesses *Sr7b* and *Sr9b*; Renown W3125 used as a differential for *Sr17* also carries *Sr7b* and *Sr9d*; W1656 = C.I.12632 carries *Sr7b* as well as *Sr36*). The presence of *Sr7b* was of no consequence when these lines were chosen as differentials prior to the 1970s since all pathotypes were virulent on *Sr7b* at that time. During the 1970s, pathotypes with avirulence for *Sr7b* became widespread, hence current survey results must be adjusted to account for this avirulence.

When the stripe rust differentials developed in Europe are tested in Asia or North America there is no assurance that the detected resistances are identical with those described in genetic studies with European cultures. For example, the European differential, Heines VII, with *Yr2*, carries a second gene detected with certain non-European pathotypes (Johnson, 1992). Similarly, Chen and Line (1992b) detected a number of genes in differentials effective against North American cultures, but probably ineffective against European isolates.

For these reasons, other research groups have emphasised backcross-derived near-isogenic lines with single genes for resistance. However, such near-isogenic lines may have several disadvantages. As in the case of *Sr7b* discussed above, they may carry additional genes for resistance, and hence their use will be limited to arrays of cultures that are virulent on the genotype of the recurrent parent. As interest extends to geographic areas beyond those for which the lines were originally developed, or to populations of the target pathogen adapted to non-hexaploid wheats or related species, there may be an increased likelihood of detection of additional resistance genes in the recurrent parent used in the development of the near-isogenic set. With such extended dimensions in both time and space, it is extremely difficult to define, identify or develop a universally susceptible host genotype. In addition, near-isogenic lines based on extremely and widely susceptible genotypes may be more difficult to use in laboratories lacking controlled environment facilities. For example, Luig and Rajaram (1972) showed that the transference of

Sr5 to a highly susceptible genetic background resulted in a significantly higher, but still incompatible, infection type which may be difficult to detect in non-controlled environments. Genes such as *Sr15* and *Sr17* appear to require lower temperatures for the expression of resistance when present in a Chinese Spring background than when present in the backgrounds of Renown and Norka which were selected as differentials in Australia.

The least progress in the development of widely accepted sets of near-isogenic lines has been with stripe rust resistance genes. In several laboratories, presumed susceptible recurrent parents were later found to be resistant to certain cultures of local or exotic origin, or as adult plants in the field. A Federation derivative with *Yr9* from Kavkaz (RA McIntosh and CR Wellings, unpublished 1982) was used by Stubbs (1988) to demonstrate that the European differentials, Clement and Riebesel 47/51, possessed seedling-effective resistance genes additional to *Yr9*. However, Federation has not been used as a general recurrent parent because grass clump dwarfs are produced in many of its hybrids.

Another disadvantage with near-isogenic lines is that entire sets may not be agronomically adapted to some regions for seed increase. In this respect, most of the extremely valuable set of Thatcher near-isogenics for leaf rust work have a strong photoperiod response making them very late and ill-adapted for use in areas with low latitudes. Likewise, the vernalisation requirement of Chinese Spring may be prohibitive in some environments. A final concern is the morphological similarity of near-isogenics and hence the dangers of wrong labelling in harvest operations.

Progressive Increases in Virulence

Most systems of pathotype nomenclature are based on binary notation involving contrasting categories of incompatibility or compatibility, that is, resistance or susceptibility in the host, avirulence or virulence in the pathogen. In Australia, intermediate levels of avirulence/virulence expression ('progressive increases in virulence', *sensu* Watson and Luig, 1968) are indicated in pathogenicity formulae. The genetic bases for this phenomenon remain to be determined. In addition to heterozygosity (Samborski, 1963), there are possibilities for multiple allelism at the pathogenicity loci or for non-allelic ('genetic background') modification of the presumed major gene avirulence loci. Recognition of these intermediaries can be important in relation to changes in field responses of resistant cultivars, and as distinctive markers in the study of evolutionary changes and epidemiology within the pathogen populations. However, international agreement on the recognition and importance of intermediate responses appears to be difficult.

Adult Plant Tests

The major objective of rust surveys is to apply the results to the field. Thus it would be advantageous if adult plant tests with chosen differentials could be undertaken in the greenhouse or field, although few attempts have been made to do this. Samborski (1984) and Park and McIntosh (1994) reported testing of leaf rust pathotypes for adult plant virulence in greenhouse experiments. The only annual survey concerned with adult plant responses is the testing of stripe rust cultures in polythene tunnels in the field at the National Institute of Agricultural Botany, UK (Priestley *et al.*, 1984*a*; 1984*b*).

There are several difficulties in establishing international agreement for adult plant differentials. Firstly, if the lines are poorly adapted, relevant correlations to local and international scenes will be more difficult to obtain. Secondly, if there are additional genes for adult plant resistance the value of lines for such tests will be reduced. For example, the effectiveness of *Lr22b* for adult plant resistance in Thatcher against one important Australasian *P. recondita* f. sp. *tritici* pathotype makes it impossible to determine the pathogenicity of this culture in the adult Thatcher background.

Despite these difficulties, monitoring pathogenicity for resistance genes expressed at adult plant growth stages will become more important as these types of resistance are more widely exploited.

Conclusion

Ultimately the researcher is interested in whether a particular isolate of the pathogen is avirulent or virulent for a number of individual host genes or resistance sources. If the pathogenicity phenotype is encoded in a pathotype name, the information must be available for its translation to a binary code or avirulence/virulence formula. Clearly, surveys cannot include all known resistance genes, but surely if a culture is worthy of preservation as a genetic resource and of being described in the international literature, then an internationally complete description should be available. This implies that research groups should be free to routinely use an arbitrary set of testers for local survey purposes, but for published papers and international exchange of information, a full pathogenicity genotype for an agreed set of testers would be valuable. With agreed testers, a gene-based designation procedure could follow, provided there are provisions for revision at regular intervals of five or ten years. Limpert *et al.* (1993) addressed the possibility of similar designation systems for all plant pathogens based on octal notation. This proposal may prove to be a satisfactory compromise, in terms of both brevity and convenience in deciphering the information.

Genetics of Rust Resistance

The perceived role and impact of genetic studies of host resistance in applied breeding programs has been variable. The serious outbreak of southern corn leaf blight in North America in 1970 prompted a retrospective analysis which revealed that vulnerability was predisposed by a genetic uniformity evident in the common cytoplasm used for hybrid cultivars (Day, 1977). Similarly, the emergence of two closely related pathotypes of *P. graminis* f. sp. *tritici* in Australia resulted in the susceptibility of over 70% of entries in the 12th International Triticale Screening Nursery (McIntosh *et al.*, 1983). These and other events have led to a re-evaluation of breeding strategies for disease resistance. The concept of durable resistance (Johnson 1981; 1988) was proposed in part, to help identify and exploit stripe rust resistances which would be less prone to commercial failure. In the simplest case, only one source of resistance is required if there is an assurance of prolonged resistance. However, the lack of assurance of durable resistance under most circumstances requires the adoption of genetic diversity to provide protection and insurance against major epidemics.

Regardless of individual viewpoints on the development and management of disease resistance, the first requirement for the breeder is a source of resistance, or the means of developing a level of resistance that results in less disease and/or a reduction in crop losses. Following the identification of a resistance source, genetic studies are undertaken to determine the components of resistance and their phenotypic and genetic characteristics. These studies are usually independent of resistance breeding because the choice of susceptible parents is based on different criteria. Following these studies, decisions can be taken on the advisability of, and the optimal strategy for, commercial exploitation. Seedling resistances or highly effective adult plant resistances characterised by hypersensitivity may be avoided by some breeders or selected by others, after careful consideration of the perceived risks of resistance breakdown.

Genetic studies of resistance sources usually proceed in several stages, often beginning with the identification of resistance among the materials comprising a field rust nursery. The next step is to determine if the resistance is effective at the seedling stage and whether it is effective against a wide array of pathotypes. From the pedigree and multi-pathotype seedling tests it may be possible to postulate the presence of known genes in candidate lines and to decide whether a genetic study is worthwhile. Presumed novel resistances are crossed with a susceptible genotype to permit determination of the number of genes for resistance to a particular pathotype, and whether the individual genes are effective against all pathotypes. Because of the large number of resistance genes already known and the relatively large chromosome number in hexaploid wheat, potentially new genes are usually localised to chromosomes and mapped to a particular chromosome arm before being permanently named. Relevant linkage studies may then be undertaken and, in some cases, tests of allelism may be necessary to decide if a distinctive locus is involved.

Because of the likely application of genetic studies to resistance breeding, it is essential to correlate greenhouse seedling tests with field assessments. Parallel tests are best based on F2 families from backcrosses or F3 lines from direct crosses. This provides the necessary seed numbers for concurrent tests with multiple pathotypes and for replication of plants within lines. F4 families may be required for the isolation and confirmation of lines with single genes for resistance. If new genes are involved, preservation of the single gene lines is essential for use in future work. This can be accomplished by backcrossing to selected genotypes, such as a recurrent parent forming the basis of an isogenic set, or to adapted agronomic germplasm for use as a resistance source in further breeding activities. Collaborative studies among laboratories within and between geographical areas can be undertaken with the exchange of seed from structured populations such as backcross F2 families, F3 lines from direct crosses and single seed descent populations. However, the actual nature of the material exchanged and its immediate use may be determined by local phytosanitary regulations.

Gene Location
Chromosome location is an intermediate step in establishing the identity and allelic relationships of a new rust resistance gene. The procedure of location varies with the availability of suitable aneuploid stocks (McIntosh, 1987). The most common method for locating a dominant resistance gene is the F2 method whereby the resistant line, or a derivative, is crossed to a monosomic series, comprising each of the 21 chromosome pairs, in a susceptible genotype. A monosomic

plant (2n = 41) from each of the 21 chromosome stocks is selected by mitotic or meiotic chromosome observation and is pollinated with the resistant line. Monosomic F1 plants are selected, allowed to self-pollinate and the progenies (F2) are inoculated with the pathogen. When a single dominant gene is involved, segregation in the F2 generation will be 3 resistant : 1 susceptible in 20 of the 21 crosses just described. In the 21st or 'critical' cross there will be a significant excess of resistant plants and only the nullisomic (2n = 40) progeny and certain secondary aneuploids, which occur at low frequency (usually less than 10%), will be susceptible. Confirmation that the resistance gene is located in the particular chromosome can be established by further association of resistance phenotype and chromosome number, by demonstrating linkage with the centromere by means of telocentric mapping (Sears, 1966) and by linkage to genes previously known to occur in the chromosome (McIntosh, 1987; Knott, 1989). Only when these tests are completed or when a unique gene from an alien chromosome is translocated to a wheat chromosome, will a new gene or new allele symbol be allocated to the Catalogue of Gene Symbols for Wheat (McIntosh, 1988a).

Chromosome substitution lines can also be used for gene location (Sears, 1953; Knott, 1989). This procedure can be used to confirm monosomic analyses when suitable materials are available, or if the inheritance of resistance is suspected to be other than a single dominant gene in a line for which monosomics are not available. In future, molecular and cytological markers may circumvent the time-consuming process of chromosome location using aneuploidy, especially when such genes are derived from interspecific or intergeneric transfers of alien chromosome segments which are more readily identified by molecular techniques. For example, a gene in a Japanese wheat line ST-1, with leaf rust resistance derived from rye was located in chromosome 2A by C-banding (Gill et al., 1991) to identify wheat chromosomes and by genomic (Mukai and Gill, 1991) or fluorescence in situ hybridisation to detect the rye chromatin (RA McIntosh, J Jiang, B Friebe, D The and BS Gill, unpublished 1993). The long arm and much of the short arm of chromosome 2A was replaced by rye chromatin (Plate 1-8).

Genetic study of a resistance gene culminates with its placement in a genetic map showing its relative position in the chromosome. This information can then be used by breeders in planning resistance gene combinations. In the medium to long-term, detailed genetic and physical maps will be essential for effective manipulation of genetic characteristics in vivo. Significant progress in the development of physical maps in wheat was made with the discovery of the means to generate wheat chromosomes with terminal deletions (Endo, 1988). Association of the presence or absence of genetic markers in a series of deletion stocks with the C-banding maps permits the ordering and localisation of markers to chromosome segments defined by C-banding.

Transfer of Rust Resistance Genes to Wheat from Related Species

In order to stay ahead of constantly changing rust pathogens it has been necessary for wheat breeders and their supporting geneticists and pathologists to maintain genetic diversity by seeking resistance genes from sources other than common wheat. Developments in wheat breeding following the rediscovery of Mendelian inheritance have closely paralleled improving success in the achievement of wide crosses with sufficient fertility and ability to cross the progeny to permit introgression of genetic attributes. Early successes in rust resistance involved crosses of bread wheat with cultivated tetraploid wheats (durum and emmer) resulting in the important derivatives, Marquillo (Hayes et al., 1920), a parent of Thatcher, and Hope and H-44 (McFadden, 1930), which were widely used as sources of resistance to stem rust. Later, T. timopheevii and diploid and polyploid species that share chromosome homologies with wheat were used as gene donors, either in direct crosses, via bridging crosses or as amphiploids.

In 1956, ER Sears (Sears, 1956) pioneered the use of pollen irradiation to transfer a gene (Lr9) for leaf rust resistance to wheat from the related grass species T. umbellulatum, which exhibited no chromosome homology with wheat. Further success was later achieved by irradiation of dry seed (Knott, 1961).

The discovery of genetic control of diploid chromosome behaviour in wheat by Okamoto (1957) and Riley and Chapman (1958) provided possibilities for greater control of alien gene transfer via homoeologous recombination. This technique improved the expectation of obtaining derivatives with more competitive agronomic performance than those produced by irradiation. Transfers resulting from the disrupted control of the chromosome pairing gene (Ph) have been achieved through the use of the genome of T. speltoides which suppresses the diploid control mechanism (Riley et al., 1968), or by use of a mutant or deletion stocks of chromosome 5B which carries the Ph gene (Sears, 1973).

More recently it was shown that chromosome translocations may occur during culture of somatic tissues. Larkin et al. (1990) were successful in using this technique to obtain 42 chromosome wheat lines with resistance to barley yellow dwarf virus derived from Thinopyrum intermedium.

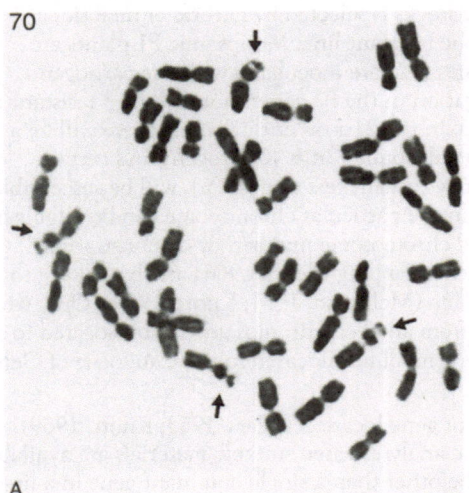

70

A

PLATE 1-8

The chromosomes of wheat.

A. A somatic cell of euploid wheat (2n = 42) prepared by the Feulgen method. Apart from two pairs of chromosomes with satellites (arrowed) wheat chromosomes prepared by this method are difficult to group into individual pairs.

B. C-banding enables the individual wheat chromosomes to be identified. This plant is known to carry leaf rust resistance from cereal rye. The translocated chromosome T2AS-2RS.2RL was recognised by loss of the typical chromosome 2RL features and gain of the terminal dark region typical of chromosome 2RL. Courtesy B Friebe and JM Jiang.

C. The wheat–rye translocation (see B) was also characterised by fluorescence *in situ* hybridisation. The yellow region indicates the presence of rye chromatin. The intense yellow fluorescence is due to a high concentration of rye specific repetitive DNA sequences. The combined information from C-banding and *in situ* hybridisation permits identification of the wheat chromosomes and the precise positions of the translocation break-point. Courtesy RA McIntosh, B Friebe, J Jiang, TT The and BS Gill.

D. The lower three images compare the C-banding patterns of wheat chromosome 2A and rye chromosome 2R with the translocated chromosome shown in B and C. Note the similarity of the long arms (right) of the 2R and translocation chromosomes. The genomic in situ hybridisation (GISH) pattern of the translocation chromosome (topmost image) shows that the rye segment extends beyond the centromere and into the short arm. C-banding permits identification of the translocated chromosome (*see* Plate 1-8B), whereas GISH provides an accurate estimate of size of the translocated segment.

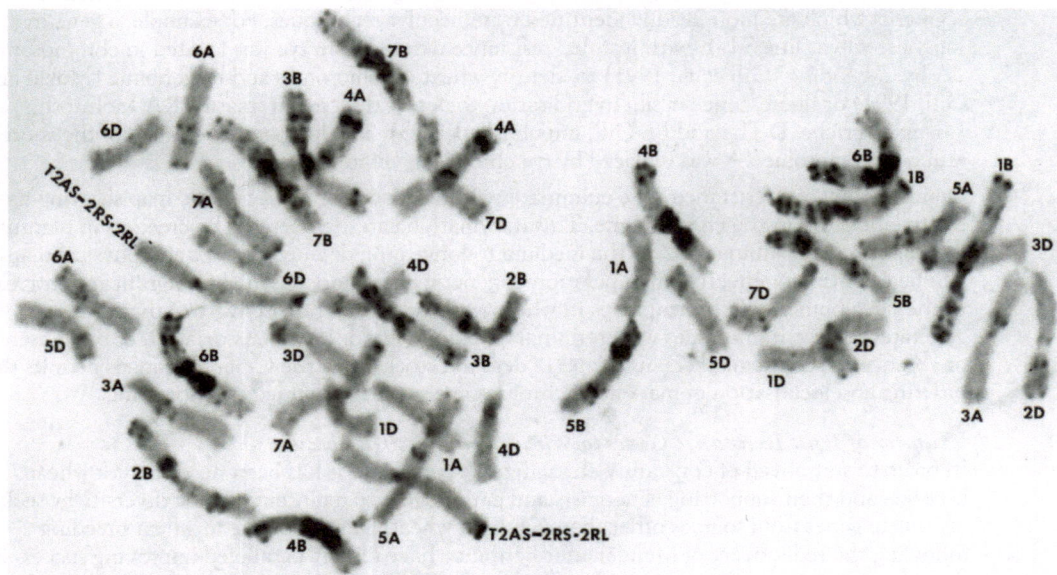

B

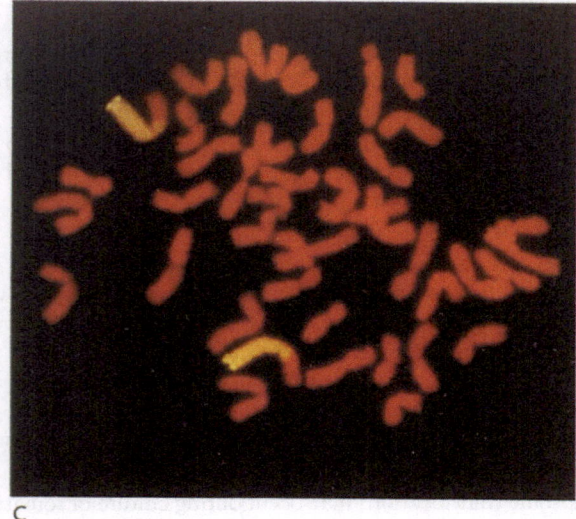

C

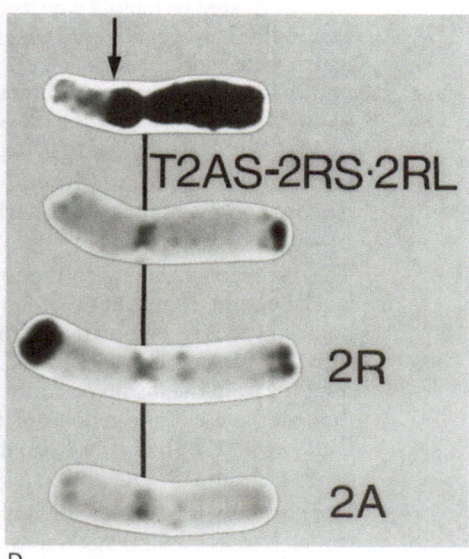

D

Success in achieving wide crosses has been improved with the embryo rescue technique which employs hormone treatment of female florets followed by physical removal and tissue culture of early embryos. Thus, the potential for spontaneous transfers of genetic material either in intact plants or in culture should not be neglected.

With the recent successful transformations of wheat (Vasil *et al.*, 1992; Weeks *et al.*, 1993; Becker *et al.*, 1994; Nehra *et al.*, 1994; RIS Brettel, *pers. comm.* 1994) there are now imminent possibilities for the transfer of genes for disease resistance beyond the limits of sexual combatibility. However, progress in the transfer of rust resistance genes will be delayed until gene products are identified or until alternative methods of gene cloning are developed.

Breeding for Rust Resistance

Several factors will determine the emphasis, and therefore the strategy best suited for rust control. Paramount among these factors will be the current experiences of rust epidemics at the local, regional and national level. Where epidemics are anticipated to be frequent and losses, therefore, significant, considerable resources can be devoted to resistance breeding. A short-term objective in these situations should be the removal of cultivars with high levels of disease. Longer-term strategies emphasise investments in the reduction of crop losses and the lowering of pathogen inoculum levels through the selection and deployment of resistant cultivars.

Ideally, breeding for rust resistance will be the outcome of the integration of studies in both pathogen and host. However, rust resistance is only one of several objectives in wheat breeding and thus strategies to achieve resistance need to be considered in the context of the whole program. Breeding is a long-term process and decisions that influence the outcome of a breeding program will be made at least five years before a cultivar is commercially released. By that time, pathogen populations could be significantly different from those present when the decisions were made. Successful breeding for disease resistance, therefore, depends on the ability of breeders and pathologists to predict the future pathotype structure of pathogen populations. If critical mutations occur locally, it can be assumed that future pathotypes will be similar to current pathotypes, except for the new pathogenic attributes. On the other hand, if variation occurs as the consequence of introduction, or possibly somatic hybridisation, the overall phenotype of such pathotypes cannot be predicted. This occurred during the early 1980s, when *P. recondita* f. sp. *tritici* pt. 53-1,(6),(7), 10,11 was introduced to Australasia (Luig *et al.*, 1985). This pathotype was favoured because it possessed virulence for *Lr13*, but it was avirulent for a wide range of resistance genes that had been overcome by the indigenous pathogen population (including *Lr1*, all *Lr2* alleles, *Lr3a*, *Lr23* and *Lr22b*). Some of the latter genes were fortuitously combined with *Lr13*, resulting in comprehensive resistance among most cultivars to all pathotypes. However, where resistance genes are deployed alone, and the corresponding avirulence genes in the pathogen have been lost or significantly reduced in effectiveness, there may be selection and rapid increase in frequencies of virulent or near-virulent variants. The greater the dependence on, and the greater the effectiveness of, any source of resistance, the greater the potential for damaging epidemics.

Where genetic vulnerability is suspected, additional sources of resistance should be incorporated. This has been the objective of the National Wheat Rust Control Program which has been active in Australia since 1975. One outcome of this national program has been the exploitation of the alien-derived genes *Lr24* and *Sr24*, which have become an important component of the suite of rust resistance genes used in eastern Australia. Although virulence for either gene has not been detected in national rust surveys, virulences to both are known to occur elsewhere. It is assumed, therefore, that virulences will occur in Australia and resistant lines are being developed to rapidly replace cultivars if this eventuates. The aim will be to manage a rapid cultivar replacement program without jeopardising the contemporary balance of industrial qualities of the crop.

Longer-term breeding strategies are aimed at achieving durability of resistance combined with the deployment of genotypic diversity to buffer potential losses. Retrospective genetic analyses of durable resistance sources either identify and characterise individual genes that have been deployed repeatedly (e.g. *Sr2* and *Sr26* for resistance to stem rust, *Lr34* and *Yr18* for resistance to leaf rust and stripe rust, respectively) or determine that resistance is based on oligogenic combinations that act additively. The use of such resistance sources is a means by which breeders can incorporate durable resistance in breeding programs.

Having decided that resistance to a particular disease is a breeding objective, the next step is to decide on the breeding strategy to achieve resistance. A detailed account of techniques of managing rust nurseries and of the common breeding methods is given by Knott (1989). The usual procedure is to identify a source of resistance and to transfer that resistance to locally adapted germplasm by available means. The facilities for selection will vary from totally uncontrolled to relatively sophisticated. At one extreme, the breeder will depend on genetically undefined resistance, the natural pathogen flora and opportunistic infections, possibly encouraged by early plantings of susceptible cultivars, frequent watering and the use of rust-prone sites for nurseries. At the other extreme, there will be the use of known resistance sources to ensure genetic variability, temperature-controlled testing facilities and specialised field sites with controlled releases of known pathotypes (Plate 1-9A). The approach of the CIMMYT group in Mexico is to identify international 'hotspots' in order to expose breeding populations to the widest possible range of pathogenic variability. This process is supplemented with disease data recorded for the International Nursery Program. Lines with low coefficients of infection over all sites are assumed to have a greater probability of durable resistance. The success of the strategy has been vindicated because many cultivars selected and released from breeding populations developed by breeders at CIMMYT and elsewhere are known to possess certain genes now associated with durable resistance (Rajaram *et al.*, 1988).

Epidemics are cyclical and the probability of an epidemic in a particular region depends upon prevailing climate during the crop season, local survival of inoculum during the non-crop season, the length of the crop season and the distribution of susceptible host genotypes throughout the entire year. Because rust pathogens may be airborne over large distances, these factors apply both locally and to large areas, often spanning several countries. A predominance of resistant cultivars decreases epidemic probability, whereas a predominance of susceptible cultivars will increase it. Indeed, a farmer who chooses to grow a susceptible cultivar when resistance predominates is likely to escape significant losses. However, if all farmers grew susceptible cultivars, the overall risk of loss may be high, and any small area of resistance would be constantly threatened with potential mutant urediospores with the capability of overcoming that resistance. With high levels of rust resistance and exposure of the wheat crop to a range of potential diseases for which there is competition for limited research funding, the perceived risk of rust epidemics is reduced, and thus scientists and farmers may become complacent and the cycle begins anew. The plea, therefore, is to maintain wheat research and breeding in order to ensure durable long-term crop protection.

Management of Resistance

Management of rust resistance in cultivars after they are released in agriculture is perhaps more important than achieving resistance in the first instance. The history of stem rust resistance in the summer rainfall areas of eastern Australia has involved the progression from cultivar cohorts that were predominantly susceptible and frequently sustained moderate to high levels of disease inoculum and crop losses, to the current cohorts that are predominantly resistant and carry very little rust. In the intermediate stages, mixtures of resistant and susceptible wheats were characterised by cyclical phases including periods of high infection and crop losses, significant

A   B

PLATE 1-9
Stripe rust in wheat.
A. Experimental plots grown for selection. A susceptible spreader is planted at either end of test rows.
B. A genetic difference between the responses of adjacent cultivars in South Australia, 1987.

pathogenic variability and poor durability of resistance. Currently, there is a single predominant pathotype of *P. graminis* f. sp. *tritici* with rare occurrence on occasional susceptible wheat plants or crops, certain grasses and barley (Park and Wellings, 1992). At this level, characterised by low pathogen populations and a policy of obligatory resistance for cultivar recommendation, the likelihood of mutation in the pathogen is greatly reduced and even single gene resistances appear to confer durable resistance.

The objective of cultivar management, regardless of epidemic probability as outlined above, is to maximise the potential durability of deployed resistances. This can be achieved in several ways, although the success of the strategies depends largely on the level of control that can be exercised in cultivar choice. Host genotype diversity has been advocated at several levels. From a regional and/or national perspective, varietal diversification schemes described for cereals in the UK (Priestley and Bayles, 1982) recommend cultivar selection to ensure a range of resistant genotypes between fields. Similar schemes have been informally adopted in areas of eastern Australia for the wheat rust diseases. At a more local level, multi-line and mixed cultivars based on components of resistant and susceptible materials, aim to increase host genotype diversity within fields (Wolfe *et al.*, 1992). In our opinion, all host genotypes must have levels of resistance adequate to prevent not only crop loss but also to significantly reduce the pathogen population. Genetic variation among or within cultivars at the field or regional level then function to buffer losses in the event of rust increase resulting from pathotype change.

However, a vital aspect in the controlled deployment of cultivars in agriculture is the level of co-operation forged with the farming community. Ballantyne *et al.* (1994) applied the term 'public risk diseases' to describe the impact of decisions to choose susceptible genotypes on the broader agricultural community. The rusts are aptly described as 'public risk diseases' and control will be achieved only at the levels of farm, community and epidemiological region if all growers become involved in control.

Several examples serve as illustrations of this principle. In 1986 there was a higher incidence of stem rust in the southern USA Great Plains than was common in the preceding years. This was apparently caused by the cultivation on a relatively small area of rust-prone material in south Texas (Roelfs *et al.*, 1987; AP Roelfs, *pers. comm.* 1988). Stem rust development on the non-recommended cultivar in a favourable environment was sufficient to threaten a large area of the North American wheat belt. In north-eastern Australia, a new pathotype of *P. graminis* f. sp. *tritici* with virulence for *Sr36* was detected and increased rapidly on several cultivars in 1984 (Zwer *et al.*, 1992). Effective extension activities and active farmer co-operation resulted in the rapid removal of susceptible cultivars, followed by a decline in the *Sr36*-virulent pathotype to undetectable levels by 1987 (Park and Wellings, 1992). The stripe rust epidemic in South Australia in 1987 demonstrated in particular locations that leaf damage could occur on moderately resistant cultivars when grown adjacent to susceptible cultivars (Plate 1-9B).

In summary, if rust control is to be sustained, the use of 'the wrong cultivar in the wrong place' must not be tolerated, even when it provides individual growers or a group of growers, with a temporary advantage. The cost of producing a successful crop from a susceptible genotype may be borne by the growers of resistant cultivars. Active extension and publicity campaigns are often necessary to remind farmers of the scourge of rust, especially during periods of complacency following the use of resistant cultivars.

CONCLUSION

In this introductory chapter we have described some of the developments that have led to the current implementation of control strategies for wheat rusts. We selected examples from an international perspective, although the description of a co-ordinated strategy for rust control draws on the scheme that was inaugurated in Australia following a severe stem rust epidemic in 1973. The outstanding success of the Australian National Wheat Rust Control Program has been largely due to the application of the sciences of wheat genetics and plant pathology in the agripolitical context of the Australian wheat industry. In addition, long-term financial commitment from the wheat industry, co-operation with regional colleagues and the dedication of research and technical staff, which has spanned 70 years at The University of Sydney, have significantly contributed to the success of the national program. Thus the illustrations in this book reflect in some measure the philosophy of rust research at the PBI, and represent the culmination of experience provided by colleagues over several generations.

GUIDE TO GENE DESCRIPTIONS AND ILLUSTRATIONS

The following chapters provide a profile for each rust resistance gene in wheat and also triticale which may be infected by the rust pathogens of wheat. Illustrations of interactions of the resistance genes with pathotypes possessing corresponding avirulence genes are given where possible. In a few instances, especially for recently designated genes and for some of those with temporary designations, it was not possible to obtain illustrations. The following notes provide guidelines regarding the content and approaches adopted in formulating the profiles.

Synonyms

Resistance genes are often given a temporary name prior to formal designation for listing in the Catalogue of Gene Symbols for Wheat. The temporary designations are listed to provide continuity with the literature and/or laboratory usage.

Chromosome Location

This section provides the chromosome location and some relevant genetic linkage information. Gene locations in chromosomes 4A and 4B have been interchanged to conform with the revised genomic groupings as resolved by delegates of the Seventh International Wheat Genetics Symposium (Miller and Koebner, 1988), that is, the chromosome previously designated 4A (Sears' original designation IV) was redesignated 4B and that previously designated 4B (VIII) became 4A.

Low Infection Type

We have adopted the 0–4 description scales of Stakman *et al.* (1962) with minor modifications. Stripe rust responses were also converted to this scale (see Tables 1-1, 1-2).

For leaf rust and stem rust infection types (IT) 3^+ and 4 are interpreted as compatible; for stripe rust ITs 3 and 4 are interpreted as compatible.

Environmental Variability

All host:pathogen interactions are likely to be affected by environment. We have attempted to identify, and sometimes illustrate, significant environmental effects that characterise individual genes and so assist in their identification in germplasm that has not been subjected to genetic analysis.

Origin

The origins of some genes trace to a single source. In other cases when they are not clear, we have listed the stocks and references when first reported.

Pathogenic Variability

Comments under this heading are general. At the global level, there are major geographical areas within which pathogen populations seem to move relatively freely. Interchange between pathogen populations in those areas is less likely than interchange within areas. Nevertheless, such interchanges do occur. For example, Watson and de Sousa (1983) provided pathological and climatic evidence to demonstrate the likely wind dispersal of inoculum of *P. graminis* f. sp. *tritici* from southern Africa to Australia.

The influence of human activities in the transportation of pathogen genotypes is underestimated because of the lack of adequate phenotypic description to distinguish putative exotic forms from the local pathotypes. However, the exotic introduction of *P. striiformis* f. sp. *tritici* into Australia was concluded to be due to the movement of spores adherent to clothing on travellers from Europe (Wellings *et al.*, 1987).

Pathogenic variation between regions within major geographic areas can also be significant. Long *et al.* (1993) showed significant differences among *P. recondita* f. sp. *tritici* populations in various regions of the USA. These differences reflect the influences of geographical barriers, local inoculum survival between seasons and wheat genotype on the pathogen populations. Because rust pathogens usually propagate asexually, selection for a single avirulence/virulence allele causes parallel changes in the frequencies of all other virulence/avirulence alleles in the particular pathogen genotype or genotype group.

Whereas there is a vast literature coverage of pathogenic variability (e.g. Luig, 1983 for stem rust; Huerta-Espino, 1992 for leaf rust and stem rust; Stubbs and Fuchs, 1992 for stripe rust in Europe) we have limited comment to general observations.

Reference Stocks

This section lists the most appropriate genetic sources for particular genes, as far as possible with full literature referencing. The pedigree system adopted is that of Purdy *et al.* (1966) where slashes separate parents in crosses and the number of recurrent crosses to a particular parent is separated from the parent by an asterisk. An accession number for a genetic stock follows the name without punctuation. A comma is inserted when the accession number applies to a complex pedigree. For example, Centenario/6*Thatcher, R.L.6003 refers to a line carrying a gene from Centenario backcrossed five times to Thatcher. The derived line carries the accession number R.L.6003. If the comma were lacking, the number would apply to Thatcher.

The sources of various accession numbers are as follows: C followed by a year suffix, for example C91, refers to The University of Sydney Cytogenetics Register; C.I. and P.I. are USA accession numbers; R.L. are Agriculture Canada Research Station, Winnipeg, numbers; W relates to The University of Sydney Wheat Accession Register; and WYR refers to accessions held in the Vavilov Institute Collection, St Petersburg, Russia.

Stock listings are divided into: near-isogenic lines (i:); chromosome substitution lines (s:); common wheat cultivar stocks (v:); tetraploid wheat (tv:); diploid species accessions (dv:); alien substitution lines (su:); and alien addition lines (ad:).

Near-isogenic lines are often produced by backcrossing after which they can be compared with the recurrent parent, or if a common recurrent parent is involved, they can be compared with each other. They can also be produced by the extraction of contrasting homozygous genotypes following several generations of self-pollination and recurrent selection of heterozygotes, or may similarly be produced by induced mutation.

Intercultivar substitution lines are generated by the replacement of particular chromosome pairs in one cultivar by their homologues from another. This is achieved through the use of monosomy to prevent recombination of the target chromosome during backcrossing, followed by selfing, and selection of euploid lines.

Cultivars are listed from simple genotypes to complex genotypes in ascending allele order and gene number.

Alien chromosome substitution lines involve the replacement of related wheat chromosome pairs by alien chromosomes. Disomic alien substitution lines have 2n = 42.

Alien addition lines involve the addition of chromosomes from related species to the wheat genome. Disomic alien addition lines have 2n = 44.

Alien translocation stocks are given in two ways. 3D/3Ag represents a translocation involving a wheat 3D chromosome and an uncharacterised 3Ag homoeologue from *Thinopyrum*. Where the translocation was cytologically characterised it may be written 1DS.1DL–7Ai#2L with the point representing the centromere and the translocation breakpoint in the long arm represented by a dash. In this example, a segment of the long arm of the 7Ai#2 chromosome from *Th. intermedium* is distally located. Where additional catalogued genes in a particular stock are known to be present, these are also listed, but the referencing relates only to the gene under discussion. Catalogued genes of no practical interest, such as *Sr18* and genes not formally catalogued, are usually excluded in genotype lists. A number of genes with temporary designations and genetic stocks with uncatalogued genes are listed at the end of each disease section.

Source Stocks

Source stocks cover a wider range of genetic material that may be of more interest to breeders than geneticists and pathologists. Where the numbers of stocks are large, they are listed under known or presumed countries of origin in similar order to reference stocks. Literature citations for this section are intentionally incomplete in order to avoid unnecessary complexity and the fact that much of the information is common knowledge among laboratories specialising in cereal rust research and rust resistance breeding. References covering several, but not necessarily all, genotypes in a particular section may be given before the genotype listings.

Use in Agriculture

This section attempts to identify how, when and where particular genes were used in agriculture, and may include important historic information that is not noted in other sections.

Plate

The plates for each gene attempt to give insight to certain features associated with the responses of lines possessing the particular gene. The identification of genes in germplasm collections and breeding populations is based not only on the manipulations of avirulence and virulence in the pathogen, but also on the actual infection type expression, including its stability or variability over host and pathogen genotypes and over environments.

Plates are constructed to show contrasts between low and high infection types, between avirulent and virulent pathotypes and between environments that permit the expression of resistance and those that do not. Low infection types and avirulent cultures are presented firstly in each plate. Host genotypes within plates are listed left to right (L to R) (and centre, C, where appropriate) and pathotypes are described by the Australian designation systems except when photographs were supplied from other laboratories when local nomenclature is used. Actual cultures used to produce the photographs are provided only when the pathogenicity phenotype is not available, or to avoid ambiguity. A listing of culture accession numbers is given in Appendix II. In many instances the relevant pathogen phenotype is listed to assist with interpretation of the plate. For example (*P1*) and (*p1*) describe avirulence and virulence, respectively, for seedlings with gene *1* for the relevant disease.

Where adult plant reactions are illustrated, the leaves chosen were either the uppermost (flag = F) leaf, or one (F-1) or two (F-2) leaves below it.

CHAPTER 2

The Genes for Resistance to Leaf Rust in Wheat and Triticale:

CATALOGUED LEAF RUST RESISTANCE GENES

Lr1 (Ausemus *et al.*, 1946) (Plate 2-1)

Chromosome Location

5D (McIntosh *et al.*,1965); 5DL (McIntosh and Baker, 1970*a*). In one study *Lr1* mapped 47 cM from the centromere; in another study it was genetically independent of the centromere (Knott and McIntosh, 1978).

Low Infection Type

Predominantly 0;⁼ to ;. Huerta-Espino (1992) recorded some cultures producing IT 2.

Environmental Variability

Low (Dyck and Johnson, 1983). Browder and Eversmeyer (1986) reported high infection types at very low temperatures, that is, 5°C. Two days at 20°C prior to 5°C conditions were sufficient to produce full resistance.

Origin

Common wheat; present in the Mains and Jackson (1926) differential Malakof. *Lr1* was designated by Ausemus *et al.* (1946) based on the earlier work of Mains *et al.* (1926).

Pathogenic Variability

P. recondita tritici populations are globally polymorphic with respect to host lines possessing *Lr1* (Huerta-Espino, 1992).

Reference Stocks

i: Centenario/6*Thatcher, R.L.6003 (Dyck and Samborski, 1968*b*); Malakof/6*Prelude, R.L.6028 (Dyck and Samborski, 1968*b*); Witchita*4/Malakof (Johnston and Heyne, 1964).
v: Centenario (Dyck and Samborski, 1968*b*); Malakof C.I.4898 (Johnston and Levine, 1955).

Source Stocks

Many (McIntosh, 1988*a*). Some examples:
Australia: Tarsa. Miling *Lr3a*. Hartog *Lr10 Lr13* (=Pavon 'S'). Suneca *Lr13*. Bindawarra *Lr20*.
CIMMYT: (See Singh and Rajaram, 1991; Singh, 1993). Sonora 64; Chicora 'S'. Roque *Lr3a Lr13*. Nuri 70 *Lr13*; Tobari 66 *Lr13*; Yecora 70 *Lr13*. Mahome 81 *Lr13 Lr16*. Noroeste 66 *Lr13 Lr17*. Torim 73 *Lr13 Lr17 Lr34*. Tonichi 81 *Lr13 Lr27 + Lr31*. Yecora 70 R *Lr13 Lr34*.
Europe: Halle 9H37.
Indian Subcontinent: Khushal 69; Moti; Sharbati Sonora. Pak-20 *Lr3a*. Lu26 *Lr13*; UP301 *Lr13*. NP846 *Lr13 Lr17*. See Singh and Gupta (1991).
North America: Daws; Dirkwin; Hyslop; McDermid; Norco. Coker 9323 *Lr2a Lr9*. Coker 9766 *Lr2a Lr9 Lr11*. Profit 75 *Lr2c*. Oslo *Lr2c Lr10*. Newton *Lr3a*; Plainsman V *Lr3a*. Rosen *Lr3a Lr10*. Victory *Lr3a Lr10 Lr14a*. Karl *Lr3a Lr10 Lr24*. Mit *Lr10*. Thunderbird *Lr24*.

Use in Agriculture

Lr1 probably continues to play a role in gene combinations. It is highly effective against avirulent pathotypes but, due to the occurrence of virulence in most wheat growing areas, is of little use when deployed alone.

Lr2 (Ausemus *et al.*, 1946)

Lr2 is a complex locus comprising at least three resistance alleles which can be distinguished with different pathotypes and/or by different infection types. Inheritance of pathogenicity in the pathogen relating to alleles *Lr2a*, *Lr2b* and *Lr2c* appears to involve more than a single gene

A   B

PLATE 2-1. Lr1

Seedling leaves of (L to R): Malakof, Tc + *Lr1*, Pavon 76, Norka and Thatcher; infected with A. pt. 162-1,2 [*P1*] and B. pt. 104-2,3,6,(7) [*p1*]. Virulence for *Lr1* [pt. 104-2,3,6,(7)] exposes the low responses conferred by genes *Lr13* in Pavon 76 and *Lr20* in Norka.

(Dyck and Samborski, 1974; Kolmer, 1992*a*). The gene currently designated *Lr15* may be a further allele (McIntosh and Baker, 1968; RA McIntosh, unpublished 1976). *Lr2a* is the most important of the designated *Lr2* alleles because it confers the widest array of resistance and the lowest infection types to avirulent cultures.

Chromosome Location

2DS (Luig and McIntosh, 1968). *Lr2* was mapped 38 cM from the centromere (McIntosh and Baker, 1968), 10 cM from and probably distal to gene *C* for compact spikes (Luig and McIntosh, 1968) and 1 cM from *Sr6* (McIntosh and Baker, 1968).

Lr2a (Dyck and Samborski, 1974)   (Plate 2-2)

Synonym

Lr2 (Ausemus *et al.*, 1946; Dyck and Samborski, 1968*b*).

Low Infection Type

Depends on pathogen culture and host genetic background (Dyck and Samborski, 1974), varying from $0;^=$ to 2 or X^+.

Environmental Variability

Medium (Browder, 1980). Low infection type response varied with pathogen isolate and temperature (Dyck and Johnson, 1983). *Lr2a* is slightly more effective at higher temperatures.

Origin

Common wheat; present in Webster C.I.3780 (Johnston and Levine, 1955) and some plants of Mediterranean C.I.3332 (Singh and McIntosh, 1985). In North American studies, Mediterranean always responded similarly to Democrat (*Lr3a*), whereas in Australian and European studies, these wheats differed with some cultures. Singh and McIntosh (1985) suggested that the Australian source of Mediterranean W1728 (*Lr2a* and *Lr3a*) was probably selected to show an array of phenotypes different from either Webster (*Lr2a*) or Democrat (*Lr3a*).

Pathogenic Variability

P. recondita tritici populations tend to be polymorphic worldwide (Huerta-Espino, 1992). *Lr2a* is effective against predominant pathotypes in eastern Australia (Park and Wellings, 1992).

Reference Stocks

i: Webster/6*Thatcher, R.L.6016 (Dyck and Samborski, 1974); Prelude*6/Webster, R.L.6018 (Dyck and Samborski, 1974); Red Bobs*6/Webster, R.L.6017 (Dyck and Samborski, 1974); Wichita*4/Webster (Johnston and Heyne, 1964).

v: Webster C.I.3780 (Johnston and Levine, 1955). Some plants of Mediterranean C.I.3332 have *Lr2a* and *Lr3a*. These were accessioned as W1728 in Australia (Singh and McIntosh, 1985). In the international culture survey undertaken by Huerta-Espino (1992) the response of his Mediterranean stock was highly correlated with the predicted combination of *Lr2a* and *Lr3a*. However, there were exceptions, some of which were probably misclassifications or errors, but of particular note were four cultures from Chile recorded as avirulent for *Lr2a* and virulent on Mediterranean.

Source Stocks

With the exception of North America, this gene is not common.

Australia: Festiguay (Luig and McIntosh, 1968).

North America: Eureka C.I.17738; Voyager. Coker 9323 *Lr1 Lr9*. Coker 9766 *Lr1 Lr9 Lr11*. Guard *Lr3a Lr10*. Coker 762 *Lr9 Lr11*. Alex *Lr10*; Erik *Lr10*; Florida 303 *Lr10*; James *Lr10*; Len *Lr10*; Waldron *Lr10*. Compton *Lr10 Lr11*; Florida 302 *Lr10 Lr11*. Butte *Lr10 Lr13*. Marshall *Lr10 Lr13 Lr34*. Hickory *Lr18*. See Browder (1980) and McVey (1989).

Use in Agriculture

Limited value if used alone, but could be a useful component of multiple gene resistances.

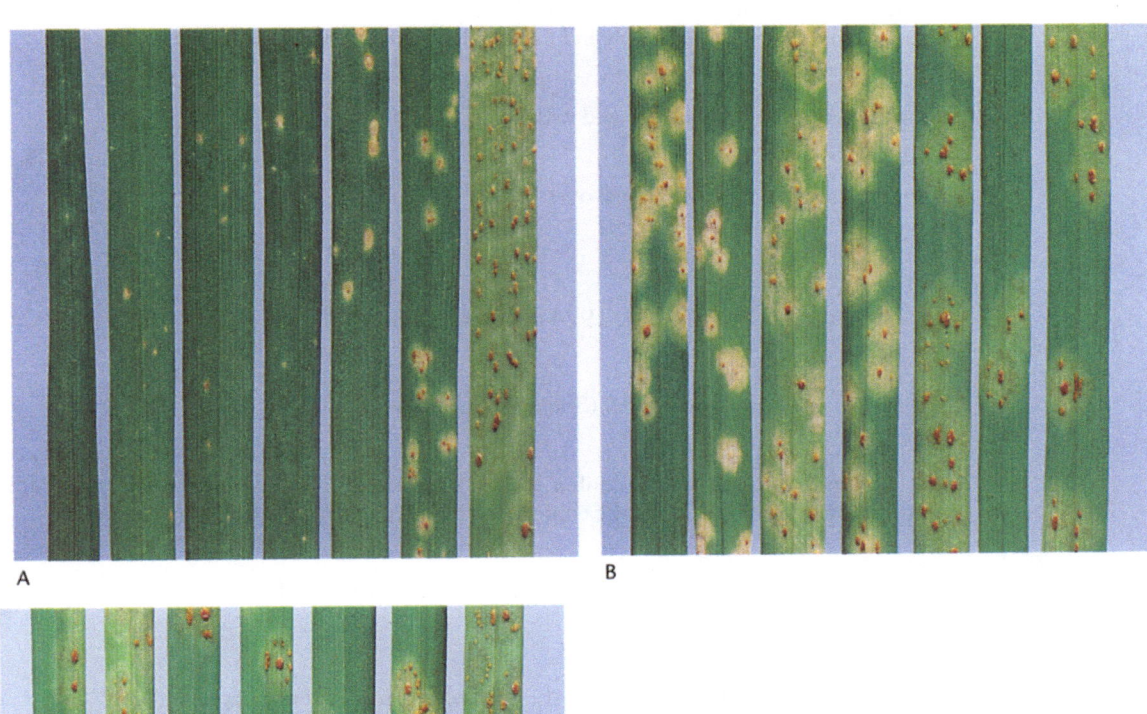

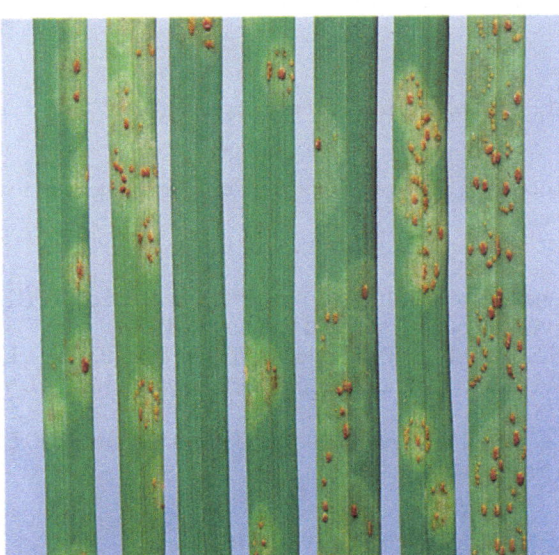

PLATE 2-2. *Lr2a, Lr2b* and *Lr2c*
Seedling leaves of (L to R): Webster, Tc + *Lr2a*, Carina, Tc + *Lr2b*, Loros, Tc + *Lr2c* and Thatcher; infected with A. pt. 53-1,(6),(7),10,11, B. pt. 104-2,3,6,(7) and C. pt. 10-1,2,3,4 and incubated at 22°C. The first culture is avirulent for all three *Lr2* alleles and produces very low infection types on wheats with *Lr2a* and *Lr2b*. The second, more typically Australian culture, produces intermediate infection types with *Lr2a* and *Lr2b*. With such cultures plants possessing *Lr2a* are usually more resistant than those with *Lr2b* and plants with *Lr2c* are susceptible. Pt. 10-1,2,3,4 is virulent for all three alleles.

Lr2b (Dyck and Samborski, 1974)   (Plate 2-2)

Synonym

Lr2² (Soliman *et al.*, 1964).

Low Infection Type

The pattern of response is similar to that conferred by *Lr2a*. Infection types range from 0; but are usually higher than those conferred by *Lr2a*.

Environmental Variability

Medium (Browder, 1980). Similar to *Lr2a* (Dyck and Johnson, 1983).

Origin

Common wheat. Carina and its derivatives are the only wheats reported to carry this gene but it could be confused with *Lr2a*.

Pathogenic Variability

Patterns of response are similar to *Lr2a*.

Reference Stocks

i: Thatcher*6/Carina, R.L.6019 (Dyck and Samborski, 1974); Prelude*6/Carina, R.L.6021(Dyck and Samborski, 1974); Red Bobs*6/Carina, R.L.6020 (Dyck and Samborski, 1974); Wichita*4/Carina (Johnston and Heyne, 1964).·
v: Carina *LrB* C.I.3756 (Johnston and Levine, 1955; Dyck and Samborski, 1968*b*).

Source Stocks

None known, but could be confused with *Lr2a*.

Use in Agriculture

Would seem to have no advantage over *Lr2a*.

Lr2c (Dyck and Samborski, 1974)   (Plate 2-2)

Synonym

Lr2³ (cv. Brevit), *Lr2⁴* (cv. Loros) (Soliman *et al.*, 1964).

Low Infection Type

Ranges from 0; to 2⁺ or X. *Lr2c* confers low ITs only when *Lr2a* and *Lr2b* are low and is often high when the others are low. *Lr2c* confers resistance to only one Australian pathotype.

Environmental Variability

Moderate (Browder, 1980). Temperatures from 10°C to 25°C had little effect on low infection type (Dyck and Johnson, 1983).

Origin

Common wheat. Brevit and Loros are the only wheats reported to carry this allele.

Pathogenic Variability

Both avirulence and virulence for *Lr2c* occurs in most geographic regions. Avirulence in Australia is limited to a single pathotype.

Reference Stocks

i: Thatcher*4/Brevit, R.L.6022 (Dyck and Samborski, 1974); Prelude*5/Brevit, R.L.6024 (Dyck and Samborski, 1974); Red Bobs*6/Brevit, R.L.6023 (Dyck and Samborski, 1974); Thatcher*6/Loros, R.L.6025 (Dyck and Samborski, 1974); Prelude*6/Loros, R.L.6027 (Dyck and Samborski, 1974); Red Bobs*6/Loros, R.L.6026 (Dyck and Samborski, 1974); Wichita*4/Loros (Johnston and Heyne, 1964).
v: Loros C.I.3779 (Johnston and Levine, 1955; Dyck and Samborski, 1968*b*). Brevit *LrB* C.I.3778 (Johnston and Levine, 1955; Dyck and Samborski, 1968*b*).

Source Stocks

North America: Profit 75 *Lr1*. Probrand 715 *Lr1 Lr10*; World Seeds 25 *Lr1 Lr10*. See Statler (1984).

Use in Agriculture

Very limited usefulness (Dyck and Samborski, 1974). *Lr2a* will always be preferred because of its wider range of effectiveness.

Lr3 (Ausemus *et al.*, 1946)

Three resistance alleles have been described at or near the *Lr3* locus. Alleles *Lr3bg* (cv. Báge) and *Lr3ka* (cv. Klein Aniversario) are temporary symbols that have not been converted to standard designations.

Chromosome Location

6B (Heyne and Livers, 1953); 6BL (RA McIntosh, unpublished 1970). Luig (1964 and unpublished, 1963) was unable to obtain recombination between *Lr3* and *Sr11*. Because Sears (1966) obtained 45% recombination between *Sr11* and the centromere, *Lr3* must also be distally located in chromosome 6BL. *Lr3a*, assigned by Browder (1980), is commonly referred to as *Lr3*.

Lr3a (Browder, 1980)   (Plate 2-3)

Synonym

Lr3 (Dyck and Johnson, 1983; McIntosh, 1988a).

Low Infection Type

0;$^=$ to X$^-$. *Lr3a* is usually incompletely dominant and may display disturbed genetic ratios (Luig, 1964).

Environmental Variability

Low (Browder, 1980). Dyck and Johnson (1983) found that the low infection type showed some variability with decreasing temperatures in respect to one pathogen culture.

Origin

Common wheat; present in the Johnston and Levine (1955) differentials Democrat C.I.3384 and Mediterranean C.I.3332.

Pathogenic Variability

Pathogen populations are polymorphic in most geographic areas, but because this gene is present in cultivated wheats in many geographic regions the frequencies of virulence are often very high. In the international survey of Huerta-Espino (1992) virulence on Mediterranean was highly correlated with predicted virulence for the combination of *Lr2a* + *Lr3a* although there were exceptions. Virulence on the Wichita + *Lr3a* (ex Mediterranean) and Thatcher + *Lr3a* (ex Democrat) lines generally corresponded with virulence on Democrat.

Reference Stocks

i: Democrat/6*Thatcher, R.L.6002 (Dyck and Samborski, 1968a); Wichita*4/Mediterranean (Johnston and Heyne, 1964); Sinvalocho/ 6*Prelude, R.L.6029 (Haggag and Dyck, 1973).
v: Democrat C.I.3384; Mediterranean C.I.3332. Some stocks of Mediterranean (e.g. W1728) also carry *Lr2a* (Singh and McIntosh, 1985). Sinvalocho C.I.12096 (Haggag and Dyck, 1973).

Source Stocks

This allele is widely dispersed in both winter and spring wheats from many countries (see McIntosh, 1988a). Selected examples are listed below. Because of similarity, some wheats reported to have *Lr3a* may carry *Lr3bg*. For example, on the basis that Mentana produced IT 23 or X with some cultures virulent on Democrat, NH Luig (*pers. comm.* 1980) believed that Mentana carried *Lr3bg*. If so, then various Mentana derivatives would be expected to carry this allele rather than *Lr3a*.
Australia: Cook; Diaz; Eradu. Gutha *Lr10*. Reeves *Lr17*. Angas *Lr20*. Wialki *Lr23*.
CIMMYT: Galvez 87 *Lr10 Lr13* (Singh and Rajaram, 1991).
Europe: Hana; Mentana; Mironovskaya 264; Mironovskaya 808; Skorospelka 3b. Solaris *Lr26* (Bartos *et al.*, 1983). Bezostaya 1 *Lr34* (Bartos *et al.*, 1969b).

Indian Subcontinent: Pak 20 *Lr1*; CPAN 1235 *Lr1* (Singh and Gupta, 1991).
Japan: Shirahada.
North America: Bennett; Gage; Homestead; Kawvale; Lancota; McDermid; Ottawa; Pawnee; Ponca; Quantum 542; Rita; Rose; Shawnee; Stallion; Turbo; Warrier. Kaizer *Lr1*; Newton *Lr1*; Plainsman V *Lr1*. Pioneer 2551 *Lr1 Lr10*. Karl *Lr1 Lr10 Lr24*. Guard *Lr2a Lr10*. Cardinal *Lr10*; Hawk *Lr10*; Len *Lr10*; Nell *Lr10*; TAM105 *Lr10*; TAM107 *Lr10*. Coker 916 *Lr10 Lr11*. Coker 983 *Lr10 Lr11 Lr18*. Vona *Lr10 Lr14a*. Wared *Lr10 Lr20*. Mustang *Lr10 Lr24*. Coker 9134 *Lr11*; Pioneer 2548 *Lr11*; Saluda *Lr11*. Pioneer 2157 *Lr11 Lr14a*. Heil *Lr14a*; Scout *Lr14a*; Scout 66 *Lr14a*. Redland *Lr16*. Payne *Lr24*.

Use in Agriculture

Widely dispersed and probably has a minor role in gene combinations.

Lr3bg (Haggag and Dyck, 1973)   (Plate 2-4)

Lr3bg was designated as a distinctive allele by Haggag and Dyck (1973) on the basis that the near-isogenic line Báge/8*Thatcher, R.L.6042 gave resistance to isolates of race 15 (IT 2), a race virulent on the comparable near-isogenic derivative of Democrat (R.L.6002). In other respects R.L.6042 responded similarly to R.L.6002 (i.e. low IT 0; to 0;1). *Lr3bg* has not been confirmed in other wheats but obviously this allele is very difficult to distinguish from *Lr3a*. See comment relating to Mentana in 'Source Stocks' section for *Lr3a*. Báge carried a second gene conferring IT2 (possibly *Lr14b*) that was inherited independently of the *Lr3* locus (Haggag and Dyck, 1973; Haggag *et al.*, 1973).

Low Infection Type

;, 1^+, $2^=$ to 2C or X. The lowest responses are obtained only when *Lr3a* is effective.

Environmental Variability

Low (Dyck and Johnson, 1983).

Origin

Common wheat, present in cv. Báge P.I.193910 (Haggag and Dyck, 1973).

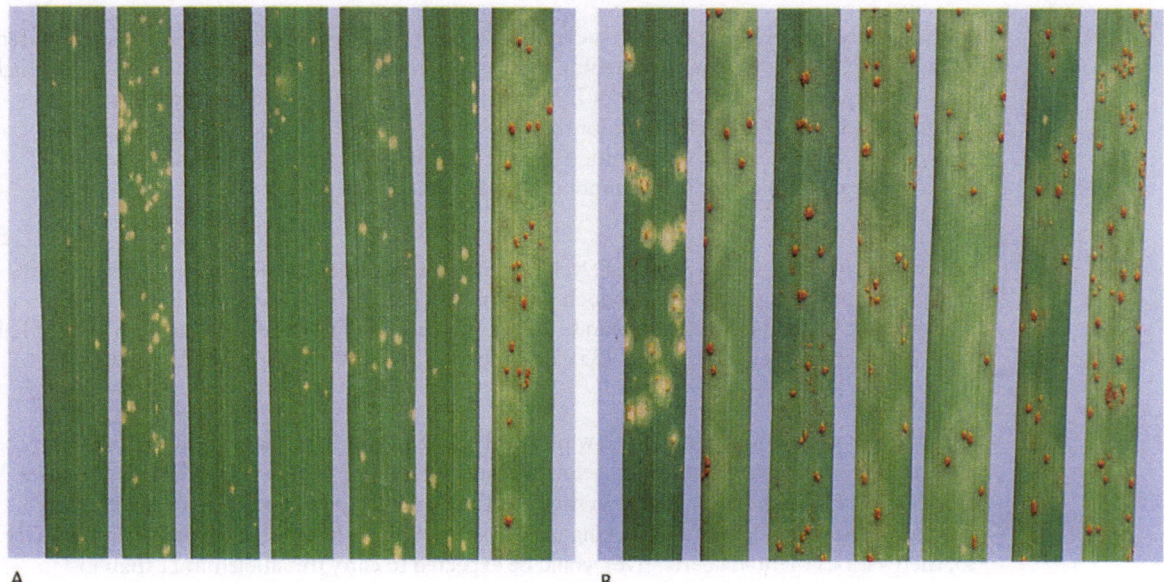

A   B

PLATE 2-3. Lr3a
Seedling leaves of (L to R): Mediterranean W1728, Mediterranean ex Canada, Mentana, Tc + *Lr3bg*, Democrat, Tc + *Lr3a* and Thatcher; infected with A. pt. 10-1,2,3,4 and B. pt. 104-2,3,6,(7). Note the resistance to pt. 104-2,3,6,(7) in Mediterranean W1728 which carries *Lr2a* and *Lr3a*. Tc + *Lr3a* and Tc + *Lr3bg* are not distinguished in this comparison.

Pathogenic Variability

This allele produces IT 0; to cultures avirulent on seedlings with *Lr3a*, and produces intermediate reactions with certain additional cultures. Huerta-Espino (1992) recorded low frequencies of cultures virulent for *Lr3a* but avirulent on Thatcher + *Lr3bg* from all parts of the world except South America.

Reference Stocks

i: Báge/8*Thatcher, R.L.6042 (Haggag and Dyck, 1973).

Source Stocks

Unknown; see note relating to Mentana under *Lr3a*.

Use in Agriculture

Probably similar to *Lr3a*. The value of the additional resistance conferred by *Lr3bg* relative to *Lr3a* is probably limited.

Lr3ka (Haggag and Dyck, 1973) (Plate 2-5)

Low Infection Type

0;1 to 2C.

Environmental Variability

Low. Assumed to be influenced by pathogen culture as Dyck and Johnson (1983) reported temperature sensitive responses with one of their cultures.

Origin

Common wheat; present in Klein Aniversario (Haggag and Dyck, 1973) and Klein Titan (RA McIntosh, unpublished, 1975).

Pathogenic Variability

Lr3ka confers a wider range of resistance than other *Lr3* alleles. Huerta-Espino (1992) found virulence in most geographic areas except South Africa and Mexico (see also Singh, 1991).

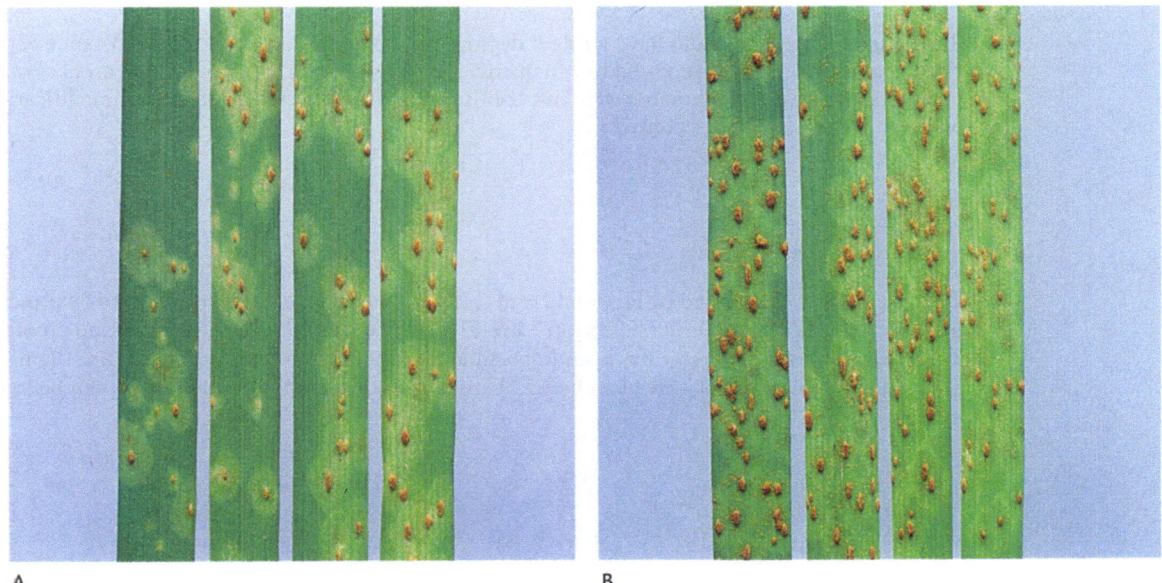

A   B

PLATE 2-4. Lr3bg

Seedling leaves of (L to R): Mentana, Tc + *Lr3bg*, Tc + *Lr3a* and Thatcher; infected with A. pt. 122-1,2,(3) and B. pt. 104-2,3,6,(7). The similar mixed infection types produced by Mentana and Tc + *Lr3bg* in A. is the response we associate with *Lr3bg* and supports the postulation that Mentana carries *Lr3bg*.

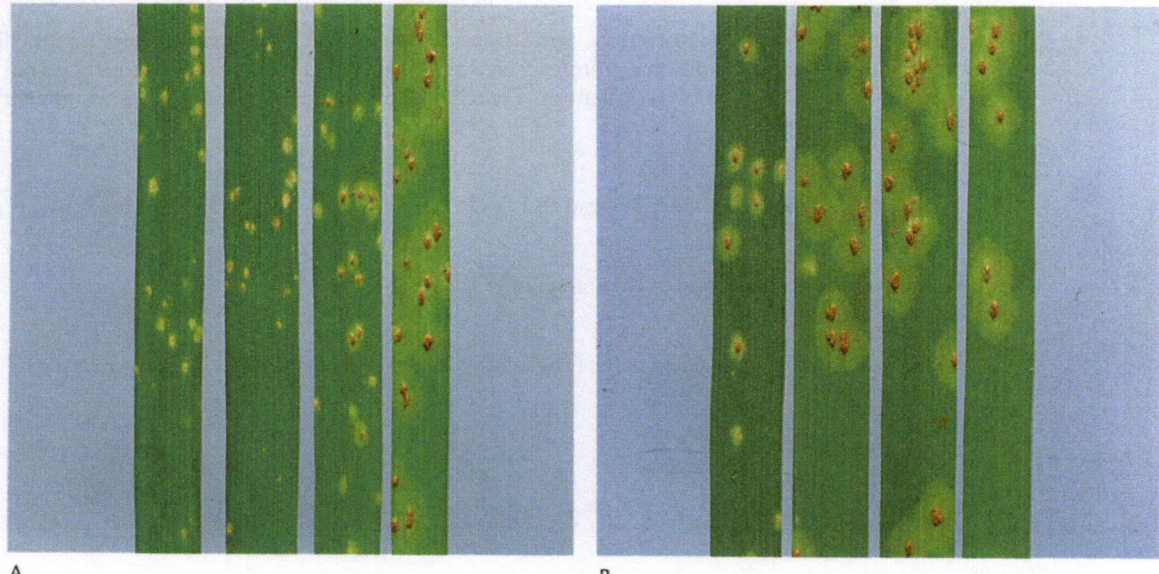

A   B

PLATE 2-5. Lr3ka

Seedling leaves of (L to R): Klein Aniversario, Klein Titan, Tc + *Lr3ka* and Thatcher; infected with A. pt. 162-1,2 [*P3kap3a*] and B. pt. 135-1,2,3,4,5 [*p3kaP3a*]. The lower infection types produced by the first two wheats in A., and probably in B., indicate the effects of additional resistance genes.

Virulence is rare in Australia. The results of Haggag and Dyck (1973, Table 1) suggest that *Lr3ka* produces lowest ITs with pathotypes which are also avirulent on *Lr3a*. However, this relationship is not apparent in the data of Huerta-Espino (1992). Our results indicate that the response conferred by *Lr3ka* is independent of avirulence or virulence for *Lr3a*.

Reference Stocks

i: Klein Aniversario/6*Thatcher, R.L.6007 (Haggag and Dyck, 1973); Klein Aniversario/6*Prelude, R.L.6030 (Haggag and Dyck, 1973).
v: Klein Aniversario and Klein Titan are the only wheats in which the presence of *Lr3ka* has been confirmed. Both wheats possess additional seedling and probably adult plant resistance factors.

Use in Agriculture

Rarely used. This gene could have a role if deployed in combination with other resistance genes. Haggag and Dyck (1973) reported low frequencies of recombination between *Lr3a* and *Lr3ka*. However, putative recombinants were not confirmed and the possibility of genetic modification by other genes was not discounted.

Lr4, Lr5, Lr6, Lr7, Lr8

These symbols were applied by Fitzgerald *et al.* (1957) to a group of genes thought to be present in Waban C.I.12992 = Purdue Selection 3369-61-1-10. Because the genes were not incorporated into separate genetic stocks it has been impossible for other researchers to characterise them. As a consequence, the symbols were abandoned. A discussion of their possible identities can be found in Browder (1980).

Lr9 (Soliman *et al.*, 1963)   (Plate 2-6)

Chromosome Location

6B (Sears, 1961; McIntosh *et al.*, 1965); 6BS (Sears, 1972*a*); 6BL (Sears, 1966). These reports presumably both relate to Translocation 47 which became known as Transfer (Sears, 1961). *Lr9* is difficult to combine with *Lr3* and *Sr11* but ER Sears recovered an *Lr9* + *Sr11* recombinant line (available as PBI Cobbitty Accession C66.10).

Low Infection Type

$0;^=$ to $0;$, occassionally 1^+. Samborski (1963) reported a culture producing IT 1^+ on plants with *Lr9*. When this culture was selfed on the alternate host, *Thalictrum flavum*, there was segregation of cultures with ITs $0;$; 1^+ and 3^+ indicating that the parent culture was heterozygous. A single pustule was found as a glasshouse mutant at The University of Sydney. This was increased and established as a reference culture (79-L-4) with an intermediate level of pathogenicity. In the absence of complete virulence this is a valuable culture for confirming the presence of *Lr9* in host lines (see Plate 2-6).

Environmental Variability

Low (Browder 1980; Dyck and Johnson, 1983).

Origin

Lr9 was transferred to Chinese Spring wheat from *Triticum umbellulatum* (Sears, 1956). Translocation 47 became known as Transfer.

Pathogenic Variability

Virulence for *Lr9* was found in the USA in 1971, four years after its use in soft red winter wheats (Shaner *et al.*, 1972). However, both Shaner *et al.* (1972) and H. Ohm (*pers. comm.* 1994) indicated that wheats with *Lr9* occupied less than 2% of the area. Increased areas of wheats with *Lr9* after 1971 probably assisted the significant increase in frequency of virulence for *Lr9* in the soft red wheat areas. Virulence was also reported in Brazil and Argentina. Huerta-Espino (1992) found virulence in isolates from Italy, Burundi and Pakistan. However, the overall frequency of virulence was very low.

Reference Stocks

i: Transfer which is near-isogenic to Chinese Spring; Thatcher*6/Transfer, R.L.6010; Wichita*4/Transfer (Johnston and Heyne, 1964).
v: Transfer.

Source Stocks

North America: Abe; Adder; Aim; Arthur 71; McNair 701; McNair 2203; Riley 67; Sullivan. Coker 9835 *Lr11*; Coker 9904 *Lr11*; Oasis *Lr11*. Terral *Lr11 Lr18 Lr24*. Coker 9733 *Lr24*; Coker 9877 *Lr24*.

Use in Agriculture

Lr9 has not been widely deployed despite its widespread effectiveness.

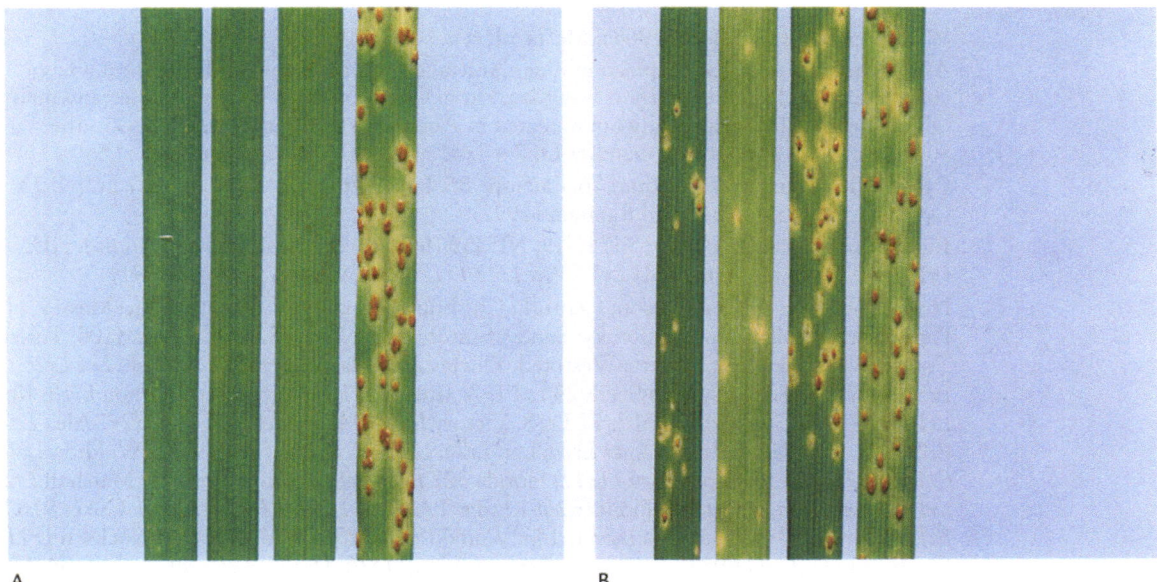

A   B

PLATE 2-6. *Lr9*
Seedling leaves of (L to R): Transfer, Arthur 71, Tc + *Lr9* and Thatcher; infected with A. pt. 162-1,2,3,4 and B. a mutant derivative of pt. 162-1,2,3,4 selected for IT 1^+ on seedlings of Transfer. Note the additional resistance of Transfer (no known additional seedling resistance gene) and Arthur 71 (probably *Lr13*) relative to the Thatcher derivative.

Lr10 (Choudhuri, 1958) (Plate 2-7)

Synonym

LrL (Anderson, 1961).

Chromosome Location

1A (Dyck and Kerber, 1971); 1AS (RA McIntosh, unpublished 1970). Howes (1986) reported linkage values of 3 cM for *Lr10 – Gli–A1* (gliadin), 4 cM for *Gli–A1 – Hg* (hairy glumes) and 6 cM for *Lr10–Hg*. In addition, *Pm3* for reaction to powdery mildew is also closely linked with *Hg*, at 1 cM (Briggle and Sears, 1966) and 5 cM (McIntosh and Baker, 1968).

Low Infection Type

; to 12. Some UK cultures (Dyck and Johnson, 1983) and some Indian cultures tend to give IT 2^{++} to 3. At least some of these responses are interpreted (probably incorrectly) in India as indicative of virulence.

Environmental Variability

Medium (Browder, 1980).

Origin

Bread wheat. *Lr10* was originally reported in Lee and Timstein. RA McIntosh and LE Browder (unpublished, 1969) found that this gene was widely distributed in older Australian wheats. It occurs frequently in USA and CIMMYT wheats.

Pathogenic Variability

Pathogenic polymorphisms for *Lr10* occur in North America and some variation is apparent in the UK, Europe, Africa, South and East Asia and South America. With one recent exception, *Lr10* has not been effective in Australasia since formal testing was initiated in the 1920s. It is possible that as a result of the earlier widespread cultivation of Federation and other wheats with *Lr10*, Australian pathogen populations evolved to uniform virulence before surveys began. The first pathotype [53-(6),(7),9,10,11] with avirulence for *Lr10* in New Zealand was found in 1987 (Park and Wellings, 1992).

Reference Stocks

i: Exchange/6*Thatcher, R.L.6004; Gabo/6*Thatcher, R.L.6143; Lee/6*Prelude, R.L.6031; Selkirk/6*Thatcher, R.L.6145; Timstein/6*Thatcher, R.L.6146. (PL Dyck, *pers. comm.* 1993). s: CS*5/Timstein 1A; CS*6/Kenya Farmer 1A. These lines can be considered near-isogenic to Chinese Spring.

Source Stocks

Widespread (see McIntosh, 1988*a*; McIntosh *et al.*, 1989, 1990).

Australia: *Lr10* is very widespread in older Australian wheats including Federation, Gabo (=Timstein) and Warden *Lr16*. It was passed from Gabo to Kenya Farmer and Lee and many other wheats. Although usually not detected in Australian work, it is present also in the Australian local differentials Gatcher *Lr27 + Lr31* and some plants of Timson *Lr17*.

CIMMYT: Nainari 60; Nacozari 76; Cucurpe 86. Pavon 76 *Lr1 Lr13*. Galvez 87 *Lr3a Lr13*. Opata 85 *Lr34*. See Singh and Rajaram (1991).

Indian Subcontinent: BW11; NI747-19; NI5439 (leaf rust differential used in India). HD2009 *Lr13*; HD2329 *Lr13*; Punjab 81 *Lr13*. Arz *Lr13 Lr17*. See Singh and Gupta (1991).

North America: Arthur; Carson; Centurk; Chisholm; Condello; Coteau; Holley; Massey; Norak; Pacer; Parker; Quantum 561; Rocky; Sandy; Sinton; Solar; Stoa; TAM101; TAM106; Tascosa; Tomahawk; Wakefield; Walera; Westbred. Bluejay *Lr1*; Klasic *Lr1*; Mit *Lr1*. Oslo *Lr1 Lr2c*; Proband 715 *Lr1 Lr2c*; World Seeds 25 *Lr1 Lr2c*. Rosen *Lr1 Lr3a*. Hunter *Lr1 Lr3a Lr18*. Karl *Lr1 Lr3a Lr24*. Coker 9803 *Lr1 Lr11 Lr18*. Victory *Lr1 Lr14a*. Blueboy II *Lr1 Lr24*. Alex *Lr2a*; Erik *Lr2a*; Florida 303 *Lr2a*; James *Lr2a*; Len *Lr2a*; Norsman *Lr2a*; Steele *Lr2a*; Waldron *Lr2a*. Guard *Lr2a Lr3a*. Compton *Lr2a Lr11*; Florida 302 *Lr2a Lr11*. Butte *Lr2a Lr13*; Marshall *Lr2a Lr13*. Hawk *Lr3a*; Nell *Lr3a*; Pioneer 2551 *Lr3a*; TAM105 *Lr3a*; TAM107 *Lr3a*. Coker 916 *Lr3a Lr11*. Gaines *Lr3a Lr13*. Vona *Lr3a Lr14a*. Wared *Lr3a Lr20*. Florida 201 *Lr11*; Jackson *Lr11*; Madison *Lr11*. Coker 9227 *Lr11 Lr18*. Exchange *Lr12 Lr16*. Houser *Lr13*. Era *Lr13 Lr34*. Selkirk *Lr14a Lr16*. Brule *Lr16*. Lee *Lr23*; Timstein *Lr23*. Norkan *Lr24*; Parker 76 *Lr24*.

Use in Agriculture

Lr10 is not widely effective on its own, but may play a role in gene combinations in most areas except Australia.

PLATE 2-7. *Lr10*
Seedling leaves of (L to R): Federation, Tc + *Lr10*, CS*5/Timstein 1A, Thatcher and Chinese Spring; infected with pt. 53-(6),(7),9,10,11 and incubated at 18°C. This is the first pathotype detected in Australasia with avirulence for *Lr10*. Because of its distinctive pathogenicity formula it is assumed to have an exotic origin.

Lr11 (Soliman *et al.*, 1958) (Plate 2-8)

Synonym

LrEG in cv. El Gaucho (Samborski and Dyck, 1976).

Chromosome Location

2A (Soliman *et al.*, 1964).

Low Infection Type

;, 1, 2, 3⁻,Y.

Environmental Variability

High (Williams and Johnston, 1965). More effective at lower temperatures (Johnston and Heyne, 1964). Background genotype also has an important influence on low infection type (Modawi *et al.*, 1985).

Origin

Common wheat; present in the Johnston and Levine (1955) differential, Hussar C.I.4843.

Pathogen Variability

P. recondita tritici populations are widely polymorphic for pathogenicity on seedlings with this gene. Virulence for *Lr11* increased in Texas from 1985 to 1990 (Marshall, 1992), in the USA generally in 1988–1990 (Long *et al.*, 1992) and in Canada from 1990 to 1991 (Kolmer, 1991, 1993). Virulence was less frequent in Mexico (Singh, 1991) and South Africa (Pretorius and Le Roux, 1988). Although Australian cultures show variability on seedlings of Hussar and Tc + *Lr11* ranging from IT 2⁻ to IT 3, tests of adult plants of Tc + *Lr11* with those cultures producing the highest seedling responses showed that such cultures were avirulent, while Thatcher remained susceptible (RF Park and RA McIntosh, unpublished 1992). However, two cultures isolated from New Zealand (pts 53-(6),(7),9,10,11 and 122-1,3,4,6,7) produced IT 3⁺ on seedlings of both Hussar and Tc + *Lr11* at both low and high temperatures. The latter culture also produced infection type 3⁺ on adult plants of Tc + *Lr11*. Adult plants of Tc + *Lr11* are resistant to pt. 53-(6),(7),9,10,11 because this culture is avirulent for *Lr22b* (RF Park, unpublished 1993).

In the survey of Huerta-Espino (1992) moderate to high frequencies of avirulence were recorded for samples collected in Pakistan, Turkey, south-western Europe, the African continent and China. High frequencies of virulence occurred among samples from Nepal, Romania, Bulgaria and much of South America. Schafer and Long (1988) described the probable role of Coker 983 in supporting pathotypes with virulence for *Lr11*.

Reference Stocks

i: Thatcher*6/El Gaucho, R.L.6048 (Samborski and Dyck, 1976); Thatcher*6/Hussar, R.L.6053 (Dyck and Johnson, 1983); Wichita*6/Hussar, KS7110704 (Browder, 1973*a*).

v: Hussar C.I.4834 (Johnston and Levine, 1955).

Source Stocks

India: HS86 *Lr10 Lr13* (Singh and Gupta, 1991).

USA: Bulgaria 88 (Browder, 1973*a*); Hart C.I.15929 (Modawi *et al.*, 1985). Oasis *Lr9* (Browder, 1980). AP Roelfs (*pers. comm.* 1993) — Beau; Beeker; Coker 68-15; Pioneer 2543; Pioneer 2550; Sawyer; Southern Belle. Coker 9803 *Lr1 Lr10 Lr18*. Coker 762 *Lr2a Lr9*; Coker 86-25 *Lr2a Lr9*; Coker 9907 *Lr2a Lr9*. Compton *Lr2a Lr10*; Florida 302 *Lr2a Lr10*. Coker 86-32 *Lr2a Lr18*. Coker 9134 *Lr3a*; Pioneer 2548 *Lr3a*; Saluda *Lr3a*. Coker 916 *Lr3a Lr10*. Coker 983 *Lr3a Lr10 Lr18*. Pioneer 2157 *Lr3a Lr14a*. Coker 9835 *Lr9*; Coker 9904 *Lr9*. Florida 201 *Lr10*; Jackson *Lr10*; Madison *Lr10*. Coker 9227 *Lr10 Lr18*.

Use in Agriculture

Apart from the soft red winter cultivars of the USA this gene is not widely used in agriculture. Browder (1973*a*) attributed the slow rusting characteristics of Bulgaria 88 (Caldwell *et al.*, 1970) to the action of *Lr11*. Because of difficulties in identification by the gene matching method, *Lr11* could be present in a wider array of wheats than has been reported. On the other hand, certain reports of the presence of *Lr11* are not correct. *Lr11* could have potential use in gene combinations.

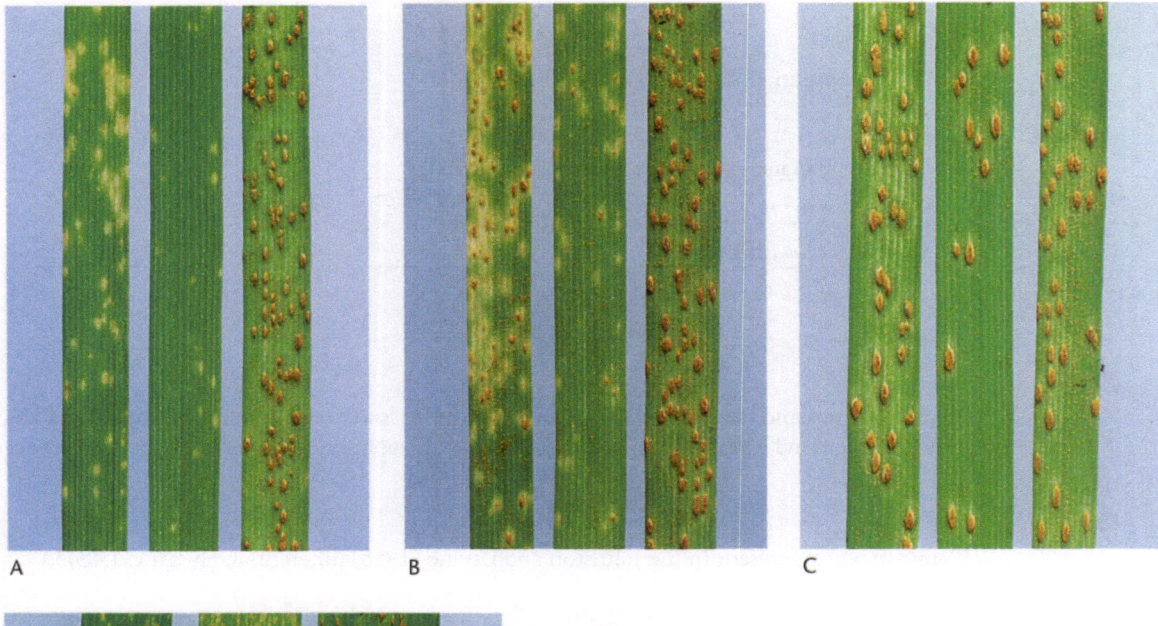

A   B   C

D

PLATE 2-8. *Lr11*

A. to C. Seedling leaves of (L to R): Hussar, Tc + *Lr11* and Thatcher; infected with A. pt. 64-11, B. pt. 53-1,(6),(7),10,11 and C. pt. 53-(6),(7),9,10,11 and incubated at 17°C. Note the three levels of response for Hussar and Tc + *Lr11*. On the basis that standard race 53 should be virulent on seedlings of Hussar (Johnston and Levine, 1955), pt. 53-1,(6),(7),10,11 is incorrectly designated.

D. Adult responses of flag-1 leaves of (L to R): Tc + *Lr11*, Tc + *LrEG* (from El Gaucho) and Thatcher; infected with pt. 76-0. Note the similar low responses of the two Thatcher derivatives.

Lr12 (Dyck *et al.*, 1966) (Plate 2-9)

Adult plant resistance

Chromosome Location

4B (McIntosh and Baker, 1966; Dyck and Kerber, 1971). This gene has not been further mapped.

Low Infection Type

Flag leaf symptoms IT 2 (Dyck *et al.*, 1966); IT ;1 (RF Park and RA McIntosh, unpublished 1991).

Environmental Variability

Presumed to be stable, but no information is available.

Origin

Common wheat cultivars Chinese Spring and Exchange.

Pathogenic Variability

Largely unknown as pathogenicity surveys rarely involve adult plant resistances. Variability probably occurs in most geographic areas. Virulence occurs in a related group of pathotypes in Australia (Park and McIntosh, 1994). Caldwell (1968) described the progressive loss of resistance over ten years in USA soft red winter derivatives of Chinese Spring C.I.6223. Singh and Gupta (1992) reported that the Thatcher near-isogenic line with *Lr12* gave no resistance at two field sites in Mexico. Virulence also occurs in India (Sawhney, 1992).

Reference Stocks

i: Exchange/6*Thatcher, R.L.6011 (Dyck, 1991).
v: Exchange *Lr10 Lr16* C.I.12635 (Dyck *et al.*, 1966).

Source Stocks

Opal (PL Dyck, *pers. comm.* 1975). Uruguay *Lr1* W1064 (Athwal and Watson, 1957). Benito *Lr1 Lr2a Lr13* (Campbell and Czarnecki, 1981). Sturdy *Lr13 Lr34* (Dyck, 1991). Chinese Spring *Lr34* (Dyck, 1991). Adult plant resistance in the *T. timopheevii* derivatives C.I.12632 and C.I.12633 is assumed to be derived from Chinese Spring via line 2666A2-2-15-6-3 (Illinois No.1/Chinese C.I.6223) (Allard and Shands, 1954).

Use in Agriculture

Chinese Spring and Exchange have been used as parents in breeding programs in North America. For example, Cox (1991) pointed out that Chinese Spring C.I.6223 appeared in the pedigrees of 40 soft red winter wheats. The number of these wheats possessing *Lr12* is not known.

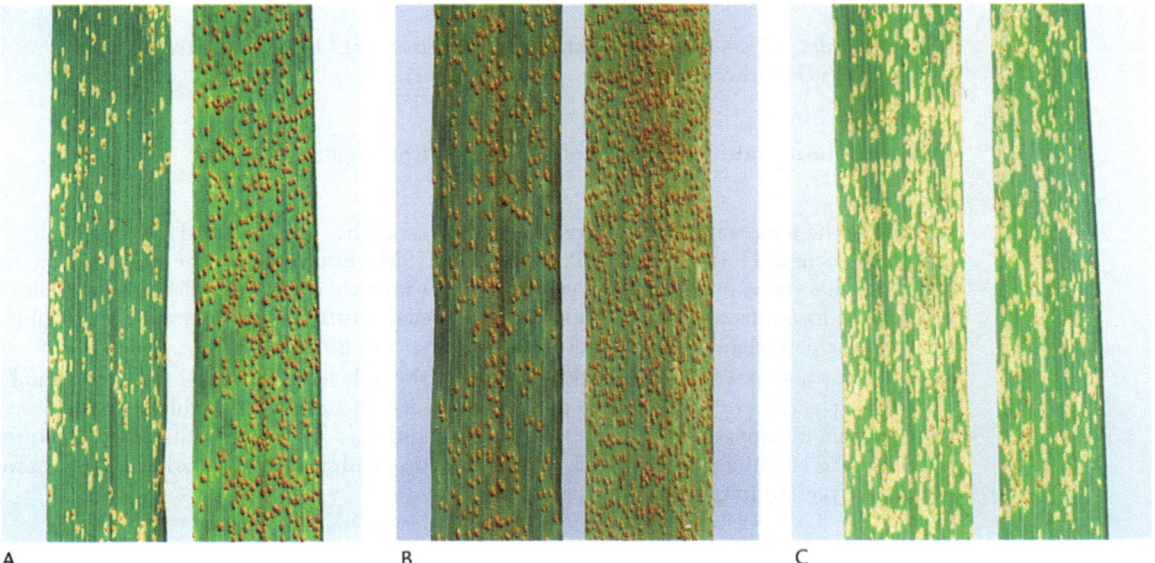

A   B   C

PLATE 2-9. *Lr12*

Adult leaves (flag-1) of (L to R): Tc + *Lr12* and Thatcher; infected with A. pt. 104-2,3,(6),(7),11, B. pt. 104-2,3,6,(7) and C. pt. 53-1,(6),(7),10,11. This shows that pt. 104-2,3,(6),(7),11 is avirulent for *Lr12* whereas pt. 104-2,3,6,(7) is virulent. The former pathotype with virulence for *Lr16* was originally assumed to be a mutant derivative of the latter. These results indicate that pt. 104-2,3,(6),(7),11 probably had a different origin. The pathogenicity of pt. 53-1,(6),(7),10,11 with respect to *Lr12* cannot be ascertained because gene *Lr22b* in Thatcher is effective and both Thatcher and Tc + *Lr12* give similar adult plant responses.

Rajaram *et al.* (1971*a*) reported that the Australian wheat line TR135, later named Timgalen, possessed a gene for adult plant resistance in common with C.I.12632, as well as possessing at least two genes for seedling resistance. The latter were subsequently identified as *Lr3a* for which Timgalen is genetically heterogeneous, and the complementary genes *Lr27 + Lr31*. Pathotypes that overcame the resistance of Timgalen in the field were virulent on seedlings with *Lr3a* and *Lr27 + Lr31*, and with such cultures no gene for adult plant resistance could be detected in Timgalen. Recent studies showed that the related pathotypes that predominated following the use of Timgalen in the north-eastern wheat-growing areas of Australia were also virulent on adult plants of Tc + *Lr12*, thus indicating that Timgalen could indeed carry *Lr12* (Park and McIntosh, 1994). The *Lr31* complementary factor is also located in chromosome 4B.

Lr13 (Dyck *et al.*, 1966)   (Plate 2-10)

Lr13 was originally reported as a gene for adult plant resistance although it was always clear that the onset of resistance occurred at a relatively early growth stage (Dyck *et al.*, 1966).

Chromosome Location

2BS (WM Hawthorn and RA McIntosh unpublished, 1981). *Lr13* is closely associated with the gene *Ne2m* for progressive necrosis. Indeed most wheats with *Ne2m* carry *Lr13*. All wheats with *Lr13* appear to carry *Ne2m*. Cultivar Kalyansona has been reported to carry *Ne2m*, but does not carry *Lr13*. Because the pedigree of Kalyansona involves Frontana, it is possible that a recombination event occurred between *Ne2m* and *Lr13*. Hawthorn (1984) reported a recombinant from a testcross, but the genotype was not confirmed. Singh and Gupta (1991) confirmed the close linkage of *Lr13* and *Ne2m*. *Lr13* is also very closely linked with *Lr23* (WM Hawthorn and RA McIntosh, unpublished 1981) and a group of Indian wheats that carry *Lr13* and *Lr23* in combination have been identified (E Gordon-Werner and RA McIntosh, unpublished 1989; Singh and Gupta, 1991). Because of the lack of pedigree information, it is not possible to determine the number of independent recombination events that were involved. Obviously, if the genes are closely linked, then once in coupling they will tend to inherit together. With the available *P. recondita* cultures in Australia it is quite difficult to study joint segregations involving *Lr13* and *Lr23*.

Low Infection Type

;1 through X to 3 on seedlings depending on pathogen culture, environmental conditions and host genetic background. Dyck *et al.* (1966) reported low infection types on adult plants of cv. Manitou ranging from 0;1$^-$ to 2^{++} depending on the particular culture.

Environmental Variability

High (Browder, 1980). Seedling resistance conferred by *Lr13* is more effective at higher temperatures (Hawthorn, 1984; Pretorius *et al.*, 1984).

Origin

Common wheats, particularly those of South American origin.

Pathogenic Variability

Because of the temperature sensitivity, pathogenic variability is uncertain. However virulence is known to occur in North America (Pretorius *et al.*, 1984), South Africa, the Indian subcontinent, and Australia where a single pathotype is virulent. Virulence can be established from adult plant tests, but in locations where field epidemics are based on pathotype mixtures, it is impossible to establish the pathogenic status of individual pathotypes or isolates.

In the survey undertaken by Huerta-Espino (1992) the only host line with *Lr13* was Columbus which also possesses *Lr16*. Columbus gave low seedling reactions with all cultures and, in many tests, gave an incompatible infection type lower than that of Tc + *Lr16*. This indicated either that *Lr13* was effective in many of the tests, interacting with *Lr16*, or that additional resistance genes were present in Columbus.

Reference Stocks

i: Thatcher*7/Frontana, R.L.4031. Being a backcross derivative, Manitou can also be considered near-isogenic relative to Thatcher. Red Bobs*6/Manitou, R.L.6067. Kolmer (1992*b*) listed 15 two-gene combinations involving *Lr13* and other genes. Based on Thatcher, most of these lines should also carry *Lr22b*.

s: CS*7/Ciano 67 2B; CS*7/Atlas 66 2B.

v: Frontana *Lr34* C.I.12470 (Dyck *et al.*, 1966).

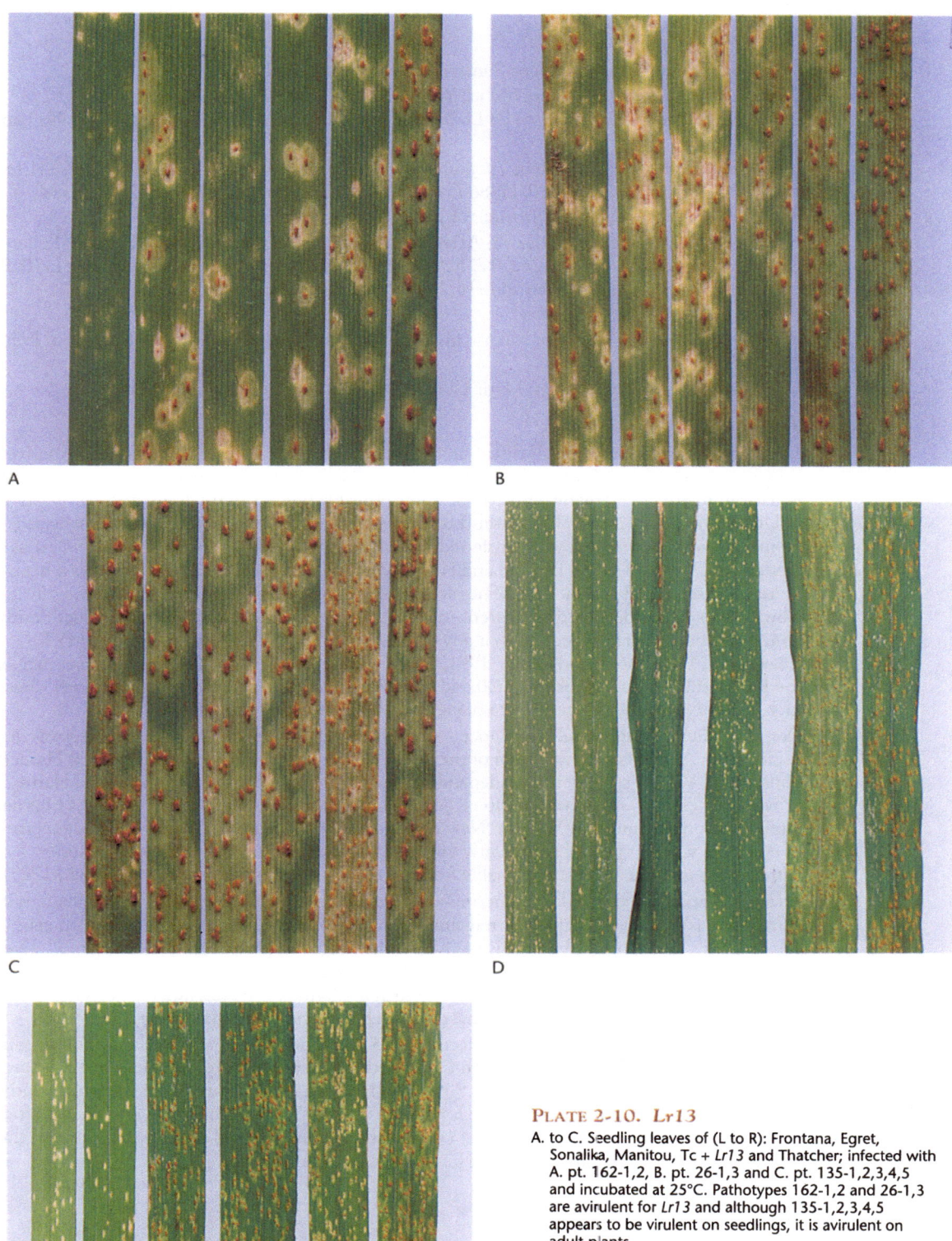

A

B

C

D

E

PLATE 2-10. *Lr13*

A. to C. Seedling leaves of (L to R): Frontana, Egret,
Sonalika, Manitou, Tc + *Lr13* and Thatcher; infected with
A. pt. 162-1,2, B. pt. 26-1,3 and C. pt. 135-1,2,3,4,5
and incubated at 25°C. Pathotypes 162-1,2 and 26-1,3
are avirulent for *Lr13* and although 135-1,2,3,4,5
appears to be virulent on seedlings, it is avirulent on
adult plants.

D. and E. Adult leaves (flag) of (L to R): Manitou, Tc + *Lr13*,
Red Bobs + *Lr13*, Sunstar, Thatcher and Red Bobs;
infected with D. pt. 104-2,3,6,(7) and E. pt. 53-
1,(6),(7),10,11. The first of these pathotypes is avirulent
and the second, virulent for *Lr13*. The first four wheats
carry *Lr13*. Resistance to pt. 53-1,(6),(7),10,11 in the
first, second and fifth wheats is conferred by *Lr22b*. Red
Bobs and Sunstar do not carry this gene. The phenotype
(;) conferred by *Lr13* is lower than that (;11$^+$) conferred
by *Lr22b*.

Source Stocks

Africa: Kenya Plume.

Australia: Avocet; Egret; Flinders; Sunstar; Vulcan. Suneca *Lr1*.

CIMMYT: Cajeme 71 *Lr1*; Nuri 70 *Lr1*; Yecora 70 *Lr1*. Pavon *Lr1 Lr10*. Tonichi 81 *Lr1 Lr27* + *Lr31*. Inia 66 *Lr14a Lr17*. Cumpas 88 *Lr26*; Genaro 81 *Lr26*. See Singh and Rajaram (1991) and Singh (1993).

Indian Subcontinent: Bahawalpur 79. Jauhar 78 *Lr1 Lr3a*. Lyallpur 73 *Lr1 Lr34*; UP115 *Lr1 Lr34*; WL2265 *Lr1 Lr34*. PWB65 *Lr10*. HS86 *Lr10 Lr11*. HD 2329 *Lr10 Lr34*; IWP72 *Lr10 Lr34*; Punjab 81 *Lr10 Lr34*. Sonalika *Lr14a*; WL711 *Lr14a*.

North America: Anderson; Atlas 50; Atlas 66; Coastal; Katalpa; Lark (USA); Napayo; Polk; Taylor; Tayland. Benito *Lr1 Lr2a Lr12*. Roblin *Lr1 Lr10 Lr34*. Butte *Lr2a Lr10*. Houser *Lr10*. Era *Lr10 Lr34*. Columbus *Lr16*; Kenyon *Lr16*. Manitou *Lr22b*; Neepawa *Lr22b*. Anza (heterogeneous) *Lr34*; Chris *Lr34*.

South America: Klein Cometa. Klein Aniversario *Lr3ka*. Frontana *Lr34*. Buck Manantial *Lr3a Lr16 Lr17*.

UK: Hustler; Kinsman; Mardler; Maris Huntsman; Norman; Virtue. Hobbit *Lr17*.

Use in Agriculture

Lr13 is probably the most widely distributed gene for leaf rust resistance in wheat, being derived from the commonly used parent cultivars Frontana, Fondoso and Frontiera as well as their South American relatives and many derivatives. It was probably present in the line Americano 25d used in Uruguay and its synonym Universal II used in Argentina to produce various Klein wheats. From these sources the gene was transferred to American winter wheats such as Atlas 50, Atlas 66, Anderson, Coastal, Taylor, Tayland and Houser, and presumably to the English winter wheats Maris Huntsman and Hobbit *Lr17*. The North American wheat, FKN (Frontana-Kenya 58-Newthatch), was widely used as a parent in spring wheat programs along with the earlier South American sources. *Lr13* is present in many wheats developed and distributed by CIMMYT (Hawthorn, 1984; Singh and Gupta, 1991; Singh and Rajaram, 1991). However, its presence on the Indian subcontinent probably predated CIMMYT and the Rockefeller program, with wheats such as Rio Negro and Klein Cometa being described in the literature in the 1950s.

Despite earlier attempts to use resistance sources such as Frontana and Rio Negro, *Lr13* was first exploited in Australia with adoption of the WW15 (=Anza) derivatives Egret, Avocet, Flinders, Sunstar and Vulcan and later the independently selected wheats Suneca, Sunkota and Hartog (Pavon 'S'), each of which has additional genes. *P. recondita* pathotype 53-1,(6),(7),10,11 became established on Karamu (=WW15) in New Zealand in 1981 (Luig *et al.*, 1985) and later appeared in Australia where it proved particularly virulent on Sunstar. This widely avirulent exotic pathotype is isozymically distinguishable from other Australasian pathotypes (Luig *et al.*, 1985).

Lr13 is an important gene for leaf rust resistance. Because it is often involved in gene combinations its current role is difficult to establish. It remains widely effective in Australia where many two-gene combinations involving *Lr13* continue to give excellent leaf rust protection. According to PL Dyck (*pers. comm.* 1988) and Kolmer (1992*b*) *Lr13* still plays a significant role in protecting Canadian wheats, but resistance in wheats with only *Lr13* such as Manitou and Neepawa has been less effective than that of other cultivars with additional genes such as Chris and Era (Kolmer *et al.*, 1991).

Kolmer (1992*b*) described interactive effects of *Lr13* and various genes which produce intermediate responses when present alone. Although *Lr13* continues to provide protection against leaf rust in the USA and Canada (Kolmer 1992*b*), Singh and Rajaram (1992) state "in Mexico and South America, this gene has been ineffective for at least 25 years". *Lr13* is ineffective in South Africa where Inia 66 (*Lr13 Lr17 Lr14a*) and T4 (=Anza) have been reported as susceptible, and in the UK where severe leaf rusting on Maris Huntsman has occurred.

Lr14 (McIntosh *et al.*, 1967)

Two distinct alleles, *Lr14a* and *Lr14b* and a 'combined' allele, *Lr14ab*, have been described. Because recombination was possible, the two genes are not true alleles.

Chromosome Location

7B (McIntosh *et al.*, 1967); 7BL (Law and Johnson, 1967). *Lr14* showed linkage of 22 to 35 cM with *Pm5* and 19 cM with *Sr17* (McIntosh *et al.*, 1967). Law and Wolfe (1966) placed *Pm5* 47.5 cM from the centromere, whereas Law and Johnson (1967) reported the gene order centromere—*Pm5*—*Lr14a*.

Lr14a (Dyck and Samborski, 1970)   (Plate 2-11)

Low Infection Type
X^- to X^+, 2^+.

Environmental Variability
High (Browder, 1980). *Lr14a* was observed to be more effective at temperatures below 20°C (Dyck and Johnson, 1983).

Origin
Lr14a was almost certainly transferred to the *T. aestivum* lines Hope and H-44 from Yaraslav emmer, along with genes for resistance to stem rust (*Sr17*) and powdery mildew (*Pm5*). *Lr14a* continues to be found in many wheats possessing the linked gene *Sr17*.

Pathogenic Variability
Lr14a is no longer widely effective in any geographical area. In the study of Huerta-Espino (1992) the highest frequencies of avirulence were found among samples from Rwanda, Tanzania, Turkey and Canada. Only certain putative exotic pathotypes are avirulent in Australasia.

A further complicating factor is that host genetic background has a significant influence on the response conferred by *Lr14a*. Law and Johnson (1967) described a gene that modified the expression of resistance attributed to *Lr14a*. Huerta-Espino (1992) included Hope, Rescue and Tc + *Lr14a* as testers for *Lr14a*. The Tc + *Lr14a* line (but not Thatcher) often gave a resistant response when one, other, or both of the other cultivars were susceptible.

Reference Stocks
i: Selkirk/6*Thatcher, R.L.6013 (Dyck and Samborski, 1970).
s: Chinese Spring*6/Hope 7B (McIntosh *et al.*, 1967).
v: Hope; Spica (McIntosh *et al.*, 1967).

Source Stocks
Lr14a is present in Hope and H-44 and many derivatives. The *Mli* gene for resistance to powdery mildew was recently considered synonymous with *Pm5* (Heun and Fischbeck, 1987a, 1987b; Hovmøller, 1989). It would be of interest to determine if wheats with *Mli* also carry *Lr14a*. RA McIntosh, NM Maseyk and RF Park (unpublished, 1991) found heterogenous responses in adult Kalyansona and related wheats tested with pathotypes avirulent for *Lr14a*. These wheats are usually scored susceptible when tested in the seedling stage with the same cultures (see 'Pathogenic Variability').

Australia: Gala; Glenwari; Hofed; Lawrence; Spica.

CIMMYT: Some derivatives of cross II8156, for example Mexipak 65 (mixed) (RA McIntosh, unpublished 1980) and Kalyansona. Pitic 62; Penjamo 62. Nadadores 63 *Lr10*; Jahuara 77 *Lr10*; Esmeralda 86 *Lr10* (Singh and Rajaram, 1991). Inia 66 *Lr13 Lr17* (RA McIntosh, unpublished 1990).

Indian Subcontinent: Sonalika *Lr13*; WL711 *Lr13*.

New Zealand: Aotea.

North America: Minter; Newthatch; Redman; Regent; Renown. Victory *Lr1 Lr3a Lr10*. Heil *Lr3a*; Scout *Lr3a*; Scout 66 *Lr3a*. Pioneer 2157 *Lr3a Lr11*. Vona *Lr10*. Selkirk *Lr10 Lr16* (Dyck and Samborski, 1970).

In most studies *Lr14a* will not be detected because the cultures used will be virulent, other genes may be epistatic and/or the infection type produced will not be sufficiently low or discerning to be recognised as *Lr14a*.

Use in Agriculture
No longer considered effective. However, wheats such as WL711 and Sonalika with *Lr13* + *Lr14a* are highly resistant in Australia in the presence of pt. 53-1,(6),(7),10,11 which is virulent on genotypes with *Lr13* alone.

Lr14b (Dyck and Samborski, 1970)   (Plate 2-12)

Low Infection Type
X^- to X, 1^-C, 2 to 3^-.

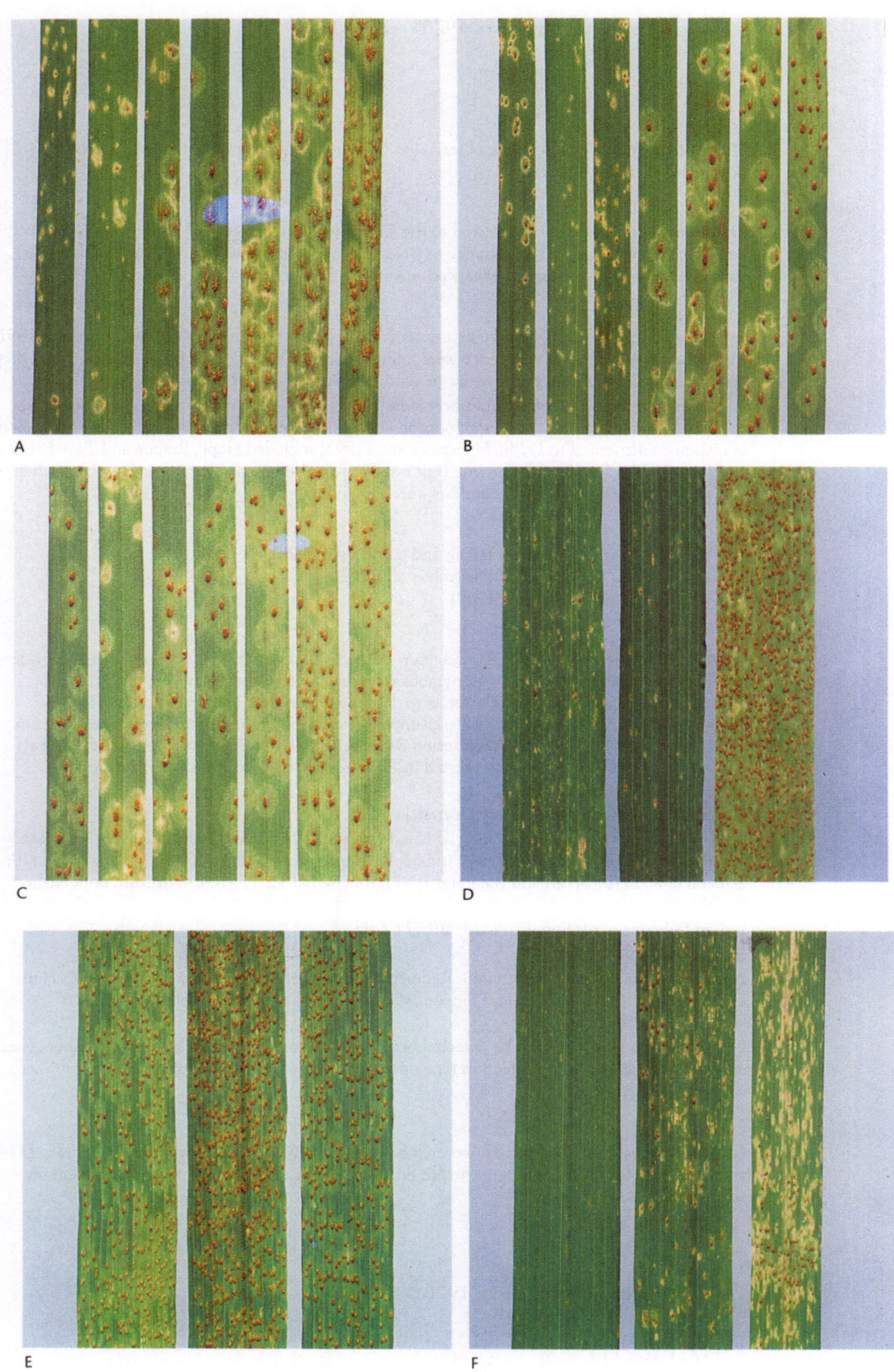

A

B

C

D

E

F

Environmental Variability

Less effective at warmer temperatures (Dyck and Johnson, 1983).

Origin

Common wheat.

Pathogenic Variability

Comparatively little is known of pathogenic variability on wheats with *Lr14b*.

In the survey of Huerta-Espino (1992) avirulence was most frequent among cultures from South America and Mexico. Low levels of avirulence were found in cultures from Africa as well as southern and eastern Asia. This author cited evidence indicating that avirulence was common in parts of the former Union of Soviet Socialist Republics (USSR).

Dyck and Samborski (1970) found that pathogen virulence was recessive for *Lr14a* and dominant for *Lr14b*. Dyck and Johnson (1983) reported incompatibility with one English culture at higher temperatures. Although susceptible at the seedling stage in both Australia (RA McIntosh, unpublished 1985) and India (Sawhney *et al.*, 1992), Thatcher derivatives with *Lr14b* display adult plant resistance to at least some pathotypes in both countries (Plate 2-12). It is not known if the adult plant resistance is conferred by *Lr14b*.

Reference Stocks

i: Maria Escobar/6*Thatcher, R.L.6006 (Dyck and Samborski, 1970). Thatcher*6/Rafaela, R.L.6056.

v: Bowie *Lr3a*. Maria Escobar *Lr17* (Dyck and Samborski, 1970).

Source Stocks

Rafaela *Lr17* (Dyck and Kerber, 1977*a*).

Use in Agriculture

This gene is not commonly used in agriculture.

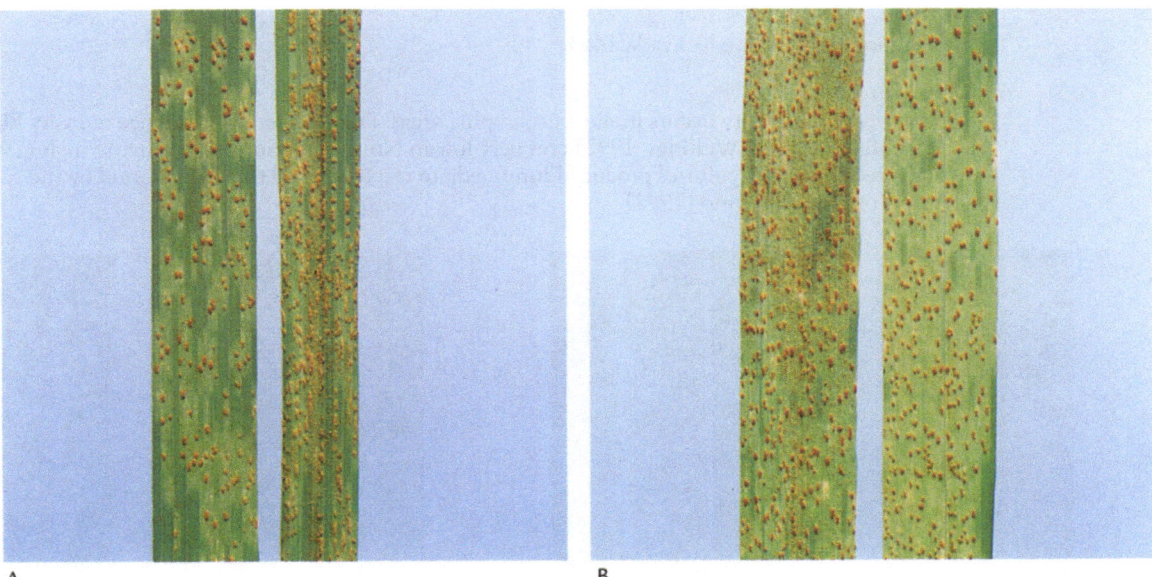

A   B

PLATE 2-12. *Lr14b*

Adult leaves (flag-2) of (L to R): Tc + *Lr14b* and Thatcher; infected with A. pt. 104-2,3,6,(7) and B. pt. 104-2,3,(6),(7),11. This difference in pathogenicity was discovered only recently and it is unknown if gene *Lr14b* is involved. Under field conditions Tc + *Lr14b* is more resistant to pt. 104-2,3,6,(7) than shown here.

◀ **PLATE 2-11. *Lr14a***

A. to C. Seedling leaves of (L to R): Super X, Sonalika, WL711, Spica, Tc + *Lr14a*, Tc + *Lr14ab* and Thatcher; infected with A. pt. 53-1,(6),(7),10,11 [*P14a*], B. pt. 76-0 [*P14a*] and C. pt. 10-1,2,3,4 [*p14a*]. The first three wheats have a gene(s) additional to *Lr14a* and effective with the first two cultures. Resistance in Sonalika and WL711 with pt. 10-1,2,3,4 is conferred by *Lr13*. *Lr14ab* is a combination of the *a* and *b* alleles — see *Lr14ab*.

D. to F. Adult leaves (flag-1) of (L to R): Tc + *Lr14a*, Tc + *Lr14ab* and Thatcher; infected with D. pt. 76-0, E. pt. 104-2,3,(6),(7),11 and F. pt. 53-1,(6),(7),10,11. Lines with *Lr14a* and *Lr14ab* give similar high levels of resistance to pt. 76-0 and no resistance to pt. 104-2,3,(6),(7),11. Resistance to pt. 53-1,(6),(7),10,11 was conferred by both the *Lr14* alleles and *Lr22b*, with greater interaction involving Tc + *Lr14a*. With this pathotype the response of Tc + *Lr14ab* was not significantly lower than that of Thatcher.

Lr14ab

The designation *Lr14ab* was given to the 'allele' in a genotype derived from the cross Selkirk/6* Thatcher//Maria Escobar/6*Thatcher. Although *Lr14a* and *Lr14b* are not true alleles this designation indicates their close linkage (0.16 ± 0.16 cM, estimate ± s.e.) (Dyck and Samborski, 1970). Tests with the recombinant stock in Australia have indicated similar seedling responses to Selkirk/6*Thatcher and similar or better adult plant resistance to Maria Escobar/6*Thatcher (RA McIntosh, unpublished 1985). The recombinant stock also displays adult resistance in India (Sawhney *et al.*, 1992).

Reference Stock

Selkirk/6*Thatcher//Maria Escobar/6*Thatcher, R.L.6039 (Dyck and Samborski, 1970).

Lr15 (Luig and McIntosh, 1968) (Plate 2-13)

Chromosome Location

2D (Luig and McIntosh, 1968); 2DS (McIntosh and Baker, 1968). *Lr15* has been mapped relative to other genes in chromosome 2D. It appears to be allelic with *Lr2*, very closely linked with *Sr6* (<1 cM) and 10 cM from gene C for compact spikes (McIntosh and Baker, 1968; RA McIntosh, unpublished 1973).

Low Infection Type

$0; ^-$ to $;1^+, 2^+$.

Environmental Variability

Low (Browder 1980; Dyck and Johnson, 1983). However, Singh (1991) conducted tests at 15–18°C due to temperature sensitivity. The host:pathogen combinations resulting in intermediate responses may be more temperature-sensitive although sensitivity has not been observed in Australia.

Origin

Common wheat stock Kenya W1483.

Pathogenic Variability

Pathogenic variability occurs in most geographic areas. Frequencies of avirulence are very high in Australia (Park and Wellings, 1992) and very low in North America and Mexico. Singh (1991) suggested that some cultures produced intermediate responses and this is supported by the survey results of Huerta-Espino (1992).

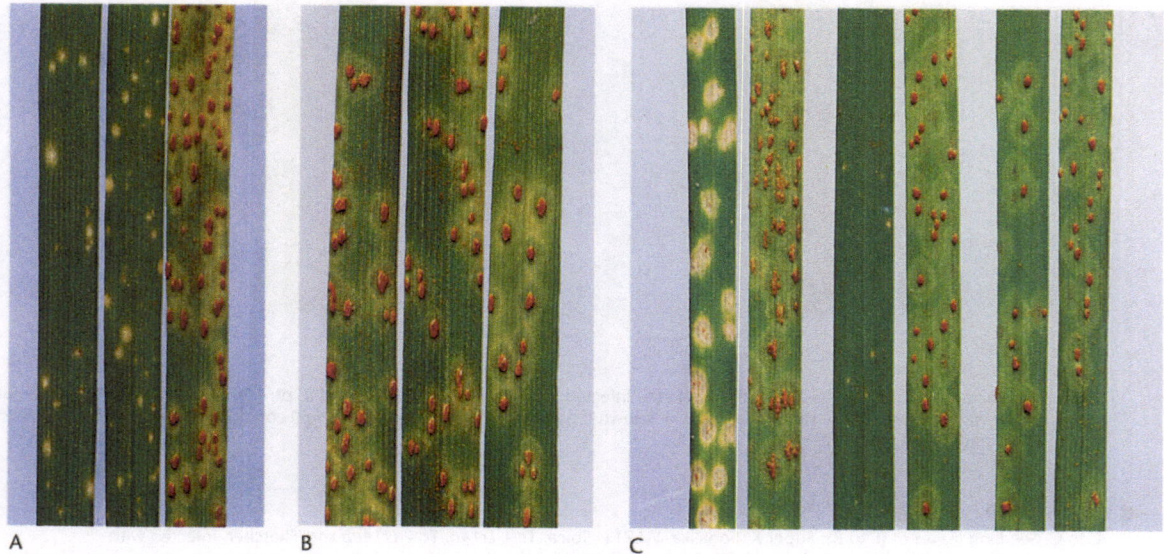

A   B   C

PLATE 2-13. *Lr15*

A. and B. Seedling leaves of (L to R): Kenya W1483, Tc + *Lr15* and Thatcher; infected with A. pt. 10-1,2,3 [*P15*] and B. pt. 10-1,2,3,4 [*p15*]. The latter pathotype is assumed to be a mutant derivative of the former with added virulence for *Lr15*.

C. Leaf pairs of: Kenya W1483 and Thatcher; infected with pt. 76-0 (left), pt. 53-1,(6),(7),10,11 (centre) and pt. 10-1,2,3,4 (right). The low infection type with pt. 76-0 is consistently higher than with other avirulent Australian pathotypes, but is significantly lower than the intermediate responses recorded by Singh (1991) and Huerta-Espino (1992).

Reference Stocks

> **i**: Thatcher*6/Kenya W1483, R.L.6052.
> **v**: Kenya W1483 = Kenya 112-E-19-J (Luig and McIntosh, 1968).

Source Stock

> Kenya W1483 is the only wheat reported with this gene.

Use in Agriculture

> Not deployed. Despite the absence of wheats with *Lr15*, virulence for *Lr15* is relatively common in most geographical areas.

Lr16 (Dyck and Samborski, 1968*a*)   (Plate 2-14)

Synonym

> *LrE* (Anderson, 1961).

Chromosome Location

> 4B (Dyck and Kerber, 1971); 2BS (RA McIntosh, unpublished 1978). These results can be reconciled given that the first authors used a Rescue monosomic series and that Rescue carries a 4B–2B reciprocal translocation. *Lr16* appears to be genetically independent of *Lr13* and *Sr36* which are proximally located in chromosome 2BS (RA McIntosh, unpublished 1980). *Lr16* is always associated with *Sr23* (McIntosh and Luig, 1973*b*).

Low Infection Type

> 1 to 3C.

Environmental Variability

> *Lr16* is more effective at higher temperatures (Dyck and Johnson, 1983). Browder (1980) considered *Lr16* to have low environmental sensitivity. Statler and Christianson (1993) reported that the environmental sensitivity varied with pathogen isolates.

Origin

> Common wheat; one source of *Lr16* (Selkirk) traces to the old Australian cultivar Warden (Quartz/Ward's White//Red Bordeaux) used as a hay wheat from 1900 to 1920 (Macindoe and Walkden Brown, 1968).

Pathogenic Variability

> Variation occurs in most geographic areas, but the frequency of virulence is usually relatively low (Browder, 1980; Huerta-Espino, 1992). Huerta-Espino (1992) reported a higher level of avirulence in samples from China than might have been predicted from the 1986 data of Hu and Roelfs (1989). The first instance of virulence in Australasia occurred with the introduction of pt. 53-1,(6),(7),10,11 in New Zealand in 1981 (Luig *et al.*, 1985). In 1984, a second pathotype [104-2,3,(6),(7),11] with increased virulence for *Lr16* was detected in Australia. This pathotype, and its derivative pt. 104-1,2,3,(6),(7),11 have become widely distributed throughout Australia (Park and Wellings, 1992) and recent pathogenicity and isozymic studies indicated that the former was not a single-step mutant of the earlier predominant pt. 104-2,3,6,(7) (Park *et al.*, 1993). As illustrated in Plate 2-14, this pathotype is not fully virulent for *Lr16* under all conditions and perhaps it would be better designated pt. 104-2,3,(6),(7),(11).

Reference Stocks

> **i**: Exchange/6*Thatcher, R.L.6005 (Dyck and Kerber, 1971; Singh and Gupta, 1991).
> **v**: Exchange *Lr10 Lr12*. Selkirk *Lr10 Lr14a* (Anderson, 1961).

Source Stocks

> Not present in current cultivars in Australia or the Indian subcontinent (Singh and Gupta, 1991). RA McIntosh (unpublished, 1993) found that *Lr16* occurred frequently among a collection of Chinese wheats.
> **Australia**: Warden *Lr10* (Anderson, 1961).
> **CIMMYT**: Ciano 79; Imuris 79; Papago 86; Topoca 89. Mahome 81 *Lr1 Lr13*. Huasteco 81 *Lr10*. See Singh and Rajaram (1991) and Singh (1993).
> **Europe**: Etoile de Choisy (Bartos *et al.*, 1969*b*).

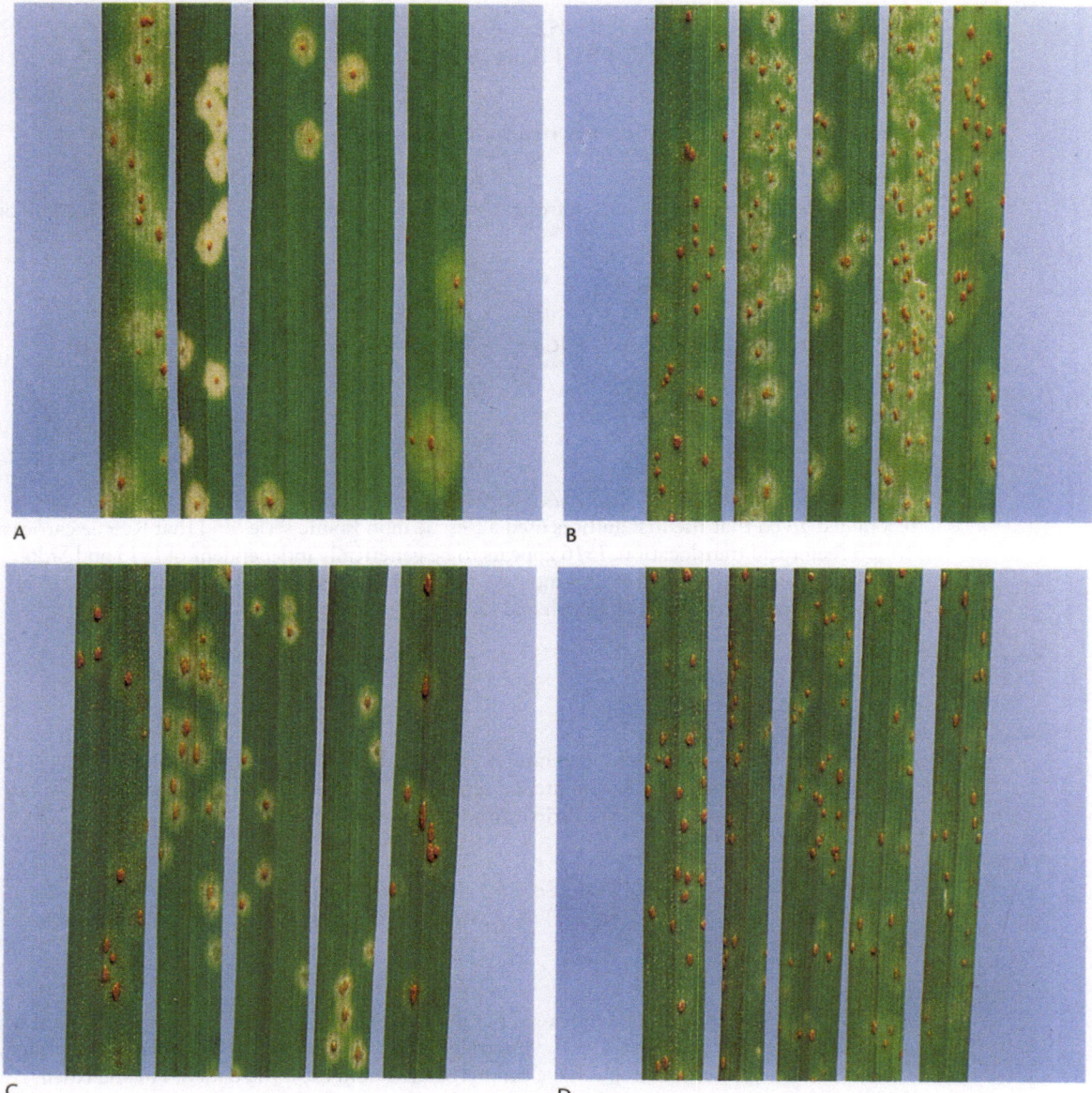

PLATE 2-14. Lr16

Seedling leaves of (L to R): Selkirk, Exchange, Ciano 79, Tc + *Lr16* and Thatcher; infected with pt. 104-2,3,6,(7) (A. and C.) and pt. 104-2,3,(6),(7),11 (B. and D.) and incubated at 22°C (A. and B.) and 15°C (C. and D.). The typical necrotic response of *Lr16* is greater at higher temperatures. In this experiment Selkirk reacted with less necrosis than the other lines with *Lr16*. Although pt. 104-2,3,(6),(7),11 is considered virulent, some interaction occurs with at least some host lines at the higher temperature. This type of host and environmental variability introduces subjectivity to the interpretation of pathogenicity data in the absence of correlated adult plant and field studies.

North America: Probrand 812; Success; Telemark. Brule *Lr10*. Arapahoe *Lr10 Lr24*. Columbus *Lr13* (heterogeneous) (Samborski and Dyck, 1982); Kenyon *Lr13*.
South America: Buck Manantial *Lr3a Lr13 Lr17 Lr34*.

Use in Agriculture

Not highly effective when present alone but may interact with some other genes to give enhanced levels of resistance (Samborski and Dyck, 1982; Kolmer, 1992*b*). The use of cultivars with resistance conferred primarily by *Lr16* such as Probrand 812 in the USA and Selkirk *Lr10 Lr14a* in Canada was rapidly followed by increased frequencies of pathotypes with virulence for *Lr16*. Schafer and Long (1988) described the likely role of Probrand 812 in influencing the pathotype distribution in Texas. On the other hand, gene combinations involving *Lr16* such as Columbus *Lr13* and Kenyon *Lr13* appeared to confer more durable resistance.

Lr17 (Dyck and Samborski, 1968*a*) (Plate 2-15)

Chromosome Location

2AS (Dyck and Kerber, 1977*a*; RA McIntosh, unpublished 1977). Bariana and McIntosh (1993) reported very close linkage between *Lr17* and *Lr37*, *Sr38* and *Yr17* present in a VPM1 derivative.

Low Infection Type

;,1,2, X, X$^+$3.

Environmental Variability

Seedlings with *Lr17* become more resistant to avirulent cultures at higher temperatures (Dyck and Johnson, 1983; Statler and Christianson, 1993). Browder (1980) considered environmental sensitivity to be low.

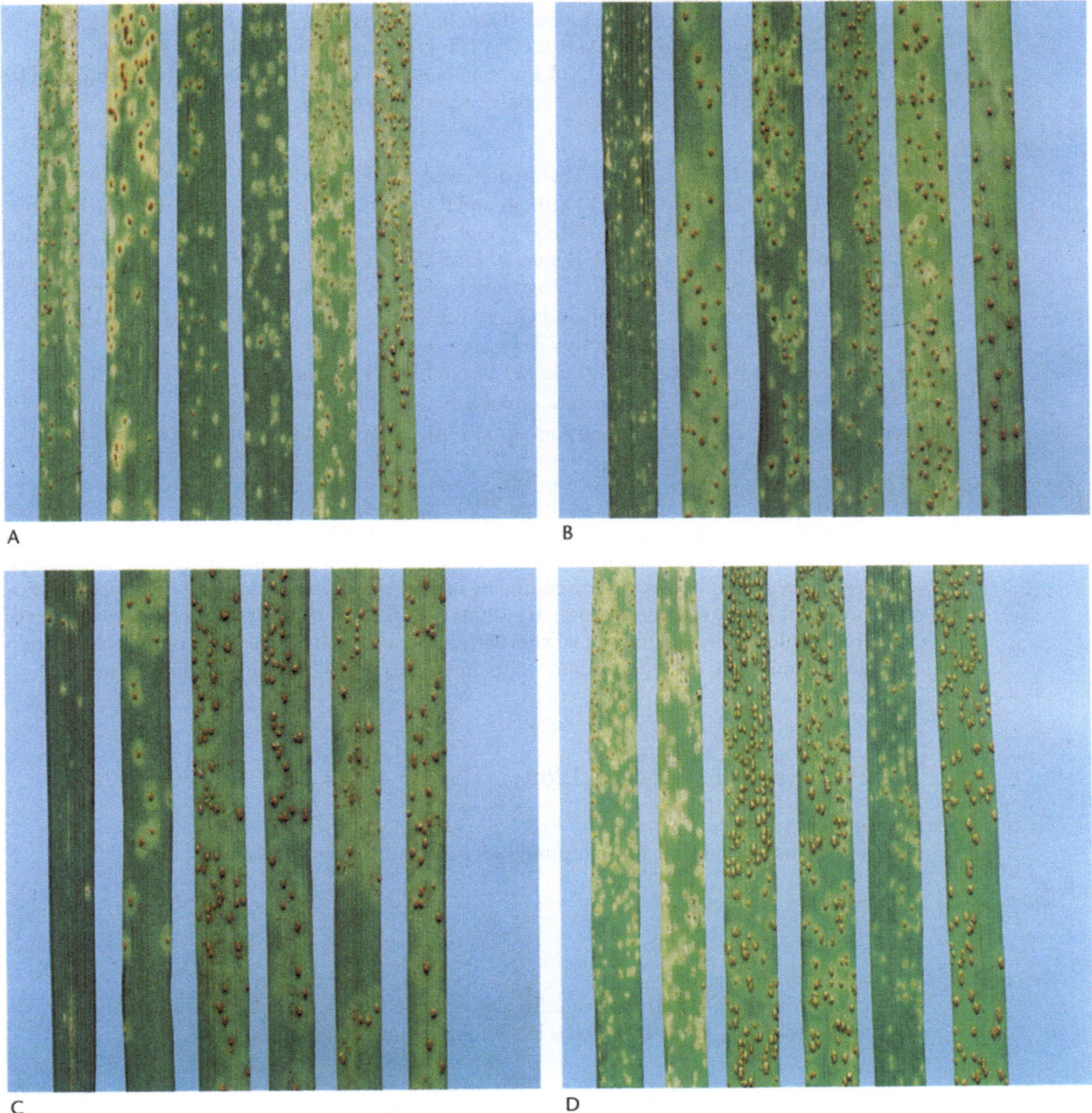

A

B

C

D

PLATE 2-15. *Lr17*

A. to C. Seedling leaves of (L to R): Sunkota, Klein Rendidor, Timson 'Resistant', Timson 'Susceptible', Tc + *Lr17* and Thatcher; infected with A. pt. 26-1,3, B. pt. 104-2,3,6,(7) and C. pt. 104-2,3,6,7 and incubated at 22/24°C. Pt. 104-2,3,6,(7) produces intermediate responses on seedlings with *Lr17*. Sunkota and Klein Rendidor have resistance genes additional to *Lr17*.

D. Paired seedling leaves of: Timson 'Resistant' and Timson 'Susceptible'; infected with pt. 76-0 [*p10*] and incubated at 22/24°C (first pair) and 18°C (second pair). The third pair of leaves was infected with pt. 53-(6),(7),9,10,11 [*P10*] and incubated at 18°C. Note the effects of temperature on the expression of response conferred by *Lr17* and the clear differentiation of response of the two selections when tested with a culture avirulent for *Lr10*, confirming the presence of *Lr10* in Timson.

Origin

Common wheat. *Lr17* appears to have a South American origin.

Pathogenic Variability

Variation occurs in most geographic areas. Researchers should ensure that greenhouse temperatures are sufficiently high to obtain resistance to avirulent cultures. One isolate of pathotype 104-2,3,6,7 [culture 791021] (NH Luig, unpublished 1979) and one of pathotype 122-1,3,4,6,7 [89172] (RF Park, unpublished 1992) with virulence for *Lr17* have been recorded in Australasia. Most of the predominant pathotypes in recent years, including those collected from cultivars such as Songlen and Timson which carry *Lr17* were incompletely virulent on seedlings with this gene and hence designated -(7). Pathogen populations in Canada were avirulent (Kolmer, 1991) whereas USA populations were variable (Long *et al.*, 1993).

Reference Stocks

i: Klein Lucero/6*Prelude, R.L.6041 (Dyck and Samborski, 1968*a*); Klein Lucero*6/Thatcher, R.L.6008 (Dyck and Samborski, 1968*a*); Thatcher*6/EAP26127, R.L.6055 (PL Dyck, *pers. comm.* 1987); Thatcher*6/Rafaela R.L.6054 (PL Dyck, *pers. comm.* 1987).
v: Klein Lucero; Maria Escobar (Dyck and Samborski, 1968*a*). The data of Huerta-Espino (1992) clearly show that Klein Lucero has a gene(s) additional to *Lr17*.

Source Stocks

Australia: Songlen. Timson *Lr10* (heterogeneous). Sunkota *Lr13*.
CIMMYT: Jupateco 73; Tanori 71 (Singh and Rajaram, 1991). Noroeste 66 *Lr1 Lr13* (Singh and Rajaram, 1991). Torim 73 *Lr1 Lr13 Lr34* (Singh, 1993). Curinda 87 *Lr3a Lr13 Lr34*. Lerma Rojo 64 *Lr13* (Singh and Rajaram, 1991); Potam 70 *Lr13* (Singh, 1993); Safed Lerma *Lr13* (Singh and Gupta, 1991). Inia 66 *Lr13 Lr14a* (Singh and Rajaram, 1991; RA McIntosh, unpublished 1990).
Indian Subcontinent: Arz (Singh and Gupta, 1991; E Gordon-Werner, unpublished 1988). NP846 *Lr1 Lr13* (Singh and Gupta, 1991).
South America: Buck Manantial *Lr3a Lr13 Lr16 Lr34* (Kolmer, 1991).
UK: Hobbit sib *Lr13* (RA McIntosh, unpublished 1991).
Unknown: EAP 26127 (Dyck and Kerber, 1977*a*). Rafaela *Lr14b* (Dyck and Kerber, 1977*a*).

Use in Agriculture

Lr17 is an important component of the multiple gene resistances of a number of wheats. It is present in the Australian wheats Songlen, Timson and Sunkota. In 1976, a culture, designated pt. 104-2,3,6,(7), produced an intermediate response on seedlings of Songlen and Timson. This change was accompanied by increased rusting on these cultivars under field conditions. However, they were not highly susceptible to such pathotypes. In 1979, a culture, 104-2,3,6,7, was identified with full virulence on seedlings of wheats with *Lr17*. This culture was shown to be fully virulent on adult plants of Tc + *Lr17* and Timson in greenhouse tests (RF Park, unpublished 1991).

Lr18 (Dyck and Samborski, 1968*a*)   (Plate 2-16)

Chromosome Location

5BL (McIntosh, 1983*a*). *Lr18* segregated independently of the centromere.

Low Infection Type

;, 12, 2, 23⁻.

Environmental Variability

High (Browder, 1980). *Lr18* becomes increasingly ineffective as temperatures increase to 25°C (Dyck and Johnson, 1983; McIntosh, 1983*a*; Statler and Christianson, 1993).

Origin

T. timopheevii. McIntosh (1983*a*) discussed the likelihood that all sources of *Lr18* were derived from this species. Y Yamamori (*pers. comm.* 1989) was able to correlate the presence of an N-banded 5BL terminal segment of several *Lr18* sources, including the near-isogenic line of Thatcher, with a chromosome of *T. timopheevii*. The source of *Lr18* in North American wheats is unknown.

Pathogenic Variability

Rust workers in most geographical regions have reported pathogenic variation on seedlings with *Lr18* with avirulence usually predominating (Kolmer, 1991; Huerta-Espino, 1992; Long *et al.*,

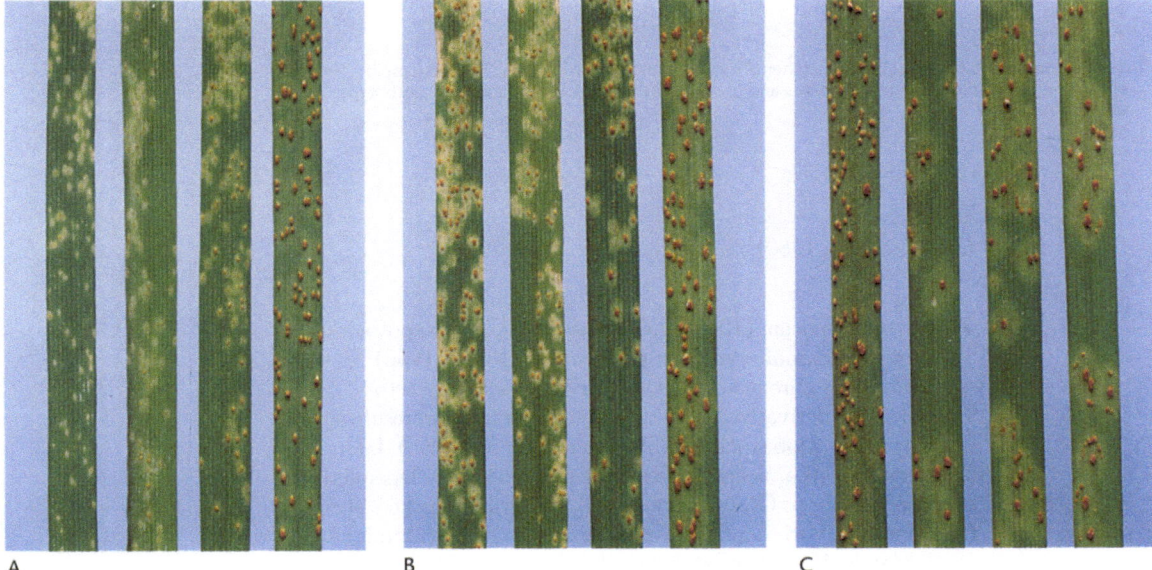

A B C

PLATE 2-16. *Lr18*
Seedling leaves of (L to R): Timvera, WYR50854, Tc + *Lr18* and Thatcher; infected with pt. 104-2,3,6,(7) and incubated at A. 15°C, B. 18°C and C. 22/24°C. *Lr18* becomes ineffective at higher greenhouse temperatures. At low temperatures *Lr18* is dominant but becomes recessive at intermediate temperatures. *Lr18* is effective in seedling and adult plants against all Australian pathotypes.

1992). High frequencies of virulence have been reported from surveys in Germany and Switzerland (discussed in Huerta-Espino, 1992). However, virulence has not been found in Australia and both seedlings at low temperatures (16–18°C) in the greenhouse and adult plants in the field from normal planting times are highly resistant. For some rust surveys based on seedling tests, greenhouse temperatures may be too high for satisfactory screening with the consequence that avirulent pathotypes might be recorded as virulent.

Reference Stocks

i: South Africa 43/7*Thatcher, R.L.6009 (Dyck and Samborski, 1968a); Thatcher*6/Sabikei 12, R.L.6090 (PL Dyck, *pers. comm.* 1989).
v: South Africa 43, P.I.159106 (Dyck and Samborski, 1968a).

Source Stocks

Red Egyptian P.I.170925 and P.I.17016-2c (Dyck and Samborski, 1968a). Timvera and Timvera Derivative W3350, Sabikei 12 (McIntosh, 1983a). Certain WYR accessions (Russian *T. timopheevii* derivatives) (McIntosh, 1983a).
USA: Coker 797. Hunter *Lr1 Lr3a Lr10*. Coker 86-32 *Lr2a Lr9*. Coker 983 *Lr3a Lr10*. Coker 9227 *Lr11*. Terral 101 *Lr11 Lr24*.

Use in Agriculture

Lr18 has not been used in agriculture, but would appear to be useful, especially in areas such as Australia. *Lr18* is the only documented gene for leaf rust resistance derived from *T. timopheevii*. Because seedlings of *T. timopheevii* are highly resistant over a range of temperatures and *P. recondita* cultures, we assume that genes additional to *Lr18* are present.

Lr19 (Browder, 1972)   Plate (2-17)

Chromosome Location

7DL (Sharma and Knott, 1966; ER Sears, *pers. comm.* 1972; Dvorak and Knott, 1977). In the translocation line 7A/Ag#12, *Lr19* is located in chromosome 7AL (Eizenga, 1987). 7AgL (ER Sears, *pers. comm.* 1972).

Low Infection Type

0, 0;$^=$ to ;, occasionally 1$^+$ or 2.

PLATE 2-17. *Lr19*

Seedling leaves of: Agatha (L) and Thatcher (R); infected with pt. 104-2,3,6,(7). This gene is effective against all Australian pathotypes and gives a consistently low response over pathotypes and environments.

Environmental Variability

Low (Browder, 1980).

Origin

Thinopyrum ponticum (10x). The winter wheat, Agrus, possessed a *Thinopyrum* chromosome pair (Sharma and Knott, 1966) substituting for chromosome 7D. Agrus was backcrossed to Thatcher and derivatives were subjected to irradiation resulting in the translocation stock T4 (Sharma and Knott, 1966), later designated Agatha. Sears (1973) made a number of independent transfers of *Lr19* to Chinese Spring. South African workers also transferred *Lr19* to Inia 66 wheat from *Th. distichum* (Marais *et al.*, 1988; Marais, 1992a) and designated the derivative Indis.

Pathogenic Variability

Virulence is rare. Huerta-Espino and Singh (1993) reported the isolation of an *Lr19*-virulent culture from cultivar Oasis 86 in Mexico. Virulent isolates reported from surveys in Switzerland and Ethiopia were not independently confirmed.

Reference Stocks

i: Agatha C.I.14048 (Sharma and Knott, 1966). Being a backcross derivative Agatha can be considered near-isogenic to Thatcher. Sears's translocations (Sears, 1973) can be considered near-isogenic to Chinese Spring.
su: Agrus C.I. 13228 (Browder, 1972).

Source Stocks

Indis (Marais, 1990); Sunnan (Knott, 1989). Oasis 86 *Lr13* (Singh and Rajaram, 1991).

Use in Agriculture

Despite the excellent level of protection provided by *Lr19* and the lack of virulent pathogen isolates this gene has not been widely utilised because the various translocations are associated with yellow coloured flours (Knott, 1980, 1989; Marais *et al.*, 1988). Knott (1980) produced mutant lines with reduced flour pigment levels relative to Agatha, but residual colour grades were still marginal. *Lr19* is always associated with *Sr25* for resistance to *P. graminis* (McIntosh *et al.*, 1976). Marais (1992a) induced chromosomal deletions in Indis with reduced flour pigment levels. Knott (1989) indicated that the Swedish cultivar Sunnan was released to take advantage of the high yellow pigment levels. In some genetic backgrounds most *Lr19* sources are preferentially transmitted (RA McIntosh, unpublished 1975; Marais, 1990). However, *Lr19* in Sears' derivative 7D/Ag#7 is not favourably transmitted in comparable genetic backgrounds (RA McIntosh, unpublished 1975).

Lr20 (Browder, 1972)   (Plate 2-18)

Chromosome Location

7AL, based on complete linkage with *Pm1* (Sears and Briggle, 1969). *Lr20* is the same gene as *Sr15* (McIntosh, 1977). The and McIntosh (1975) reported that *Lr20* was loosely linked with, and distal to, *Sr22*.

Low Infection Type

;, ;N, 1N, 3N.

Environmental Variability

Lowest responses occur at low temperatures. *Lr20* becomes completely ineffective at 30.5°C (Jones and Deverall, 1977).

Origin

Common wheat; *Lr20* is present in wheats with *Pm1* and *Sr15* (Watson and Baker, 1943; Pugsley and Carter, 1953; Watson and Luig, 1966).

Pathogenic Variability

Lr20 has provided protection to part of the Australian *P. recondita* population since the 1930s (Waterhouse, 1952; Watson, 1962). Pathogenic variation occurs on the Indian subcontinent, Africa, Europe, China and South America. North American *P. recondita* populations are uniformly virulent for *Lr20* (Browder, 1972; Singh, 1991).

Reference Stocks

i: Thatcher*6/Timmo, R.L.6092 (PL Dyck, *pers. comm.* 1990); Fedka = Federation*3/Kenya W744 (RA McIntosh, unpublished 1970). Because Federation carries *Lr10*, this gene may also be present.
s: Chinese Spring*5/Axminster 7A (Sears, 1954).
v: Thew (Browder, 1972); Kenya W744 (Watson and Baker, 1943). Norka *Lr1* (Watson and Luig, 1966).

Source Stocks

Lr20 occurs in wheats with *Pm1* and *Sr15* [see McIntosh (1988*a*) for list].
Australia: Aroona; Bonus; Festival; Kenora; Lance; Schomburgk. Bindawarra *Lr1*.
Europe: Birdproof; Maris Halberd; Normandie; Sappo; Sicco; Timmo.
North America: Axminster; Converse; Huron.

Dyck and Johnson (PL Dyck, *pers. comm.* 1990) noted that the response of Timmo is consistently higher than Maris Halberd, Sicco and Sappo at corresponding temperatures. This also occurs with Norka compared to Thew. McIntosh (1977) reported a very high induced mutation frequency in a wheat carrying *Lr20*. The mutants gave distinctly higher responses than the parental genotypes but retained the specificity of *Lr20*. All mutations of *Lr20* also affected the response conferred by *Sr15*, but were independent of mutations for *Pm1*. The *Lr20* mutants behaved as alleles of the original *Lr20* gene (McIntosh, 1983*b*).

Use in Agriculture

Very limited. The presence of wheats with *Lr20* has affected the spectrum of pathotypes in some regions of Australia. When wheats with *Lr20* are widely grown they soon become infected with *Lr20*-virulent pathotypes (Park and Wellings, 1992).

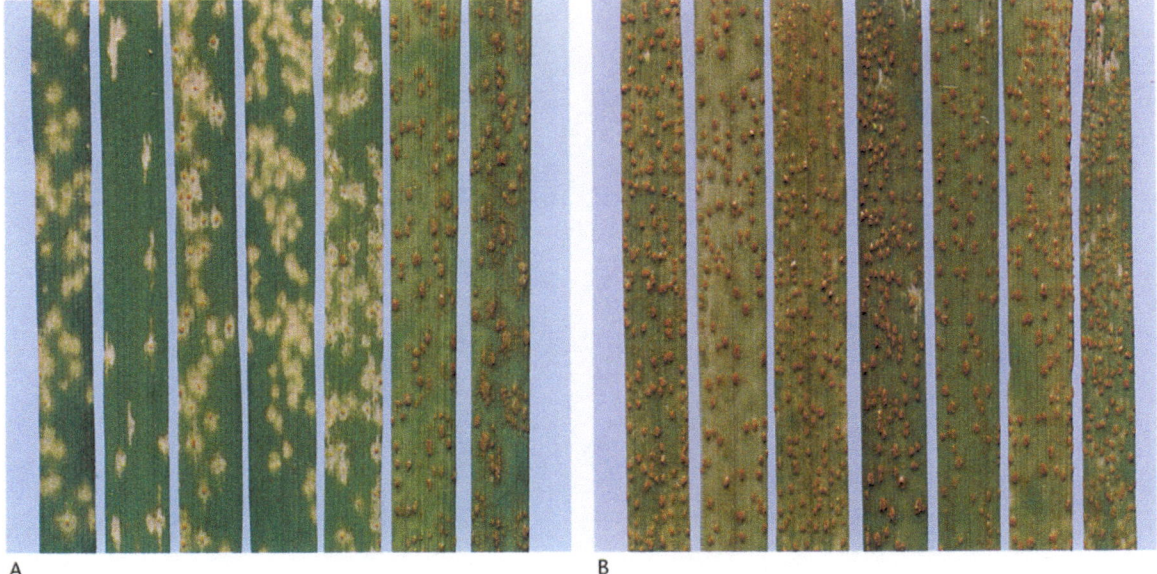

A   B

PLATE 2-18. *Lr20*
Seedling leaves of (L to R): Thew, CS*5/Axminster 7A, Sappo, Timmo, EMS-775056, EMS-775058 and Chinese Spring; infected with A. pt. 76-0 and B. pt. 76-1,2,3,6 and incubated at less than 20°C. Sappo and Timmo were not different at this temperature. 775056 (parental) and 775058 (mutant) were derived from a segregating mutant spike of an *Lr20Lr20* genotype treated with EMS (ethyl methanesulfonate). Necrosis associated with *Lr20* often extends longitudinally as with CS*5/Axminster 7A, especially at intermediate temperatures.

Lr21 (Rowland and Kerber, 1974)   (Plate 2-19)

Chromosome Location
1D (Kerber and Dyck, 1979); 1DL (Rowland and Kerber, 1974). However, Jones *et al.* (1990) placed this gene in chromosome 1DS and mapped it 4 cM distal to *Rg2* and 6 cM from *Gli-D1*. *Lr21* is also linked with a stem rust resistance gene with the same specificity as *Sr21* and with *Sr33*.

Low Infection Type
0;, 1 to 2$^+$.

Environmental Variability
Low (Dyck and Johnson, 1983).

Origin
T. tauschii var. *meyeri* R.L.5289 (Rowland and Kerber, 1974).

Pathogenic Variability
Virulent isolates have not been found in North America or Australia. Despite claims of ineffectiveness in seedlings, adult plants with *Lr21* are resistant on the Indian subcontinent. Huerta-Espino (1992) reviewed reports of virulence for *Lr21* from many countries and reported the frequencies of virulence as generally low. In some situations it is possible that avirulent responses were interpreted as virulent based on the seedling reactions.

Reference Stocks
i: Thatcher*6//R.L.5406, R.L.6043 (Browder, 1980).
v: R.L.5406 = Tetra Canthatch/*T. tauschii* var. *meyeri* R.L.5289.
dv: *T. tauschii* var. *meyeri* R.L.5289. R.L.5289 carries a gene with the same stem rust specificity as *Sr21* located in chromosome 2A of Einkorn (McIntosh, 1981). In both R.L.5289 and R.L.5406 this gene is linked with *Lr21*. The '*Sr21*' gene appears to be relatively common in *T. tauschii*. Accessions of this species with '*Sr21*' also carry leaf rust resistance, and display similar infection type responses to those conferred by *Lr21*. The two genes were separated in hexaploid backcross lines produced in the germplasm enhancement program undertaken by PBI Cobbitty.

Source Stocks
AC Cora=BW152 *Lr13* (JA Kolmer, *pers. comm.* 1993).

Use in Agriculture
Lr21 has potential for use in breeding but has not been exploited in any wheat cultivar.

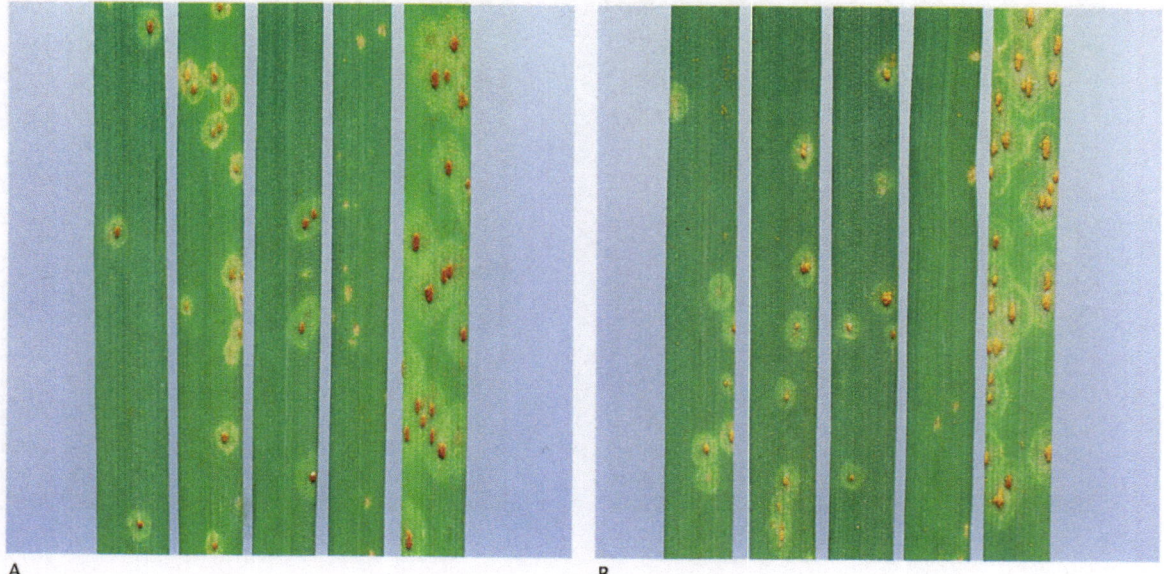

A   B

PLATE 2-19. *Lr21*
Seedling leaves of (L to R): R.L.5406, Tc + *Lr21*, R.L.5406/7*Condor, R.L.5406/2 Shortim//4*Tincurrin and Thatcher; infected with A. pt. 10-1,2,3,4 and B. pt. 104-1,2,3,6. The second pathotype has distinctive yellow uredia. The fourth host line probably carries an additional unidentified resistance gene(s).

Lr22 (Rowland and Kerber, 1974)

Two alleles for resistance have been described for this locus.

Chromosome Location

2DS (Rowland and Kerber, 1974). *Lr22* segregated independently of the centromere (Rowland and Kerber, 1974) and showed variable genetic linkage with genes for tenacious glumes (*Tg*) and wax inhibition (*W2¹*) (Dyck and Kerber, 1970; Rowland, 1972; Rowland and Kerber, 1974; Dyck, 1979).

Lr22a (Dyck, 1979)   (Plate 2-20)

Adult plant resistance

Synonym

Lr22 (Rowland and Kerber, 1974).

Low Infection Type

Low IT on flag leaves is ; to ;1.

Environmental Variability

Low.

Origin

T. tauschii var. *strangulata* R.L.5271 (Dyck and Kerber, 1970).

Pathogenic Variability

No known virulence. Observations are from field sowings.

Reference Stocks

i: Thatcher*6/R.L.5404, R.L.6044 (Dyck and Johnson, 1983).
v: Tetra Canthatch/*T. tauschii* var. *strangulata* R.L.5271, R.L.5404 (Dyck and Kerber, 1970).
dv: *T. tauschii* var. *strangulata* R.L.5271 (Dyck and Kerber, 1970).

Source Stocks

AC Minto *Lr11 Lr13* (JA Kolmer, *pers. comm.* 1994).

Use in Agriculture

The only wheat to be released which possesses *Lr22a* is the Canadian cultivar AC Minto, registered in 1993 (Townley-Smith *et al.*, 1993).

 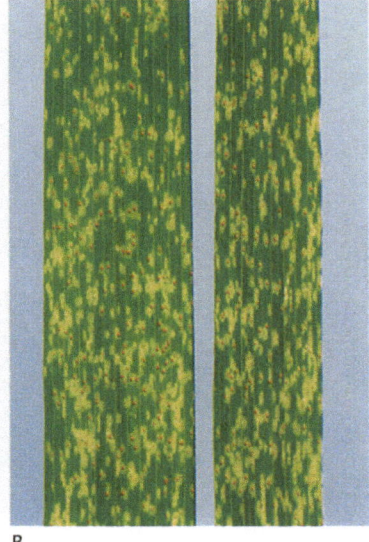

A   B

PLATE 2-20. *Lr22a*
Adult leaves (flag-1) of: Tc + *Lr22a* (left) and Thatcher (right); infected with A. pt. 104-2,3,6,(7) and B. pt. 53-1,(6),(7),10,11. Note the similarity in response of the two wheats to the latter pathotype. Because of the allelism, Tc + *Lr22a* cannot carry the *Lr22b* allele of Thatcher. Therefore, pt. 53-1,(6),(7),10,11 is considered avirulent for *Lr22a* and *Lr22b*.

Lr22b (Dyck, 1979)   (Plate 2-21)

Adult plant resistance

Low Infection Type

Low IT on adult leaves is ;1⁻ to ;1.

Low IT on adult leaves is $;1^-$ to $;1$.

Environmental Variability

Low.

Origin

Common wheat cv. Marquis (Bartos *et al.*, 1969*a*; Park and McIntosh, 1994).

Pathogenic Variability

Lr22b is effective against relatively few isolates of *P. recondita* f. sp. *tritici*. Consequently *Lr22b* will not be recognised unless controlled single pathotype tests on adult plants are undertaken. Because Thatcher was used as a standard susceptible line for leaf rust work all near-isogenic lines based on Thatcher, except that for *Lr22a*, are likely to carry *Lr22b*. *Lr22b* was first identified in Australia when Manitou, a backcross derivative of Thatcher with *Lr13*, was found to be resistant to a pathotype virulent on certain wheats with *Lr13*. In controlled tests, Thatcher was resistant to this pathotype (RA McIntosh and E Gordon-Werner, unpublished 1988). Park and McIntosh (1994) isolated a second pathotype with avirulence for *Lr22b*, in samples from Queensland and northern New South Wales. Identified as pt. 64-11 this pathotype is different from those found previously.

Reference Stocks

v: Canthatch (Dyck, 1979); Marquis (Bartos *et al.*, 1969*a*); Thatcher (Dyck, 1979).

Source Stocks

v: Manitou *Lr13* (RA McIntosh and E Gordon-Werner, unpublished 1988). Because of the widespread use of Thatcher as a parent in wheat breeding, *Lr22b* is likely to be present, but difficult to detect, in many wheats.

Use in Agriculture

Extremely limited use. In Australia *Lr22b* can be used in combination with *Lr13* to provide resistance to all current pathotypes. Because pt. 53-1,(6),(7),10,11 with avirulence for *Lr22b* is an introduction to Australasia (Luig *et al.*, 1985), avirulence is presumably present in another geographic area. The source of this pathotype is unlikely to be North America. L Broers (*pers. comm.* 1992) suggested that Thatcher displayed high levels of resistance in the field to five European pathotypes.

PLATE 2-21. *Lr22b*
Adult leaves (flag-1) of (L to R): Thatcher, Marquis and Condor; infected with A. pt. 53-1,(6),(7),10,11 [*P22b*] and B. pt. 104-2,3, 6,(7) [*p22b*]. The similar responses of Thatcher and Marquis with the first pathotype indicate that both carry the same gene, that is, *Lr22b*, and the intermediate response of Marquis with the second pathotype shows the presence of an undocumented gene for resistance.

Lr23 (McIntosh and Dyck, 1975)   (Plate 2-22)

Synonym

LrG (McIntosh and Luig, 1973*a*).

Chromosome Location

2BS (McIntosh and Dyck, 1975). *Lr23* was mapped 4 cM from the centromere (McIntosh and Dyck, 1975). It is situated very close to *Lr13* (RA McIntosh and WM Hawthorn, unpublished 1981) and linkage estimates with the *Sr9* locus in the long arm varied from 24 to 36 cM (McIntosh and Dyck, 1975; McIntosh, 1978; McIntosh *et al.*, 1981).

Low Infection Type

; to 3.

Environmental Variability

High; *Lr23* is more effective at temperatures above 20°C (Dyck and Johnson, 1983).

Origin

T. *turgidum* var. *durum* cv. Gaza (Watson and Luig, 1961). *Lr23* was transferred to the hexaploid level via Bobin W39*2/Gaza, selections of which became Gabo and Timstein (Watson and Stewart, 1956).

Pathogenic Variability

Lr23 is no longer widely effective in most geographic areas except Europe. Virulence for *Lr23* is common in Australia (Park and Wellings, 1992) and Mexico (Singh, 1991). Pathogenicity for *Lr23* is not routinely surveyed in North America (Long and Kolmer, 1989) presumably due to consistent difficulties in scoring. In most areas pathogenicity for *Lr23* shows a wide range of variation over space and time.

Reference Stocks

i: Lee FL310/6*Thatcher, R.L.6012 (McIntosh and Dyck, 1975).
s: CS*7/Kenya Farmer 2B (McIntosh and Dyck, 1975); CS*6/Timstein 2B (McIntosh and Dyck, 1975).
v: Gabo *Lr10* (Watson and Luig, 1961); Lee *Lr10* (Watson and Luig, 1961), Canadian workers found Lee to be genetically heterogenous for *Lr23* and selected FL310 as a carrier of the gene; Timstein *Lr10* (Watson and Luig, 1961).
tv: Gaza (Watson and Luig, 1961).

Source Stocks

v: See McIntosh (1988*a*).
Africa: Kenya Farmer *Lr10* (Watson and Luig, 1961).
Australia: Canna; Cranbrook; Madden; Rosella. Kulin *Lr1*. Warigal *Lr1 Lr20*. Gamenya *Lr3a*; Wialki *Lr3a*.
Europe: Rocta and various numbered accessions (Odintsova and Peusha, 1982).
Indian Subcontinent: HI977 *Lr10*; Hyb 65 *Lr10*. HD2135 *Lr10 Lr13*; HD2270 *Lr10 Lr13*; HD2278 *Lr10 Lr13*. HD2204 *Lr13*; HD2258 *Lr13*; HD2281 *Lr13*; HD2285 *Lr13*; HUW 213 *Lr13*; UP262 *Lr13*. DL153-2 *Lr34*; Girija *Lr34* (Singh and Gupta, 1991). The *Lr13 Lr23* combination in some of the above wheats was also demonstrated in genetic studies by E Gordon-Werner and RA McIntosh (unpublished, 1989). Because *Lr13* and *Lr23* are closely linked it would be interesting to know the relationships of wheats with these genes in order to establish whether the combination has occurred repeatedly. Once combined, the two genes would tend to remain together.
Mexico: (Singh and Rajaram, 1991; Singh, 1993). Mexico 82. Glennson 81 *Lr26*; Seri 82 *Lr26*. Curinda 87 *Lr26 Lr34*.
North America: Crim (Odintsova and Peusha, 1982).
tv: Many durum wheats show seedling responses that are similar to those of Gaza indicating that *Lr23* may occur widely in tetraploid wheats.

Use in Agriculture

Lr23 is no longer widely effective in Australia, thus limiting its use to gene combinations. However, it is known to confer resistance to some important Indian pathotypes and probably plays a role in gene combinations. The significance of the *Lr23Lr13* combinations in India is unknown. In other regions it could contribute to multiple gene resistance.

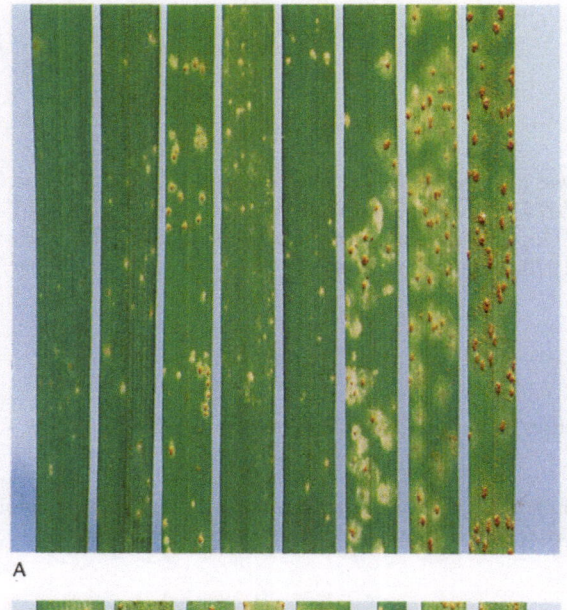

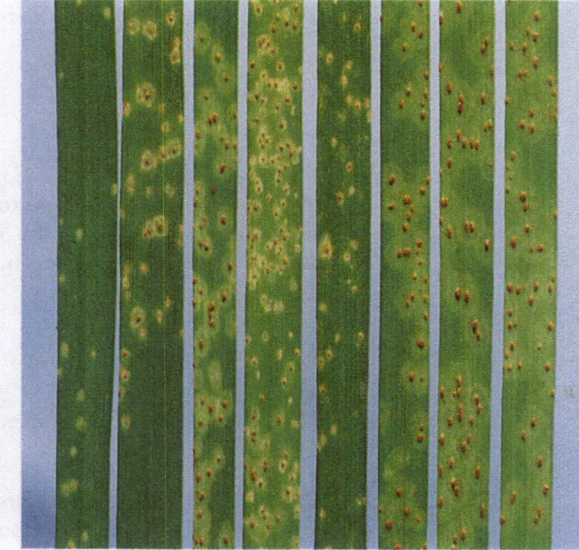

PLATE 2-22. Lr23
Seedling leaves of (L to R): Gaza durum, Lee FL310, Tc+*Lr23*, HD2278, HD2270, HD2204, Manitou and Thatcher; infected with A. pt. 26-1,3 [*P13P23*], B. pt. 53-1,(6),(7),10,11 [*p13P23*] and C. pt. 104-2,3,6,(7) [*P13p23*] and incubated at 21°C. The first five wheats carry *Lr23*. The fourth to seventh wheats carry *Lr13*. HD2278 (fourth) and HD2270 (fifth) have both *Lr23* and *Lr13* and are thus resistant to all three pathotypes.

Lr24 (McIntosh *et al.*, 1976)   (Plate 2-23)

Synonym

> *LrAg* (Browder, 1973*b*)

Chromosome Location

> 3D (Smith *et al.*, 1968; McIntosh *et al.*, 1976; Sears, 1977); 3DL (Hart *et al.*, 1976). Two translocations produced by Sears involved chromosome 3BL (Sears, 1977). A further translocation is present in cv. Amigo (The *et al.*, 1992). D The (*pers. comm.* 1992) crossed an Amigo derivative with a line possessing *Lr24* derived from one of ER Sears' derivatives. He obtained segregation indicating location of the two translocations in different chromosomes. In a recent study, J Jiang (*pers. comm.* 1993) located the *Thinopyrum* chromatin in a rust resistant derivative of Amigo to the satellite region of chromosome 1B using *in situ* hybridisation. *Lr24* is always completely associated with *Sr24* (McIntosh *et al.*, 1976).

Low Infection Type

> ; to ;1, sometimes 2C.

Environmental Variability

Moderate (Browder, 1980; Dyck and Johnson, 1983). RF Park (unpublished, 1990) observed higher infection types (;12$^=$) at low temperatures.

Origin

Th. ponticum. Leaf rust resistance was originally detected in a bread wheat line possessing a spontaneous translocation involving chromosome 3D and a chromosome from the donor species. Agent was selected from a cross of this line and cv. Triumph (Gough and Merkle, 1971). Sears (1972*b*, 1973, 1977) reported 21 independently derived translocations with *Lr24*, and subsequently modified the alien segment in two of the stocks to derive interstitial translocation stocks with reduced alien chromatin (*pers. comm.* 1989).

Pathogenic Variability

Virulence for *Lr24* occurs in North America (Gough and Merkle, 1971; Long and Kolmer, 1989), South America (Singh, 1991) and South Africa (Pretorius *et al.*, 1990) probably reflecting selection of virulent mutants as a consequence of the use of *Lr24* in commercial cultivars. Virulence for *Lr24* occurs at low frequencies in most other geographical areas (Huerta-Espino, 1992) but has not been found in Australia or India.

Reference Stocks

i: Thatcher*6/Agent, R.L.6064 (Dyck and Johnson, 1983); various Chinese Spring translocation stocks produced by Sears (1972*b*, 1973, 1977).
s: TAP67; CS/*Th. ponticum* 3Ag(3D) (ER Sears, *pers. comm.* 1975).
v: Agent (Smith *et al.*, 1968; McIntosh *et al.*, 1976); Amigo (The *et al.*, 1992).

Source Stocks

Many (see McIntosh, 1988*a*). Some examples:
Australia: Janz; Sunelg; Torres; Vasco. Sunco *Lr3a*. These white-seeded wheats are derived from translocation lines produced by ER Sears.
North America: Amigo (The *et al.*, 1992); Arkan; Centura; Century; Cimmaron; Cloud; Cody; Collin; Fox; Mesa; Osage; Rio Blanco; Sage; TAM 200; Timpaw; Trailblazer; Twain; Wanken. Thunderbird *Lr1*. Blueboy II *Lr1 Lr10*. Karl *Lr1 Lr3a Lr10*. Butte 86 *Lr2a Lr10*. Payne *Lr3a*. Siouxland *Lr3a Lr2b*. Coker 9733 *Lr9*. Terral 101 *Lr9 Lr11 Lr18*. Abilene *Lr10*; Colt *Lr10*; Norkan *Lr10*; Parker 76 *Lr10*. Arapahoe *Lr10 Lr16*. Jasper *Lr26*; Longhorn *Lr26*.
South Africa: Gamka; Kinko; Palmiet; SST25; SST44; SST102.

Use in Agriculture

Although *Lr24* is widely ineffective in North and South America and in South Africa, it remains effective in Australia and the Indian subcontinent. Several cultivars with *Lr24* are in use in Australia. The initial impediment to the use of *Lr24* from Agent was its association with red grain colour, but RA McIntosh and M Partridge (unpublished, 1974) were able to recover white-seeded recombinants from two translocation stocks (CS 3D/3Ag#3 and CS 3D/3Ag#14) from ER Sears. The *et al.* (1992) found no association between red grain colour and *Lr24* in Amigo.

A   B

PLATE 2-23. Lr24
Seedling leaves of (L to R): Agent, CS 3D/3Ag#3, Sunco, Cook and Chinese Spring; infected with A. pt. 26-1,3 and B. pt. 104-2,3,6,(7). Sunco was derived from a backcross program which incorporated *Lr24* into the cultivar Cook. Sunco and Cook carry *Lr3a* (or possibly *Lr3bg*) which is not effective against pt. 104-2,3,6,(7). Sunco is more resistant than Agent and CS 3D/3Ag#3 to pt. 104-2,3,6,(7). Cook and Sunco may also carry *Lr34*.

Lr25 (McIntosh, 1988*a*) (Plate 2-24)

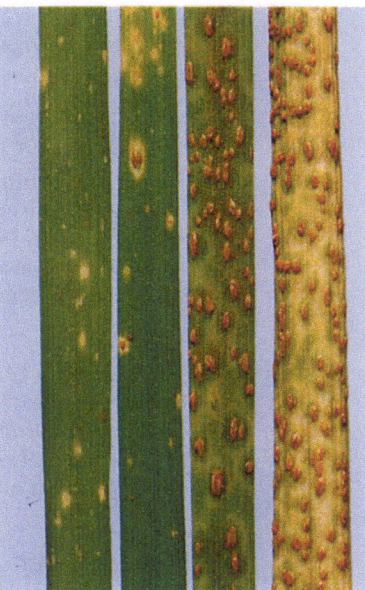

PLATE 2-24. *Lr25*
Seedling leaves of (L to R): Transec, Transfed, Chinese Spring and Federation; infected with pt. 104-2,3,6,(7).

Chromosome Location

4BS (Driscoll and Anderson, 1967; Driscoll and Bielig, 1968). *Lr25* is always associated with gene *Pm7* for resistance to powdery mildew (Driscoll and Anderson, 1967) and both are located in a segment of a cereal rye chromosome. The segment maps 1 cM from the 4B centromere (Driscoll and Bielig, 1968).

Low Infection Type

0;, ;1, 2$^+$.

Environmental Variability

Low (Browder, 1980).

Origin

Leaf rust resistance in *Secale cereale* cv. Rosen was transferred to a common wheat line, later designated Transec (Driscoll and Jensen, 1964).

Pathogenic Variability

Although *Lr25* has not been widely tested, virulence has been detected in Canada (PL Dyck, *pers. comm.* 1993), Algeria, China, Bulgaria, Egypt, Ethiopia, Hungary, Israel and Pakistan (Huerta-Espino, 1992).

Reference Stocks

i: Thatcher*7/Transec, R.L.6084 (PL Dyck, *pers. comm.* 1986).
v: Transec; Transfed (CJ Driscoll, *pers. comm.* 1970). It is not known if Transfed also carries *Lr10* from its Federation parent.

Source Stocks

None known.

Use in Agriculture

Lr25 has not been successfully introduced to agriculture. Lines with *Lr25* are agronomically inferior to comparable near-isogenic lines lacking the gene.

Lr26 (McIntosh, 1988*a*) (Plate 2-25)

Chromosome Location

1B (1BL.1RS) or 1R(1B) (Mettin *et al.*, 1973; Zeller, 1973). Some wheats consist of both substitution and translocation biotypes (Zeller, 1973). *Lr26* is completely linked with *Sr31* and *Yr9*.

Low Infection Type

0;$^=$, ;1, X2, 3. Isolates of *P. recondita tritici* can be distinguished by distinctive responses on host lines with *Lr26*. The mechanism of variability is not understood, although the variation may represent an example of 'progressive increases in virulence' (Watson and Luig, 1968).

Environmental Variability

There are no reports on the effects of environment. RF Park (unpublished, 1992) observed significant effects of temperature on intermediate responses produced by *Lr26*, but there were no effects when responses were very low (see Plate 2-25 D. and E.).

Origin

S. cereale cv. Petkus. Most wheats with *Lr26* are derived from wheat x rye hybrid derivatives produced in Germany in the 1930s (see Mettin *et al.*, 1973; Zeller, 1973).

Pathogenic Variability

Although variation in the avirulence phenotype is common, full virulence has emerged in most wheat-growing areas; for example, Europe (Bartos *et al.*, 1984), the Indian subcontinent (Nayar *et al.*, 1991), North America (Kolmer, 1991; Singh, 1991; Long *et al.*, 1992), South Africa

A

B

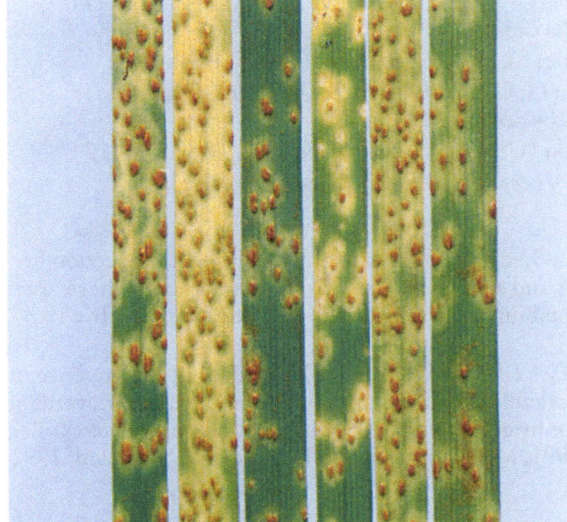

C

PLATE 2-25. *Lr26*

A. to C. Seedling leaves of (L to R): Kavkaz, Aurora, Veery#1, Veery#2, Federation*4/Kavkaz and Federation; infected with A. pt. 10-1,2,3,4 [*P3aP26*], B. pt. 104-2,3,6,(7) [*p3aP26*] and C. pt. 104-2,3,6,9 [*p3ap26*]. Aurora may carry *Lr3a* (from its Bezostaja parent) hence the lower response with pt. 10-1,2,3,4. Veery#2 possesses *Lr13* whereas Veery#1 does not, hence the different responses with pt. 104-2,3,6,9. The large pustule in the centre of the Veery#1 leaf inoculated with pt. 104-2,3,6,(7) is a likely contamination. Additional genes probably occur in the Veery lines and the Federation derivative may carry *Lr10*.

▼ D. and E. Seedling leaves of (L to R): Mildress, Federation*4/Kavkaz and Federation; infected with two cultures of pt. 104-2,3,(6),(7),11, that is, [84045] (L) and [91026] (R) and incubated at D. 22–25°C and E. 18°C. The low response of Mildress and Federation*4/Kavkaz with culture 84045 was not affected by temperature, whereas the intermediate responses with culture 91026 was significantly affected.

D

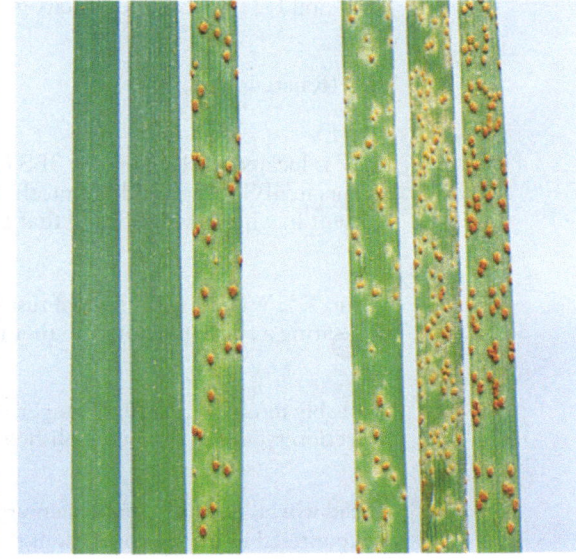

E

(Pretorius *et al.*, 1990), and South America and China (Huerta-Espino, 1992). Virulence is less common in Australasia where *Lr26* is not widely deployed (Park and Wellings, 1992).

Reference Stocks

i: Thatcher*6/ST-1.25, R.L.6078 (PL Dyck, *pers. comm.* 1990). Federation*4/Kavkaz (RA McIntosh, unpublished 1986); this line may have *Lr10*.
v: Aurora (Zeller, 1973); Kavkaz (Zeller, 1973).

Source Stocks

Many wheats have *Lr26*. In addition to the publications of Mettin *et al.* (1973) and Zeller (1973) with emphasis on European wheats, further genotypes including *Lr26* can be found in Merker (1982), Bartos *et al.* (1983, 1984), Schmidt *et al.* (1985), Hu and Roelfs (1986), Bartos and Valkoun (1988), Knott (1989), Singh and Gupta (1991), Singh and Rajaram (1991) and Singh (1993); see also McIntosh (1988*a*) and annual supplements to the gene symbols catalogue for wheat. Because *Lr26* is also associated with *Yr9* and *Sr31*, perusal of lists for these genes will provide genotypes likely to have *Lr26*. A further linked gene, *Pm8*, is not always expressed in wheats with the three rust resistance genes (Friebe *et al.*, 1989; Lutz *et al.*, 1992).

Africa: Gamtoos (=Veery #3); HAR800; HAR1023; Loeric (=Veery #5); Seri 82; Seric (=Veery #5); Vee; Viri (=Veery #5).

Australia: Grebe; Warbler.

China: Feng Kang 13; Ji Mai 23. *Lr26* was found to be present in a large number of Chinese wheats (RA McIntosh, unpublished 1993).

Europe: Clement; Hornet; Lima 1 (=Veery #3); Lovrin 10; Lovrin 13; Mildress; Selekta; Skorospelka; Slejpner. Agra *Lr3a*; Branka *Lr3a*.

Indian Subcontinent: HUW206; Pakistan 81 (=Veery #5); Sarhad 83 (=Bobwhite).

Mexico: Bobwhite; Pamir. Cumpas 83 *Lr13*; Genaro 81 (=Veery #3) *Lr13*; Ures 81 *Lr13*. Glennson 81 (=Veery #1) *Lr23*; Seri 82 (=Veery #5 'S') *Lr23*.

North America: Excel; Freedom; Salmon (USA). Siouxland *Lr3a Lr24*. Longhorn *Lr24*.

South America: Alondra; Cordillera (=Veery #3); Millaleau Inia (=Veery #3).

Use in Agriculture

This alien source of disease-resistance has had a major impact on global wheat production as indicated by its presence in many winter and spring wheats. This impact appears to have arisen with the high yield and widespread adaptability frequently found in wheats carrying the 1RS chromosome (Zeller and Hsam, 1983).

Researchers in Germany and Australia (see Dhaliwal *et al.*, 1988) showed that the presence of 1RS results in flour which produces undesirable dough stickiness that is aggravated by overmixing. The presence of 1RS in wheat can be confirmed by various disease responses as well as cytological and biochemical tests (Javornik *et al.*, 1991; May and Wray, 1991; Gupta and Shepherd, 1992).

Lr27 and *Lr31* (*Lr27* + *Lr31*) (Singh and McIntosh, 1984*a*)   (Plate 2-26)

Lr27 and *Lr31* are complementary genes. Both must be present for the expression of resistance.

Synonym

LrT (Rajaram *et al.*, 1971).

Chromosome Location

Lr27 is located in chromosome 3BS (Singh and McIntosh, 1984*b*); the complementary gene *Lr31* occurs in 4BS (Singh and McIntosh, 1984*b*). Singh and McIntosh (1984*a*) showed that *Lr27* was present in wheats with *Sr2* and that *Lr31* occurred in Chinese Spring.

Low Infection Type

$X^=$ to X^+. Wheats with this leaf rust specificity are designated as genotype *Lr27* + *Lr31*, the + indicating a single phenotype rather than different phenotypes associated with each gene.

Environmental Variability

Probably moderate. At high temperatures, chlorosis associated with *Sr2* can obscure the low infection type (RF Park, unpublished 1990).

Origin

Bread wheat. Because complementary genes are involved, the resistance phenotype may be encountered in hybrid populations derived from intercrosses of wheats possessing the individual genes. This resistance has been used mainly in Australia and was encountered by chance in the CS (Hope 3B) substitution line produced by ER Sears and in the CS (Ciano 67 3B) and CS (Ciano 67 5B) substitution lines generated at Plant Breeding Institute, Cambridge.

Pathogenic Variability

Pathogenic variability occurs in Australia (Rajaram *et al.*, 1971*a*) and Mexico (Singh, 1991) and most other geographic regions (Huerta-Espino, 1992).

Reference Stocks

s: CS*6/Hope 3B, CS*7/Ciano 67 3B, CS*7/Ciano 67 5B (Singh and McIntosh, 1984*b*). The presence of *Lr27* in the last line is assumed to be a chance occurrence.
v: Gatcher *Lr10* (Singh and McIntosh, 1984*b*).

Source Stocks

See lists in McIntosh (1988*a*), Singh and Rajaram (1991), Singh and Gupta (1991).
Australia: SUN27A *Lr1 Lr2a*. Timgalen *Lr3a* (heterogeneous) *Lr10*.
Indian Subcontinent: HW517 *Lr13*.
Mexico: Ocoroni 86. Tonichi 81 *Lr1 Lr13*. Zaragosa 75 *Lr3bg*. Cocoraque 75 *Lr13 Lr17 Lr34*. Jupateco 73 *Lr17 Lr34* (heterogeneous).

Use in Agriculture

Cultivars Timgalen and Gatcher were released in Australia with leaf rust resistance conferred by *Lr27 + Lr31*. Their release was rapidly followed by an increase in the incidence of virulent pathotypes such as 162-1,2,3,6 and 76-1,2,3,6. Resistance in Gatcher was derived from the 1959 International Spring Wheat Rust Nursery Entry No. 111 (Singh and McIntosh, 1984*a*) whereas the source of resistance in Timgalen is unknown. Timgalen is of further interest because it is considered not to carry *Sr2* whereas all other sources of *Lr27* appear to be associated with this gene. The seedling response of Timgalen is not as low as that of Gatcher.

Lr28 (McIntosh *et al.*, 1982)   (Plate 2-27)

Chromosome Location

4AL (McIntosh *et al.*, 1982). The chromosome segment containing *Lr28* mapped 53 cM from the centromere.

Low Infection Type

0;$^=$, sometimes 1$^+$ to 2$^+$ especially with South American isolates (Huerta-Espino, 1992).

Environmental Variability

Low.

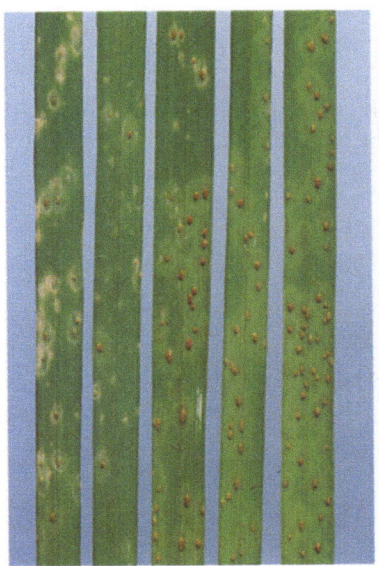
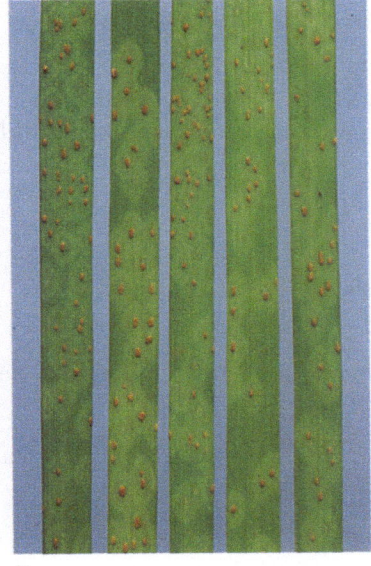

PLATE 2-26. *Lr27 + Lr31*
Seedling leaves of (L to R): Gatcher, CS/Hope 3B, CS/Ciano 67 5B, CS/Ciano 67 3B and Chinese Spring; infected with A. pt. 122-1,2,3 and B. pt. 76-1,2,3,6 and incubated at 18°C. CS/Ciano 67 3B is usually higher in response to avirulent pathotypes than the other three carriers.

A

B

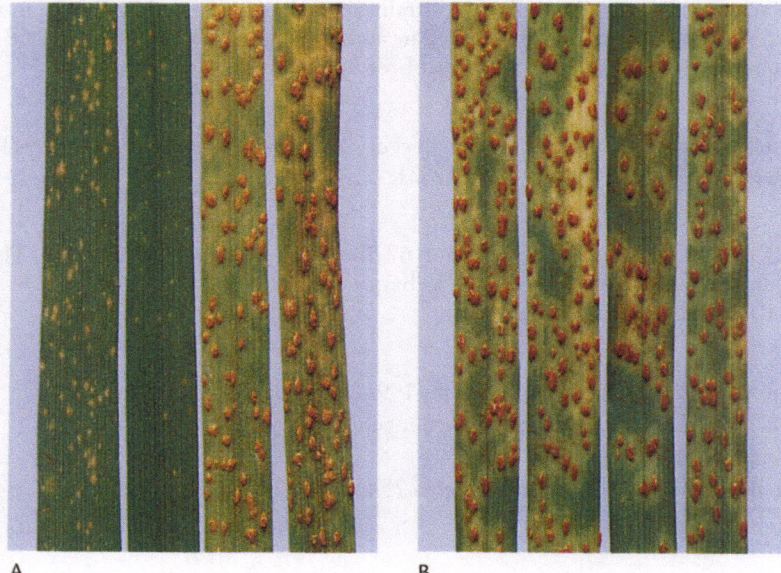

PLATE 2-27. *Lr28*
Seedling leaves of (L to R): CS 2A/2M 4/2, CS 2D/2M 3/8, Compair and Chinese Spring; infected with A. pt. 104-2,3,6,(7) and B. pt. 104-2,3,6,(7),8. The latter culture was generated from a single pustule observed on a seedling with *Lr28*. Although Compair carries *Yr8* in common with the first and second lines, it does not have *Lr28*.

Origin

T. *speltoides* (McIntosh *et al.*, 1982). This gene was inadvertently transferred during the induction of translocations between T. *comosum* chromosome 2M and group 2 wheat chromosomes using T. *speltoides* to disrupt the chromosome pairing control in wheat. RA McIntosh (unpublished, 1991) noted that *Lr28* is not present in the hexaploid wheat derivatives produced by Dvorak and Knott (1980; 1990).

Pathogenic Variability

Several spontaneous mutants with virulence for *Lr28* have been isolated in Australia. The frequency of virulence for this gene is extremely high in North America (Singh, 1991). Virulence has not been found in India but Huerta-Espino (1992) found virulence among isolates from Pakistan and from most other major geographic areas.

Reference Stocks

i: CS 2A/2M 4/2 (C77.2); CS 2D/2M 3/8 (C77.1) (McIntosh *et al.*, 1982). Thatcher*6/C77.1, R.L.6079.

Source Stocks

v: Sunland *Lr3a* (GN Brown, *pers. comm.* 1992).

Use in Agriculture

Cultivar Sunland was registered in New South Wales, Australia, as a prime hard cultivar in 1992. Although no detrimental effects appear associated with the presence of *Lr28*, the durability of resistance is likely to be low.

Lr29 (McIntosh, 1988*a*)   (Plate 2-28)

Chromosome Location

7DS (7D/Ag translocation), 7AgS (Sears, 1977; ER Sears, *pers. comm.* 1980).

Low Infection Type

;1N (Knott, 1989), ; to 2[+] (Huerta-Espino, 1992). These responses are usually higher and distinguishable from that conferred by *Lr19*.

Environmental Variability

Low.

Origin

Th. ponticum. ER Sears initially isolated this line as a leaf rust resistant 7D/Ag translocation stock. He later distinguished it from the 7D/Ag translocations with *Lr19* because it lacked the yellow flour pigment associated with *Lr19* (Knott, 1989; see discussion in Sears, 1973). Tests on an array of 7D/Ag transfers by RA McIntosh (unpublished, 1977) established that the low infection type for CS 7D/Ag#11 was distinctly different from Agatha and other CS7D/Ag lines. Moreover #11 lacked the stem rust resistance conferred by *Sr25*. ER Sears then supplied lines with the derived 7AgL and 7AgS telocentric chromosomes. The *Lr19/Sr25* phenotypes were displayed by the 7AgL line and the *Lr29* phenotype was expressed by the 7AgS line. Thus the original 7Ag chromosome carried *Lr19* and *Lr29*, but in rust tests the resistance conferred by *Lr19* was always epistatic.

Pathogenic Variability

In a global survey Huerta-Espino (1992) found only three isolates with virulence on seedlings with *Lr29*; one from Pakistan and two from Turkey.

Reference Stocks

i: CS 7D/Ag #11 (Sears, 1973; 1977). Thatcher*6//CS 7D/Ag#11, R.L.6080 (PL Dyck, *pers. comm.* 1988).

Source Stocks

None.

Use in Agriculture

Lr29 has not been exploited.

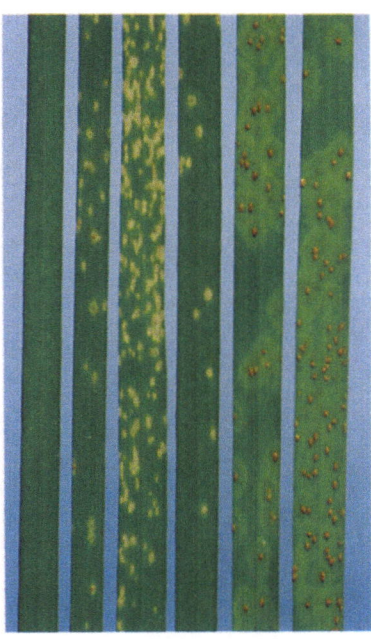

PLATE 2-28. *Lr29*
Seedling leaves of (L to R): CS + 7AgL ditelosomic addition, CS + 7AgS ditelosomic addition, CS 7D/Ag#11, Tc + *Lr29*, Thatcher and Chinese Spring; infected with pt. 104-2,3,6,(7) and incubated at 18°C. Note the distinctly lower response of the first line, which carries *Lr19*, compared with the following three wheats with *Lr29*.

Lr30 (Dyck and Kerber, 1981)   (Plate 2-29)

Synonym

LrT (Dyck and Kerber, 1981).

Chromosome Location

4AL. Inheritance tends to be recessive. *Lr30* mapped 3 cM from the centromere (Dyck and Kerber, 1981).

Low Infection Type

1^+ to 2 (Dyck and Kerber, 1981); ;1^- to 2^+ (Dyck and Johnson, 1983); 23^- (Singh and Gupta, 1991); ; to 2^+ (Huerta-Espino, 1992); 2^- to 3^- (RA McIntosh, unpublished 1985).

Environmental Variability

The usual experience in Australia has been that higher responses to avirulent pathotypes are obtained at higher temperatures. However, lower responses (IT ;12^-) are obtained using adult plants. Temperature sensitivity of the low infection type appears to be influenced by pathogen isolate (Dyck and Johnson, 1983).

Origin

Common wheat cv. Terenzio (Dyck and Kerber, 1981).

Pathogenic Variability

Virulence for *Lr30* is infrequent in Australia, Mexico (Singh, 1991; Huerta-Espino, 1992) and Canada (Kolmer, 1991). Long *et al.* (1993) found 24% and 72% virulence in two agroecological zones of north-eastern USA. Levels of virulence in other regions of the USA were low. In the international survey of Huerta-Espino (1992) virulence levels were generally low but in certain regions, for example Bulgaria and Turkey, they were about 50%. At various times in different surveys high virulence levels were reported. Because *Lr30* is unlikely to be present in the wheats under cultivation in the areas involved, variation in the frequencies of virulence for *Lr30* must be under cultivation in the areas involved, variation in the frequencies of virulence for *Lr30* must be independent of factors associated with pathogenicity for this gene.

The inheritance of pathogenicity on the near-isogenic Thatcher derivative R.L.6049, was studied by Samborski and Dyck (1976). Their culture of race 11 was heterozygous at the locus corresponding to *Lr30*. Kolmer (1992*a*) showed that virulence for *Lr30* was much higher in aecial populations than in natural uredial collections indicating high levels of heterozygosity in the latter.

Reference Stocks

i: Thatcher*6/Terenzio, R.L.6049 (Dyck and Kerber, 1981).
v: Terenzio P.I.269250 *Lr34* (Dyck and Kerber, 1981).

Source Stocks

None known.

Use in Agriculture

Current use is probably very limited. R.L.6049 shows a degree of reduced rusting relative to Thatcher under Australian (RA McIntosh, unpublished 1985) and Indian field (Sawhney *et al.*, 1992) conditions. This indicates there may be a role for *Lr30* in gene combinations. Because more effective alternative sources of resistance are available to breeders, these will be given preference in selection based on either seedling or adult plant response.

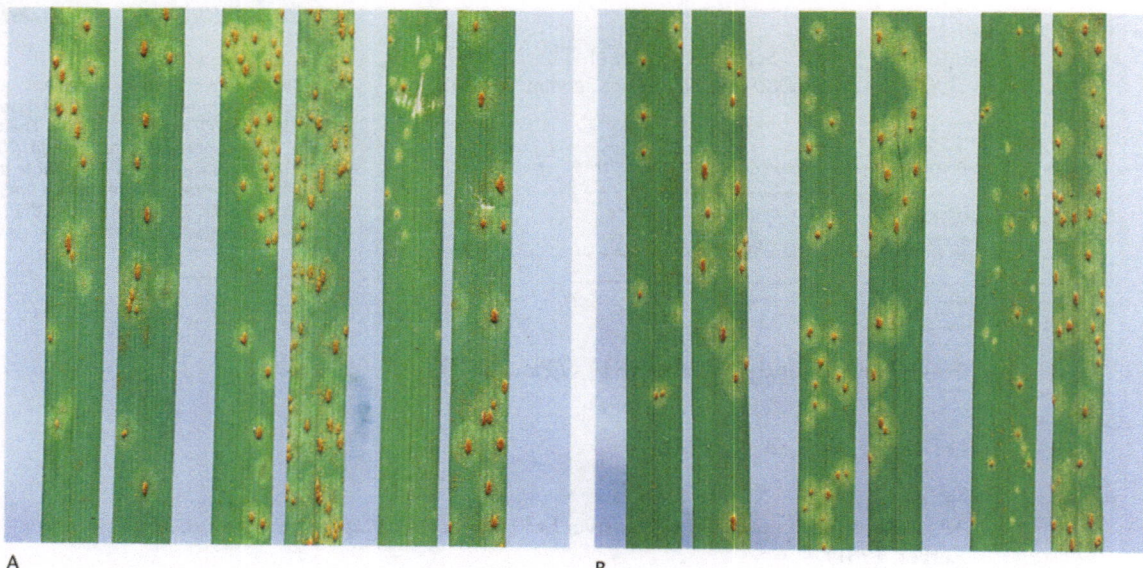

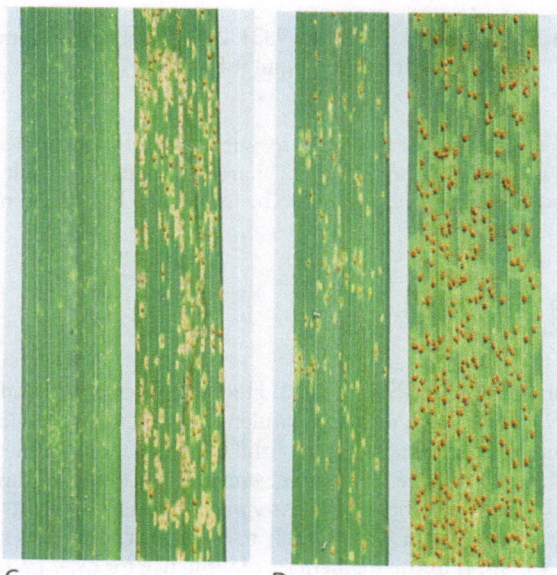

PLATE 2-29. *Lr30*

A. and B. Seedling leaf pairs of: Tc + *Lr30* (L) and Thatcher (R); infected with pt. 10-1,2,3,4 (L) pt. 104-2,3,6,(7) (C) and pt. 53-1,(6),(7),10,11 (R) and incubated at A. 15°C and B. 22/25°C. In this comparison *Lr30* appeared to be less effective at the higher temperature. Pt. 53-1,(6),(7),10,11 is clearly less virulent than the other two and pt. 104-2,3,6,(7) may be more virulent than the others. However, this latter pathotype is avirulent on adult plants with *Lr30*.

C. and D. Adult leaves (flag-1) of: Tc + *Lr30* (L) and Thatcher (R); infected with C. pt. 53-1,(6),(7),10,11 and D. pt. 104-2,3,6,(7),11. Under Australian conditions *Lr30* is much more effective at the adult plant stage. The high level of resistance that Tc + *Lr30* expresses to pt. 53-1,(6),(7),10,11 presumably results either from interaction of *Lr30* and *Lr22b* which is present in Thatcher, or from the higher level of avirulence of this pathotype as shown by the seedling test.

Lr31 (Singh and McIntosh, 1984*a*)   (Plate 2-26)

Lr31 acts as a complementary gene with *Lr27*. See *Lr27*.

Chromosome Location
4BS (Singh and McIntosh, 1984*b*).

Origin

Common wheat. Singh and McIntosh (1984*a*) showed that resistance segregated from crosses of Chinese Spring and various wheats with *Sr2*. Because *Lr31* alone confers no resistance the identification of sources is not possible unless *Sr27* is present, or unless appropriate test crosses are performed. See *Lr27* for further information.

Lr32 (Kerber, 1987)   (Plate 2-30)

Chromosome Location
3D (Kerber, 1987); 3DS (Kerber, 1988). *Lr32* mapped 30 cM from the centromere (Kerber, 1988).

Low Infection Type
; to 2 (Huerta-Espino, 1992).
1^+2^- in hexaploid derivatives (Kerber, 1987). Low infection types produced on the *T. tauschii* parent were 0; to 1.

Environmental Variability
Unknown; presumed to be stable.

Origin
T. tauschii R.L.5497-1 (Kerber, 1987).

Pathogenic Variability
Appears to be effective against all pathotypes in Australia (RA McIntosh, unpublished 1986) and Mexico (Singh, 1991) and most isolates in the USA (Long and Kolmer, 1989). Huerta-Espino (1992) identified single cultures with virulence originating from Bulgaria, Israel and Turkey; all other cultures were avirulent.

Reference Stocks
i: Thatcher*7//R.L.5497-1/Marquis-K, R.L.6086 (PL Dyck, *pers. comm.* 1988).
v: Tetra Canthatch/*T. tauschii* R.L.5497-1, R.L.5713; R.L.5713/Marquis-K (Kerber, 1987).
dv: *T. tauschii* R.L.5497-1 (Kerber, 1987).

Source Stocks
None.

Use in Agriculture
Not exploited.

PLATE 2-30. *Lr32*
Seedling leaves of: R.L.5713/Marquis-K and Thatcher; infected with pt. 104-2,3,6,(7).

Lr33 (Dyck *et al.*, 1987)   (Plate 2-31)

Chromosome Location
1BL (Dyck *et al.*, 1987). *Lr33* showed linkage of 3 cM with the centromere and with *Lr26* which is assumed to map at the centromere (Dyck *et al.*, 1987).

Low Infection Type
1^+2 (Dyck, 1977); 22^+ (Singh, 1991).

Environmental Variability
Unknown.

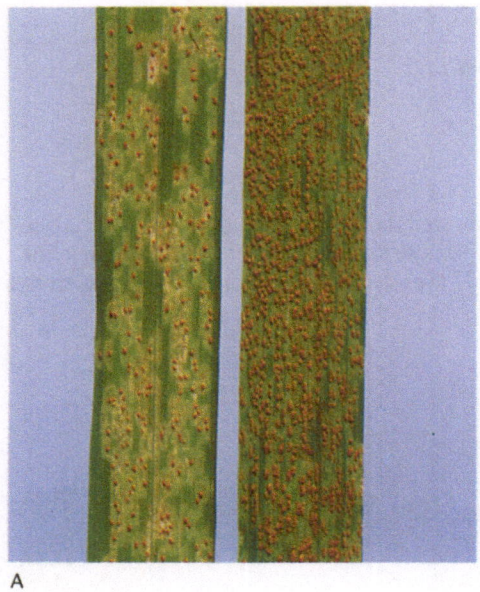

A

B

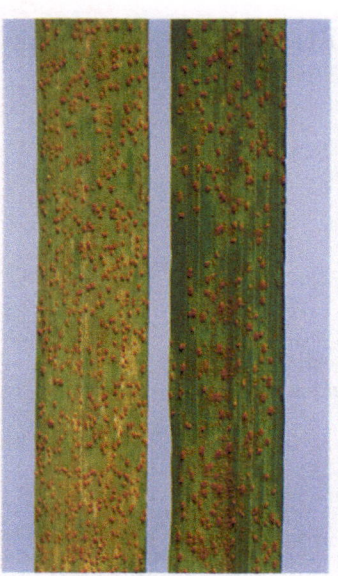

C

D

PLATE 2-31. *Lr33*
Adult leaves (flag-1) of: Tc +
Lr33 (L) and Thatcher (R);
infected with A. and B. pt. 104-
2,3,6,(7), upper (A.) and lower
(B.) leaf surfaces. C. and D.
pt. 104-2,3,(6),(7),11, upper (C.)
and lower (D.) leaf surfaces.
These greenhouse infections
indicate that the two pathotypes
differ in level of pathogenicity
and that a very low level of
avirulence occurs in pt. 104-
2,3,(6),(7),11. In these tests the
differences between upper and
lower leaf surfaces were very
obvious.

Origin

Common wheat. The original accession in which *Lr33* was identified came from China, but it was also present in wheats from Afghanistan and Iran (Dyck, 1977).

Pathogenic Variability

Dyck *et al.* (1987) state that resistance is conferred "to many races". Mexican isolates were noted to be avirulent (Singh, 1991). All Australian isolates appear virulent on seedlings with *Lr33*, but a degree of resistance was expressed on adult plants in the greenhouse (see Plate 2-31). Huerta-Espino (1992) used two Thatcher derivatives for *Lr33*; one possessed only *Lr33* whereas the second carried *Lr33* and *Lr34*. Pathogenic variability on seedlings of the line with *Lr33* occurred in most geographical areas. In some instances low responses were recorded for the *Lr33* + *Lr34* stock, whereas the respective *Lr33* and *Lr34* single gene lines gave high responses.

Reference Stocks

i: Thatcher*6/P.I.58548, R.L.6057 (Dyck, 1977).
v: P.I.268454a. P.I.124349 *Lr1 Lr34*. P.I.125387 *Lr2a*. P.I.268316 *Lr2c Lr34*. C.I.8749 *Lr34*;
P.I.58548 *Lr34*. See Dyck (1977) and Shang *et al.* (1986).

Source Stocks

None known.

Use in Agriculture

Probably limited. *Lr33* is not effective under field conditions in Australia where the predominant pathotype has been pt. 104-2,3,6,(7). In North America it provides effective resistance when in combination with other genes. For example, R.L.6059 (*Lr33 Lr34*) was noted to be resistant to all USA cultures (Long and Kolmer, 1989).

Lr34 (Dyck, 1987) (Plate 2-32)

Adult plant resistance

Synonym

LrT2 (Dyck and Samborski, 1982).

Chromosome Location

7D (Dyck, 1987). 7DS (PL Dyck, *pers. comm.* 1994). *Lr34* is completely linked with *Yr18* (Singh, 1992*a*) and a gene for leaf tip necrosis (Singh, 1992*b*).

Low Infection Type

Although designated as a gene for adult plant resistance, *Lr34* has been detected at the seedling growth stage, conferring ITs 3⁻ to 3 (Singh and Gupta, 1991), ; to 3 (Singh and Gupta, 1992), or 2 to 3 (Huerta-Espino, 1992). Seedlings of Line 897 were rated IT 2⁺ to 3, and adult plants were scored 5M by Dyck and Samborski (1982). Resistance at both growth stages increased when a second gene, *LrT3*, was also present (Dyck and Samborski, 1982). German and Kolmer (1992) described the enhancing effects of *Lr34* on the seedling and adult plant responses of some leaf rust resistance genes (see also Plate 2-32).

Environmental Variability

Infection types can vary from 0;1⁻ to 3 depending on temperature and light (Dyck and Samborski, 1982). According to these authors infection type is lowest under conditions of "cool temperature and low light intensity". Singh and Rajaram (1991) reported significantly lower responses at 18–22°C compared with 24–27°C. Resistance to three South African cultures was detected in a Thatcher line with *Lr34* under post-inoculation conditions of 7°C/low light (Drijepondt *et al.*, 1991). RF Park (unpublished, 1993) also detected resistance in a Thatcher line with *Lr34*, plus several Australian wheats when plants were kept post inoculation at 5–7°C and at low light intensity. Among these Australian wheats was Condor which Singh (1992*a*) considered to have *Lr34*. Seedling resistance has been very difficult to detect under the normal range of Australian greenhouse conditions (RA McIntosh, unpublished 1990).

Origin

Common wheat; many pedigrees of wheats with *Lr34* can be traced back to Frontana and other wheats from South America. Dyck (1991) discussed evidence suggesting that Chinese Spring may have been a contributor of *Lr34* following its introduction to South America around 1900. Frontana was originally developed in Brazil to combat a serious outbreak of stripe rust (see *Yr18*).

Pathogenic Variability

Presumably not common. Dyck and Samborski (1982) reported a unique culture (10-77) isolated from cv. Tobari 66 (*Lr1 Lr13 Lr34*) and virulent on seedlings of R.L.6050 (*Lr34*). This culture was also virulent on seedlings of Frontana and four additional lines produced at CIMMYT, but was not virulent on adult plants with *Lr34* (PL Dyck, *pers. comm.* 1993).

Reference Stocks

i: *Lr34* — Line 897 (Thatcher*6/Terenzio); Line 920 (Thatcher*6/Lageadinho); Thatcher*6/P.I.58548, R.L.6058 (Dyck and Samborski, 1982; Dyck, 1987).
Lr34 + *LrT3* — Thatcher*6/Terenzio, R.L.6050; Thatcher*6/Lageadinho, R.L.6069. German and Kolmer (1992) constructed 13 two-gene combinations of *Lr34* and other genes based on Thatcher.
v: Glenlea *Lr1* (Dyck *et al.*, 1985). Lageadinho P.I.197660 *LrT3*. Terenzio P.I.269250 *Lr3a Lr30 LrT3*. Frontana *Lr13 LrT3* (Dyck and Samborski, 1982).

Source Stocks

Lists of genotypes can be found in Shang *et al.* (1986), Roelfs (1988), Dyck (1991), Singh and Gupta (1991) and Singh (1992*a*; 1993).

Australia: Overseas workers suggested that some Australian wheats may possess *Lr34* (Singh, 1992*a*; RP Singh, *pers. comm.* 1994). These include cultivars Cook, Oxley, Egret and some selections of Condor. These wheats are derivatives of WW15 = Anza. Intercrosses among them failed to segregate for high levels of leaf rust and stripe rust susceptibility. F1 plants in the cross Oxley mono-7D/Jupateco 73 'S' selection segregated for leaf rust response with disomics less rusted than monosomics. In a corresponding cross with Jupateco 'R' selection all F1 plants responded similarly to disomic plants in the first cross (RA McIntosh, unpublished 1993).

China: Chinese Spring *Lr12*.

CIMMYT: Chapingo 53; Pima 77. Yecorato 77 *Lr1*. Trap *Lr1 Lr3a Lr10 Lr13*. Bajio 67 *Lr1 Lr13*; Jaral 66 *Lr1 Lr13*; Nortino 67 *Lr1 Lr13*; Sonoita 81 *Lr1 Lr13*; Yecora 70R *Lr1 Lr13*; Tobari 66 *Lr1 Lr13*. Torim 73 *Lr1 Lr13 Lr17*. Mango *Lr1 Lr13 Lr26*. Zacatecos 74 *Lr3a Lr13*. Victoria 81 *Lr3a Lr13 Lr17*. Cucurpe 86 *Lr10*; Nacozari 76 *Lr10*; Opata 85 *Lr10*. Esmeralda 86 *Lr10 Lr14a*. Parula *Lr13*; Rayon 89 *Lr13*; Salamanca 75 *Lr13*; Tesopaco 76 *Lr13*. Cocoraque *Lr13 Lr17 Lr27 + Lr31*. Cumpas 88 *Lr13 Lr26*. Penjamo 62 *Lr14a*; Yaqui 50 *Lr14a*. Jupateco 73R *Lr17 Lr27 + Lr31*. Curinda 87 *Lr23 Lr26*. Bucanora 88 *Lr26*. Ocoroni 86 *Lr27 + Lr31*.

Europe: Bezostaya *Lr3a*.

Indian Subcontinent: (Singh and Gupta, 1991) VL404; VL616. CPAN1235 *Lr1 Lr3a*. Lyallpur 73 *Lr1 Lr13*; UP115 *Lr1 Lr13*; WL2265 *Lr1 Lr13*. V84016-3 *Lr1 Lr13 Lr17*. BW11 *Lr10*. HD2182 *Lr10 Lr13*; HD2329 *Lr10 Lr13*; IWP72 *Lr10 Lr13*; Punjab 81 *Lr10 Lr13*; RAJ2535 *Lr10 Lr13*. DL153-2 = Kundin *Lr23*; Girija *Lr23*.

North America: Sturdy *Lr10 Lr12 Lr13*. Era *Lr10 Lr13*. Anza *Lr13* (heterogeneous); Chris *Lr13*.

South America: La Prevision *Lr13*. Frontana *Lr13 T3*.

Use in Agriculture

Because Frontana and its relatives and derivatives are considered to be sources of durable resistance to leaf rust, they have been used extensively in breeding programs. Recent studies are increasingly implicating the presence and value of *Lr34* as a source of resistance to leaf rust and only recently has it been understood in a way that will enable its direct manipulation in breeding programs. Because of its widespread effectiveness as a source of resistance under field conditions and its interactive effects (German and Kolmer, 1992) it has been selected in many breeding programs directed at leaf rust resistance. McIntosh (1992*a*) recently found that all near-isogenic lines of Thatcher known to have *Lr34* were also significantly more resistant than Thatcher to stripe rust in the field. He postulated that this genetic association of resistances occurred in cvv. Chinese Spring, Bezostaya, Bersee and Cappelle Desprez. More recent work showed that *Lr34* is not present in Bersee (RA McIntosh, unpublished 1993). The association was independently discovered by Singh (1992*a*) in certain Mexican wheats. *Lr34* was also genetically associated with leaf tip necrosis (Singh, 1992*b*).

Dyck (1987) reported that Thatcher lines possessing *Lr34* were more resistant to stem rust than Thatcher. Recent studies in Australia have sought reasons why some seedling-susceptible wheats are incapable of carrying heavy infections of leaf rust through the winter. These cultivars, including those known to carry *Lr34*, fail to develop significant levels of infection until the post-anthesis growth stages and disease seldom reaches damaging levels prior to harvest ripeness. By contrast the genotypes that carry high levels of infection throughout the (mild) winter develop early and severe levels of leaf rust. The low temperature resistance conferred by plants with *Lr34* under greenhouse and growth chamber conditions thus seems to be effective in the field where average daily temperatures range from 0°C to 20°C. On the farm the effect of such resistance in reducing disease progress would be very significant and would be influenced by disease levels in neighbouring fields.

Under some conditions, wheats with *Lr34* alone permit excessive levels of infection. D Marshall (*pers. comm.* 1992) observed relatively high levels of rusting in Texas and similar results have been found in Australia. AP Roelfs (*pers. comm.* 1993) has stated that *Lr34* when present alone is "rather ineffective" with terminal disease levels of 20–60MS. The results of Singh and Rajaram (1991) indicated higher infection levels at El Batan compared with Obregon in Mexico indicating that environment probably has a significant influence on terminal disease levels.

PLATE 2-32. Lr34

Seedling leaves of (L to R): A. and C. Tc + *Lr2a* + *Lr34*, Tc + *Lr2a*, Tc + *Lr3ka* + *Lr34*, Tc + *Lr3ka*, Tc + *Lr16* + *Lr34*, Tc + *Lr16* and Thatcher; and B. and D. Tc + *Lr18* + *Lr34*, Tc + *Lr18*, Tc + *Lr21* + *Lr34*, Tc + *Lr21*, Tc + *Lr26* + *Lr34*, Tc + *Lr26* and Thatcher; infected with pt. 104-2,3,6,(7) and incubated at 15°C (A. and B.) and 22/24°C (C. and D.). The interactive effects of the various genes and *Lr34* are expressed by a lowering and increased heterogeneity of infection type. At the higher temperature there was negligible interaction with *Lr18* which is itself, temperature-sensitive (see *Lr18*). Lines with *Lr34* alone have no clearly discernable seedling resistance using Australian cultures and normal greenhouse conditions.

Lr35 (Kerber and Dyck, 1990)   (Plate 2-33)

Adult plant resistance

Chromosome Location

2B. *Lr35* showed genetic linkage of 3.0 cM with a gene for seedling resistance to stem rust. It is not known if the latter was *Sr32*. Both genes were inherited with distorted genetic ratios (Kerber and Dyck, 1990).

Low Infection Type

Flag leaves in the greenhouse, IT ; to ;1$^+$ (Kerber and Dyck, 1990).

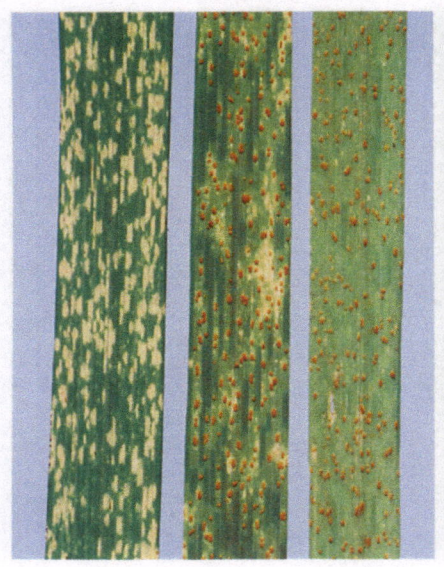 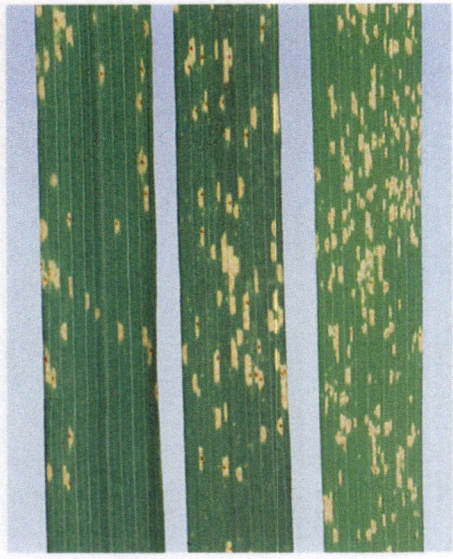

A   B

PLATE 2-33. Lr35
Adult leaves (flag-2) of
(L to R): R.L.5711,
Marquis and Thatcher;
infected with A. pt. 104-
2,3,(6),(7),11 and
B. pt. 53-1,(6),(7),10,11.
Note the resistant
response of Marquis
relative to Thatcher
with the first pathotype.
The pathogenicity of
pt. 53-1,(6),(7),10,11
with respect to R.L.5711
is uncertain because all
three lines probably
carry *Lr22b* and respond
similarly.

Environmental Variability
Unknown.

Origin
T. speltoides R.L.5344 (Kerber and Dyck, 1990).

Pathogenic Variability
Not reported, but not widely tested.

Reference Stocks
i: Marquis-K*8/R.L.5347, R.L.5711 (Kerber and Dyck, 1990). Thatcher*6/R.L.5711, R.L.6083
(PL Dyck, *pers. comm.* 1992).

Source Stocks
al: *T. speltoides* x *T. monococcum*, R.L.5347, (2n = 28).
dv: *T. speltoides* R.L.5344, parent of R.L.5347.

Use in Agriculture
This resistance has not been exploited.

Lr36 (Dvorak and Knott, 1990)   (Plate 2-34)

Chromosome Location
6BS; *Lr36* mapped 26.0 ± 7.9 cM from the centromere and was situated in the 6B satellite region
(Dvorak and Knott, 1990).

Low Infection Type
0; to 1^+N (Dvorak and Knott, 1990).

Environmental Variability
None reported.

Origin
T. speltoides, Population 2 (Dvorak and Knott, 1990).

Pathogenic Variability
None reported.

Reference Stocks
i: Line 2-9-2 = Neepawa*5/*T. speltoides* 2-9 (Dvorak, 1977).
v: E84018 = Neepawa/Line 2-9-2//3*Manitou (Dvorak and Knott, 1990).

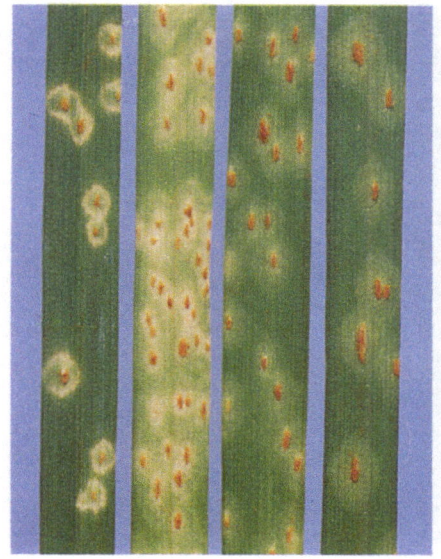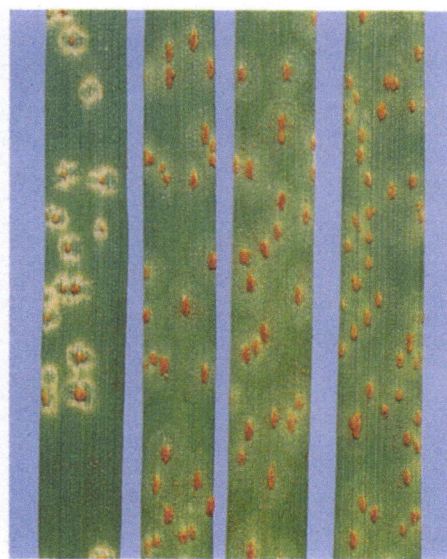

A   B

PLATE 2-34. Lr36
Seedling leaves of (L to R): Neepawa*5/*T. speltoides* 2-9 (C78.10), Manitou, Neepawa and Condor; infected with A. pt. 104-2,3,6,(7) [*P13 P36*] and B. pt. 53-1,(6),(7),10,11 [*p13 P36*]. Both Manitou and Neepawa carry *Lr13* which is effective against pt. 104-2,3,6,(7), but more so in Manitou.

Source Stocks

None known.

Use in Agriculture

Not exploited.

Lr37 (Bariana, 1991)   (Plate 2-35)

Usually expressed as adult plant resistance

Chromosome Location

2AS (Bariana and McIntosh, 1993). In wheat *Lr37* is closely linked in coupling with *Yr17* and *Sr38* and in repulsion with *Lr17*.

Low Infection Type

Low responses are difficult to achieve in seedling tests. Bariana (1991) reported ;12⁻N to 3 with a tendency for greater compatibility towards the leaf tips.

Lr37 is more effective at the adult plant stage.

Environmental Variability

Seedling resistance is more effective at lower temperatures, for example 17°C; temperatures above 20°C resulted in infection types ranging from X^+ to 3^+ (Bariana, 1991).

Origin

T. ventricosum.

Pathogenic Variability

None known.

Reference Stocks

i: Thatcher*8/VPM1, R.L.6081 (PL Dyck, *pers. comm.* 1992)
v: Rendezvous (Bariana and McIntosh, 1994); Hyak (Allan *et al.*, 1990); Madsen (Allan *et al.*, 1989); VPM1 (Doussinault *et al.*, 1981; 1988).

Source Stocks

Australia: Trident. Sunbri *Lr3a*. Various backcross derivatives produced at PBI Cobbitty.

Use in Agriculture

Interest in the French winter wheat VPM1 has been directed mainly at resistance to strawbreaker disease, or eyespot, caused by *Pseudocercosporella herpotrichoides*, located on chromosome 7D.

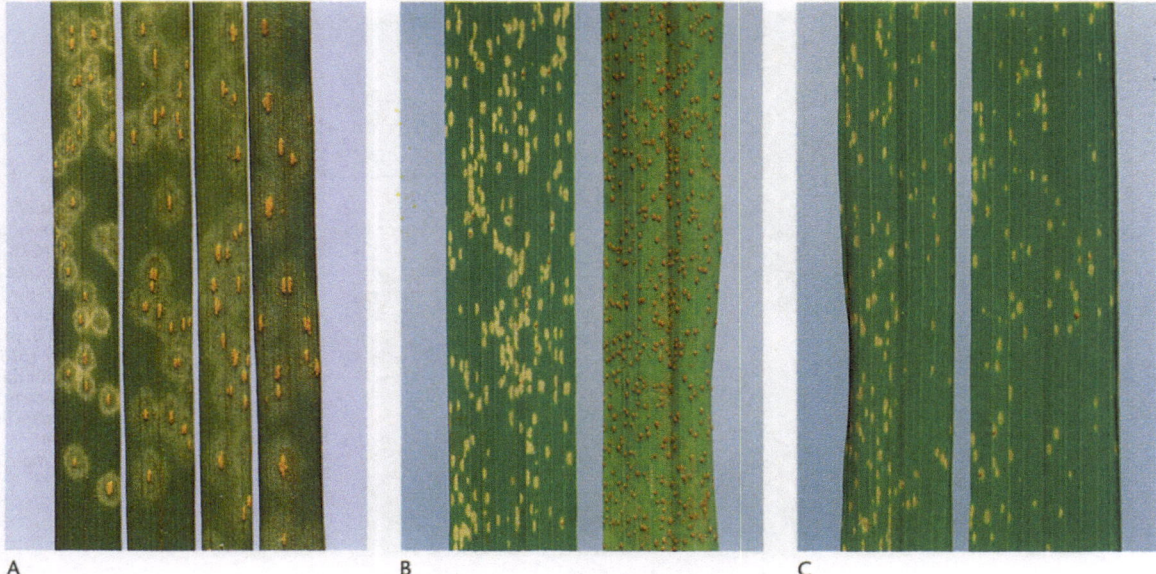

A   B   C

PLATE 2-35. Lr37

A. Seedling leaves of (L to R): VPM1, Tc + VPM1, Thatcher and W195; infected with pt. 104-1,2,3,6 and incubated at 17°C. Courtesy of HS Bariana.

B. and C. Adult (flag-2) leaves of: Tc + *Lr37* and Thatcher; infected with A. pt. 104-2,3,(6),(7),11 and B. pt. 53-1,(6),(7),10,11. The pathogenic status of the second pathotype for *Lr37* cannot be established from this comparison due to the effectiveness of *Lr22b* which is assumed to be present in Tc + VPM1 as well as Thatcher.

Lr37, Yr17 and *Sr38* in VPM1 were also derived from *T. ventricosum* (Bariana and McIntosh, 1994). Although Rendezvous, Hyak and Madsen were selected primarily for resistance to eyespot, the rust resistance genes were also obtained, presumably resulting from additional selection for resistance to stripe rust.

In Australia, selection has been directed at rust resistance. *Lr37* is highly effective under field conditions (Bariana, 1991).

Lr38 (Friebe *et al.*, 1992)   (Plate 2-36)

Chromosome Location

Translocation stocks:
1DL (T 1DS.1DL–7Ai#2L): T25 (Friebe *et al.*, 1993).
2AL (T 2AS.2AL–7Ai#2L): W49 (Friebe *et al.*, 1992); T33 (Friebe *et al.*, 1993).
3DS (T 7Ai#2L–3DS.3DL): T4 (Friebe *et al.*, 1993).
5AS (T 7Ai#2L–5AS.5AL): T24 (Friebe *et al.*, 1993).
6DL (T 6DS.6DL–7Ai#2L): T7 (Friebe *et al.*, 1993).

Low Infection Type

0;⁼.

Environmental Variability

Low.

Origin

Th. intermedium.

Pathogenic Variability

None known.

Reference Stocks

i: Thatcher*6/T7, R.L.6097 (PL Dyck, *pers. comm.* 1992).

PLATE 2-36. Lr38
Seedling leaves of (L to R): W44 [7Ai#2(7D)], W49 (T2A/7Ai#2L), W51 (a related line lacking *Lr38*) and three distinctive segregates from the F2 population of Chinese Spring/W49; infected with pt. 104-2,3,6,(7). The first segregate carried *Lr38*, the second displayed IT X⁺ similar to W51 and the third responded similarly to Chinese Spring. Based on the phenotype and knowledge of the test culture, it can be postulated that the gene conferring the intermediate response is *Lr13*.

v: W49 (Friebe *et al.*, 1992). T25, T33, T4, T24 and T7 (Friebe *et al.*, 1993).
su: W44 [7Ai#2(7D); W52 (7Ai#2(7A)].
ad: T2.

Source Stocks

None.

Use in Agriculture

Not exploited. Although no information is available, the fact that the above lines are non-homoeologous translocations reduces the likelihood of agronomic equality or superiority to current cultivars. Some Group 7 *Thinopyrum* transfers were reported to have problems of high flour pigment. However, appropriate breeding lines must be generated to allow adequate assessment of the value of these translocations. W44 with the entire 7Ai#2 chromosome is also resistant to stem rust (IT 2⁻) and stripe rust (IT 0;). W49 carries only the leaf rust resistance gene (Friebe *et al.*, 1992).

Lr39

This designation was allocated to a resistance gene in *T. tauschii* accession TA5006. Its uniqueness was questioned and remains under review.

Low Infection Type

12⁼

Reference Stock

v: KS86WGRC02.

Lr40   Plate (2-37)

This designation was allocated to a resistance gene in *T. tauschii* accession TA5017. Its uniqueness was questioned and remains under review.

Low Infection Type

2⁻.

Reference Stock

v: KS89WGRC07.

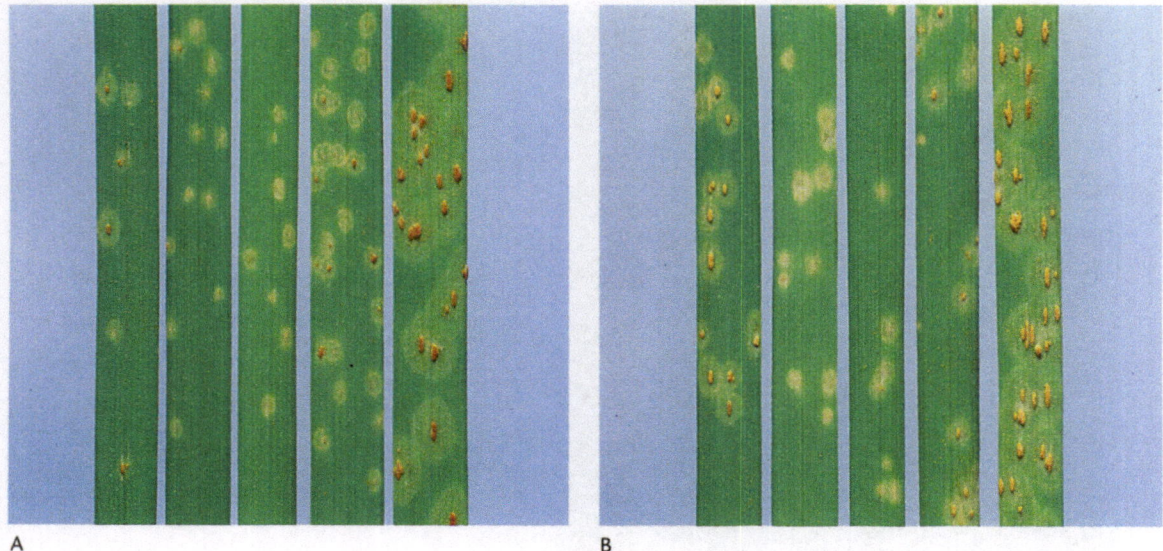

A   B

PLATE 2-37. Lr40

Seedling leaves of (L to R): R.L.5406 (*Lr21*), two leaves of KS86WGRC02, KS89WGRC07 and Thatcher; infected with A. pt. 10-1,2,3,4, and B. pt. 104-1,2,3,6. Genetic tests are required to determine if the resistance genes present in the KS lines are different from *Lr21* and different from each other. In these comparisons KS86WGRC02 produces a lower infection type than either R.L.5406 or KS89WGRC07.

Lr41 (Cox *et al.*, 1994)

A resistance gene transferred to wheat from *T. tauschii* accession TA2460.

Chromosome Location
1D (Cox *et al.*, 1994). Segregated independently of other genes derived from *Ae. squarrosa*.

Low Infection Type
;.

Reference Stock
v: KS90WGRC10 = P.I.549278 = TAM107*3/*T. tauschii* TA2460 (Cox *et al.*, 1992, 1994).

Lr42 (Cox *et al.*, 1994)

A resistance gene transferred to wheat from *T. tauschii* accession TA2450.

Chromosome Location
1D; *Lr42* showed recombination of 0.29 with *Lr21* (Cox *et al.*, 1994).

Low Infection Type
; to ;1.

Reference Stock
v: KS91WGRC11 = Century*3/*T. tauschii* TA2450 (Cox *et al.*, 1994). Century carries *Lr24*.

Lr43 (Cox *et al.*, 1994)

A resistance gene transferred to wheat from *T. tauschii* accession TA2470.

Chromosome Location
Unknown. *Lr43* segregated independently of other genes derived from *T. tauschii* (Cox *et al.*, 1994).

Low Infection Type
; to 1[+].

Reference Stock

    v: KS91WGRC16 = Triumph 64/3/KS8010-71/TA2470//TAM200 (Cox *et al.*, 1994).

Lr44 (PL Dyck, *pers. comm.* 1993)

Chromosome Location

    1B (PL Dyck, *pers. comm.* 1993).

Reference Stocks

    i: Thatcher*6/*T. spelta* 7831 (PL Dyck, *pers. comm.* 1993).
    v: *T. spelta* 7831; *T. spelta* 7839.

TEMPORARILY DESIGNATED AND MISCELLANEOUS LEAF RUST RESISTANCE GENES

In this section we list leaf rust resistances which have not been given formal gene designations but have been sufficiently characterised to warrant inclusion. Three sections detailed below include temporary designations, resistances derived from *T. speltoides* and triticale resistances.

Temporary Designations

LrB (Dyck and Samborski, 1968*b*)

Dyck and Samborski (1968*b*) identified this gene in Brevit and Carina, two of the original differential genotypes selected by Mains and Jackson (1926).

Low Infection Type

    2 (Dyck and Samborski, 1968*b*), ; to 2^+ (Huerta-Espino, 1992).

Pathogenic Variability

    Although *LrB* is ineffective in Australia, the Thatcher near-isogenic line with *LrB* provided differential responses in most geographic areas (Huerta-Espino, 1992).

Reference Stocks

    i: Thatcher*6/Carina, R.L.6051 (Dyck, 1977).
    v: Carina *Lr2b* (Dyck and Samborski, 1968*b*). Brevit *Lr2c* (Dyck and Samborski, 1968*b*).
    V336 *Lr33 LrW*.

Source Stocks

    v: P.I.268316 (Dyck, 1977).

LrEch (Samborski and Dyck, 1976)

Samborski and Dyck (1976) listed a near-isogenic line of Thatcher possessing gene *LrEch* derived from Exchange.

Low Infection Type

    ;1^+.

Pathogenic Variability

    Samborski and Dyck (1976) reported resistance to a single pathotype. Little information is available on this gene. It is ineffective in Australia.

Reference Stocks

    i: Thatcher*6/Exchange, R.L.3718.
    v: Exchange *Lr10 Lr12 Lr16*.

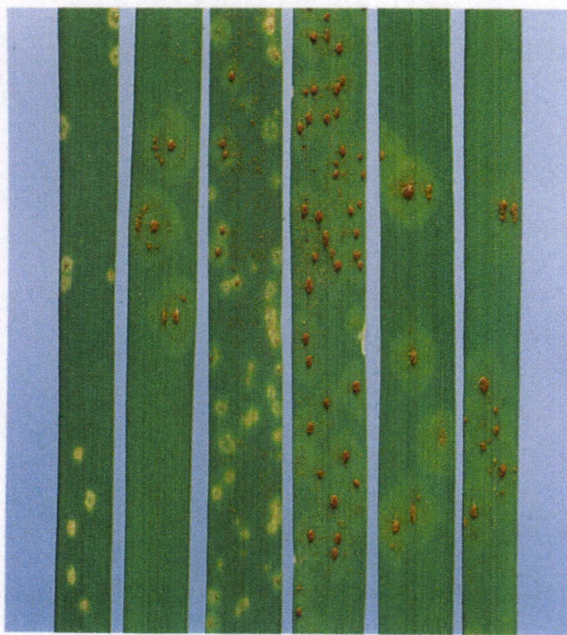

PLATE 2-38. *LrH*
Seedling leaves of: Harrier and Kalyansona; infected with pt. 53-1,(6),(7),10,11 (L pair), pt. 104-2,3,(6),(7),11 (C pair) and pt. 26-1,3 (R pair). The first two pathotypes are relatively recent putative introductions to Australasia. The Kalyansona plant infected with pt. 53-1,(6),(7),10,11 [*P14a*] did not carry *Lr14a*.

LrH (RF Park, unpublished 1992)   (Plate 2-38)

The Australian cv. Harrier carries a gene for seedling resistance to certain pathotypes that are believed to be recent introductions (RF Park, unpublished 1992). This gene is not present in any other Australian cultivar. The pedigree of Harrier is Norin 10-Brevor, 14/Kite Sib//Kite.

Low Infection Type

   ;1 to $2^=$.

Reference Stock

   v: Harrier.

LrT3 (Dyck and Samborski, 1982)

LrT3 was allocated to a gene that has no phenotypic effect on its own, but had an enhancing effect on resistance conferred by *Lr34* (= *LrT2*) (Dyck and Samborski, 1982).

LrVPM (Worland *et al.*, 1988)

Worland *et al.* (1988) reported a gene for adult plant resistance in chromosome 7DL of VPM1. This gene showed recombination of 0.19 with the completely linked *Pch1* and *Ep-D1* genes and recombination of 0.42 with *a-Amy-D2*.

LrW (Dyck and Jedel, 1989)

Allocated to a resistance gene in two wheat accessions (Dyck and Jedel, 1989).

Low Infection Type

Reference Stocks

   v: V618 *Lr33*. V336 *Lr33 LrB* (Dyck and Jedel, 1989).

Resistances Derived from *Triticum speltoides* (Plate 2-39)

Dvorak (1977) transferred resistance to leaf rust from five accessions of *T. speltoides* to either Manitou or Neepawa, both of which carry *Lr13* and presumably *Lr22b*. The gene in one of the derivatives was designated *Lr36* (Dvorak and Knott, 1980). The other derivatives are listed below and illustrated in Plate 2-39.

Neepawa*6/*Triticum speltoides* F-7 (C78.5 & .6)

Chromosome Location
1BL. Recombination with centromere was 41% (Dvorak and Knott, 1980).

Low Infection Type
0; to 2.

Manitou*6/*Triticum speltoides* E-11 (C78.7, .8 & .9)

Chromosome Location
1BL. Recombination with the centromere was 4% (Dvorak and Knott, 1980).

Low Infection Type
0; to 1.

Neepawa*6/*Triticum speltoides* H-9 (C78.11 & .12)

Chromosome Location
Not known.

Low Infection Type
; to 1^+.

Neepawa*5/*Triticum speltoides* 0-1 (C78.13)

Chromosome Location
Not known.

Low Infection Type
$0;1^-$.

Leaf Rust Resistance Genes in Triticale (Plate 2-40)

Hexaploid and octoploid triticales are generally resistant or moderately resistant to leaf rust although highly susceptible cultivars and segregates from crosses may be found. Relatively few genetic studies have been undertaken. Quinones *et al.* (1972) reported monogenic resistances in each of five triticales; two of these genes were allelic. They found no evidence for pathogenic specialisation among *P. recondita* f. sp. *tritici* collections at that time. Wilson and Shaner (1989) identified resistance genes controlling both hypersensitive resistance and slow rusting in triticale crosses. Singh and Saari (1990) identified three genotypes that differentiated components of the *P. recondita* f. sp. *tritici* population in Mexico. These differentials were Esel S, Siskiyou and Tapir/Yogui 1//2*Mus S. On the basis of infection type data, they postulated four resistance genes in these differential lines and at least two additional genes conferring resistance to Mexican pathotypes. Singh and McIntosh (1990) identified gene *LrSatu* that occurred at relatively high

PLATE 2-39
Resistances derived from *Triticum speltoides*
Seedling leaves of (L to R): A. C78.5, C78.6, C78.7, C78.8, C78.9. B. C78.11, C78.12, C78.13, Manitou and Neepawa; infected with pt. 53-1,(6),(7),10,11. This culture was chosen because it is virulent on seedlings with *Lr13* which is present in Manitou and Neepawa. C78.5 and C78.6 are selections from Neepawa*6/ *T. speltoides* F-7; C78.7, C78.8 and C78.9 are selections from Manitou*6/*T. speltoides* E-11; C78.11 and C78.12 are selections from Neepawa*6/*T. speltoides* H-9; C78.13 is a selection from Neepawa*5/*T. speltoides* 0-1.

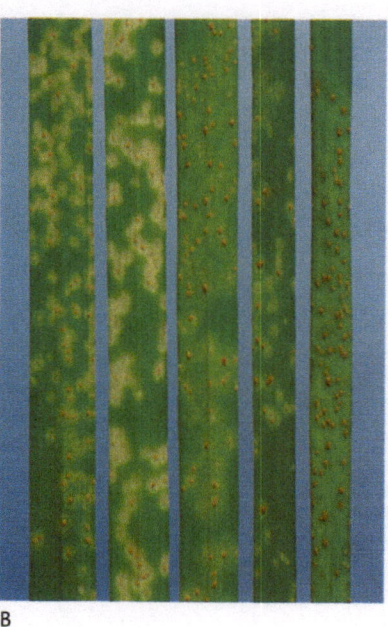

PLATE 2-40
Leaf rust resistance genes in Triticale
Seedling leaves of ten Australian and one European triticale (L to R): A. Tahara, Bejon, Satu, Ningadhu, Currency, Tiga; B. Muir, Empat Selection (IT 2⁼), Empat selection (IT 3), Coorong and Lasko; infected with pt. 26-1,3 and held at 18°C. The first eight genotypes may carry a common gene, *LrSatu*.

frequency in CIMMYT triticale populations. This gene was linked with the *Sr27/SrSatu* locus in chromosome 3R. Genotypes with *LrSatu* were more readily distinguished from some others when tests were carried out at 25°C. RF Park (unpublished, 1993) found that Lasko possessed a temperature-sensitive resistance to certain recent putative exotic Australasian pathotypes. Jones *et al.* (1992) reported that 70% of *P. recondita* f. sp. *tritici* isolates analysed from the UK in 1991 possessed virulence for Lasko triticale.

The pathogen isolated from triticale in the field is usually identified as *P. recondita* f. sp. *tritici*. This implies that the crop is generally resistant to f. sp. *secalis* in common with wheat. Because wheat is not infected with f. sp. *secalis*, Quinones *et al.* (1972) surmised that resistance in triticale was conferred only by genes derived from the wheat parent. While this appears true in at least some instances, rye genes operate in others. For example, *LrSatu* is located in a rye chromosome (Singh and McIntosh, 1990). Because the rye-derived genes, *Lr25* and *Lr26* are effective in hexaploid wheat backgrounds they ought to be effective in triticale.

The Genes for Resistance to Stem Rust in Wheat and Triticale

CATALOGUED STEM RUST RESISTANCE GENES

Sr1

Deleted — see *Sr9d*.

Sr2 (Ausemus *et al.*, 1946; Knott, 1968) (Plate 3-1)

This gene for adult plant resistance shows recessive inheritance.

Chromosome Location

3BS (Hare and McIntosh, 1979). *Sr2* is closely associated with *Lr27*, one of two complementary genes involved in conferring resistance to leaf rust (Singh and McIntosh, 1984*a*) and with pseudo-black chaff which involves melanin pigmentation of the glumes and stem, particularly below the uppermost node.

Low Infection Type

Usually characterised by slow rust development and low terminal rust responses on field-grown adult plants. More severe rusting may occur above the nodal regions and on the spike.

Environmental Variability

May be less effective under wet and overcast conditions.

Origin

T. turgidum var. dicoccum cv. Yaroslav. Resistance was transferred to Hope and H44-24 by McFadden (1930).

Pathogenic Variability

Virulence is not common and would be difficult to detect on field-grown adult plants. Knott (1977) was unable to detect resistance in lines with *Sr2* using North American race 15B-1.

Reference Stocks

s: Chinese Spring*6/Ciano 3B (Singh and McIntosh, 1984*a*); Chinese Spring*6/Ciano 5B (Singh and McIntosh, 1984*a*); Chinese Spring*6/Hope 3B (Hare and McIntosh, 1979).
v: Newthatch *Sr5 Sr7b Sr12 Sr17*. Hope *Sr7b Sr9d Sr17*; H44-24 *Sr7b Sr9d Sr17* (Hare and McIntosh, 1979).

Source Stocks

Sr2 occurs in many wheats developed in areas where stem rust has been a problem. These cultivars were developed in Australia, Canada, Kenya, USA, Mexico and the Indian subcontinent (see Luig, 1983; Roelfs, 1988).
Australia: Songlen *Sr5 Sr6 Sr8a Sr36* (Luig, 1983). Warigo *Sr7b Sr17* (Hare and McIntosh, 1979). Suneca *Sr8a Sr17* (Gyarfas, 1983). Hopps *Sr9d* (Hare and McIntosh, 1979). Hofed *Sr17* (Luig, 1983).
Canada: Pembina *Sr5 Sr6 Sr12*. Rescue *Sr5 Sr9g*. Selkirk *Sr6 Sr7b Sr9d Sr17* (Hare and McIntosh, 1979). Redman *Sr7b Sr9d Sr17*; Regent *Sr7b Sr9d Sr17* (Hare and McIntosh, 1979); Renown *Sr7b Sr9d Sr17* (Luig, 1983).
CIMMYT: Present in many CIMMYT selections. Bluebird series, for example Nuri 70 *Sr5 Sr6 Sr8a*. Lerma Rojo 64 *Sr6 Sr7b Sr9e*. Pavon *Sr8a Sr9g Sr30* (McIntosh, 1988*b*).
Indian Subcontinent: Sonalika (McIntosh, 1988*b*).
Kenya: Kenya Plume *Sr5 Sr6 Sr7a Sr8a Sr12 Sr17* (Singh and McIntosh, 1986*b*). Kenya Page *Sr7b Sr17*.
USA: Eagle (USA); Kaw. Arthur *Sr5 Sr8a Sr36*; Arthur 71 *Sr5 Sr8a Sr36*. Ottawa *Sr9d*. Lancer *Sr9d Sr17*; Scout *Sr9d Sr17*. Karl *Sr9d Sr24*. Scoutland *Sr17*.

Use in Agriculture

Sr2 is arguably the most important gene for stem rust resistance and one of the most important disease resistance genes to be deployed in modern plant breeding (McIntosh, 1988*b*; Rajaram *et al.*, 1988; Roelfs, 1988). With the possible exception of Canada, *Sr2* has provided durable resistance since its introduction to hexaploid wheat in the 1920s (McFadden, 1930). In the North American epidemics of the 1950s cultivars such as Regent, Renown and Redman became moderately susceptible. Many wheats with *Sr2*, including Hope, possess additional genes for resistance such as *Sr7b, Sr9d* and *Sr17*. When the latter genes were effective, they conferred very high levels of resistance. However, the presence of virulence for *Sr7b, Sr9d* and *Sr17* often revealed that the residual resistance conferred by *Sr2* was much less effective.

In these situations, *Sr2* was overlooked because of the apparent increased disease levels. Hare and McIntosh (1979) described phenotypic effects of *Sr2* suggestive of attributes often associated with non-hypersensitive and/or non-specific resistances, thus supporting the conclusion that *Sr2* was a source of durable resistance.

The degree of interaction involving *Sr2* and other genes is unclear. The high degree of seedling resistance of Hope and several other wheats (IT ;) to North American race 56 was apparently caused by the interaction of *Sr2* and *Sr9d* (Knott, 1968). When present alone, the latter gene confers IT 11$^+$ (Knott, 1968; 1990). The association of *Sr2* and distinctive stem and spike blackening (pseudo-black chaff) is well known (see Hare and McIntosh, 1979). Brown (1993) has shown a close association between *Sr2* and a seedling leaf chlorosis that is evident at greenhouse temperatures above 22°C.

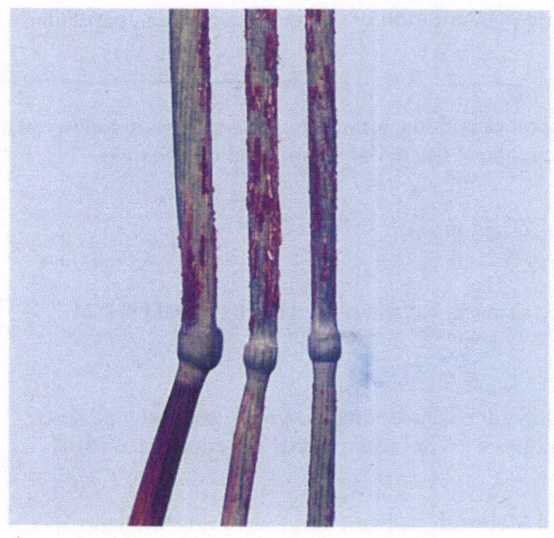

A

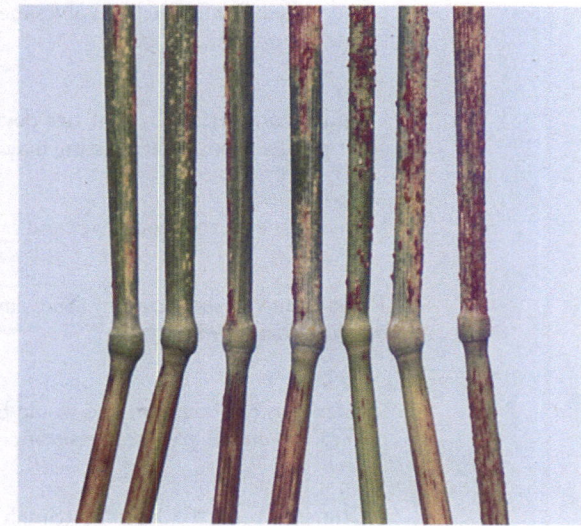

B

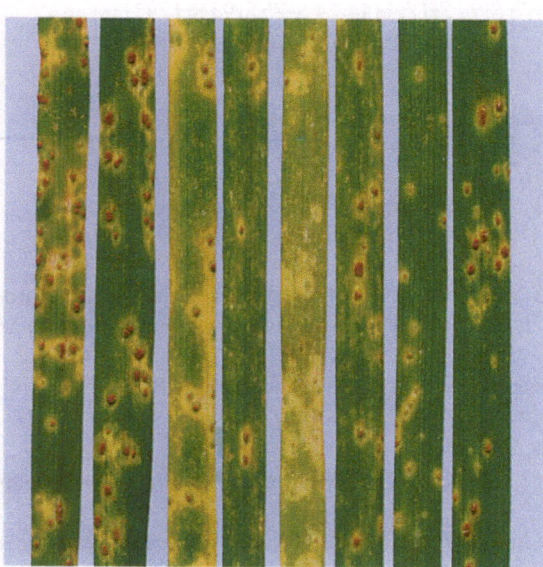

C

PLATE 3-1. Sr2

A. Stem rusted field-grown plants of (L to R): Chinese Spring/Hope 3B substitution line, the F1 of CS/Hope 3B and Chinese Spring, and Chinese Spring. CS/Hope 3B carries *Sr2* which is recessive and shown by the susceptible F1 plant. Courtesy RA Hare.

B. Stem rusted field-grown plants of (L to R): Renown, five homozygous F3 lines of the cross Renown/Line E and Line E. The first three F3 lines are resistant and the last two susceptible. Plants with *Sr2* may develop large pustules, particularly just above the node. Cultivars differ in the amount of disease that develops, for example compare CS/Hope 3B in A. with Renown. Note the melanin pigmentation below the nodes of plants with *Sr2*. Blackening of the glumes and peduncles may also occur. Courtesy RA Hare.

C. First and second seedling leaves, respectively, of (L to R): Sunstar, Hartog and two F2 seedlings of Sunstar/Hartog; infected with stem rust and incubated at 25°C. Hartog and the first F2 seedling possess *Sr2* as indicated by the chlorotic leaf symptoms. Sunstar and the second F2 plant do not carry *Sr2*. The chlorosis seems enhanced by infection with either stem rust or leaf rust and is independent of pathotype. However, seedlings of plants with *Sr2* frequently develop less sporulating uredinia than related lines lacking *Sr2*. Courtesy GN Brown.

Sr3 and Sr4

These gene symbols were allocated by Ausemus *et al.* (1946) to genes in Marquillo but the designation was later abandoned because no single line possessing either gene could be identified (Knott, 1989).

Sr5 (Ausemus *et al.*, 1946) (Plate 3-2)

Synonym

Srrl 1 (Rondon *et al.*, 1966).

Chromosome Location

6D (Sears *et al.*, 1957); 6DS (RA McIntosh, unpublished 1984).

Low Infection Type

Usually 0 to ;. Sr5 is often cited as the 'immunity' gene from wheat stem rust research. The low infection type is usually very low but can be significantly higher (IT X) when Sr5 is transferred to a highly susceptible genetic background (Luig and Rajaram, 1972).

Environmental Variability

Low (Luig and Rajaram, 1972).

Origin

Common wheat. Sr5 is present in the Stakman *et al.* (1962) differential, Reliance C.I.7370, having been inherited from Kanred, a Crimean introduction to the USA.

Pathogenic Variability

Virulence occurs in all geographic areas (Luig, 1983; Huerta-Espino, 1992).

Reference Stocks

i: ISr5-Ra C.I.14159 (Loegering and Harmon, 1969); ISr5-Rb C.I.14161 (Loegering and Harmon, 1969); Line B = W2691*7/Reliance (Luig, 1983); Sr5/7*LMPG (Knott, 1990); Thatcher/10*Marquis (Knott, 1965). Although Marquis is susceptible to many pathotypes of economic significance, lines near-isogenic to Marquis may carry the genes Sr7b, Sr18, Sr19 and Sr20, which are present in this cultivar.

s: Chinese Spring*6/Thatcher 6D (Sears *et al.*, 1957); Chinese Spring*7/Marquis 6D (Sheen and

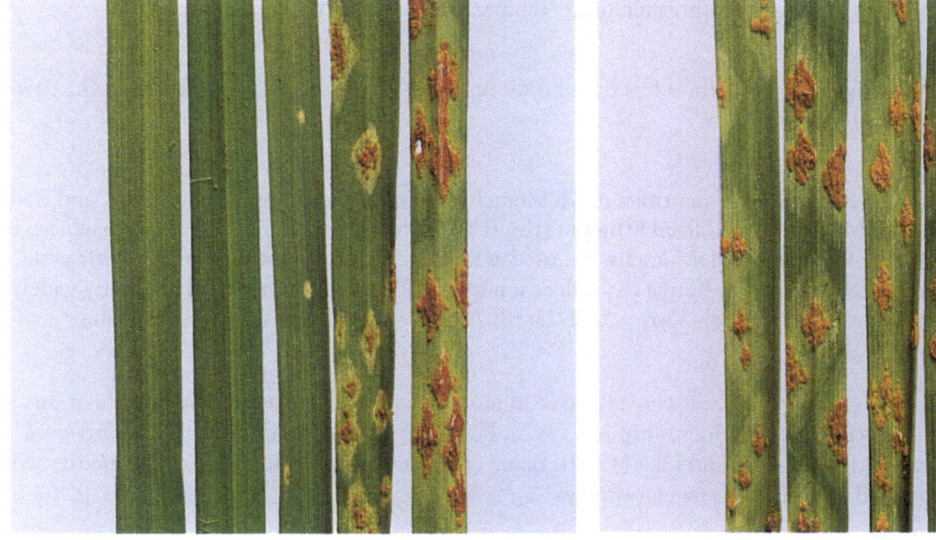

A   B

PLATE 3-2. Sr5

Seedlings of (L to R): Reliance, Zenith, ISr5-Ra, Line B (W2691*7/Reliance) and Chinese Spring; infected with A. pt. 194-2 [*P5*] and B. pt. 34-2 [*p5*]. With the avirulent culture, ISr5-Ra (genetic background of Chinese Spring) sometimes gives small necrotic pustules but in the W2691 background, Sr5 confers IT X. With the virulent culture, Reliance usually develops slightly smaller pustules than the other lines. This slightly retarded rusting has been attributed to Sr16.

s: Chinese Spring*6/Thatcher 6D (Sears *et al.*, 1957); Chinese Spring*7/Marquis 6D (Sheen and Snyder, 1964). The 'Marquis' line used by Sheen and Snyder was probably Thatcher or a closely related derivative.

v: Kanred (Sears *et al.*, 1957). Thatcher *Sr9g Sr12 Sr16* (Luig, 1983). Reliance *Sr16 Sr18 Sr20* (Luig, 1983).

Source Stocks

Many wheats possess *Sr5* (see Luig, 1983; McIntosh, 1988*a*).

Australia: Zenith. Gatcher *Sr2 Sr6 Sr8a Sr9g Sr12*. Bindawarra *Sr6 Sr9b Sr15*. Aroona *Sr8a Sr9b Sr15*. Avocet *Sr26*; Hybrid Titan *Sr26*.

China: Dong-Fang-Hong 2; Ke-Fang 1. Dong-Xie 3 *Sr31*.

Europe: Panis; Primus. Hochzücht *Sr9g Sr12*. Orlando *Sr31*.

Indian Subcontinent: NP789 *Sr11*.

North America: Cheyenne; Chisholm; Justin; Winoka. Arthur *Sr2 Sr8a Sr36*. Fortuna *Sr6 Sr7a Sr17*. Era *Sr6 Sr8a Sr17*. Centurk *Sr6 Sr8a Sr9a Sr17*. Chris *Sr7a Sr8a Sr9g Sr12*. Oasis *Sr8a Sr36*. Apex *Sr9g Sr12*; Kitt *Sr9g Sr12*. Thatcher *Sr9g Sr12 Sr16*. Agatha *Sr9g Sr12 Sr16 Sr25*.

Use in Agriculture

At various times *Sr5* has been a useful and very effective source of resistance. However, resistance proved transient in all geographic areas because of selection and rapid increase of virulent pathotypes. The dominance of Thatcher and its derivatives in the northern Great Plains of North America undoubtedly influenced the increase in pathotypes virulent for *Sr5*.

Sr6 (Knott and Anderson, 1956)   (Plate 3-3)

Synonym

SrKa$_1$ (Athwal and Watson, 1954).

Chromosome Location

2D (Sears, 1954; Wiggin, 1955; Sears *et al.*, 1957); 2DS (McIntosh and Baker, 1968). *Sr6* showed close linkage with *Lr2* and *Lr15* and linkage of 19–40 cM with gene C for compact spike (McIntosh, 1988*a*).

Low Infection Type

0; through X to 3$^+$, varying with pathogen culture and temperature. *Sr6* may be dominant or recessive depending on pathogen culture, temperature and genetic background.

Environmental Variability

Most effective at temperatures less than 20°C. Becomes ineffective at 24–27°C (Forsyth, 1956; Watson and Luig, 1968).

Origin

Common wheat. *Sr6* was identified in McMurachy, a farmer's selection from Canada, and Red Egyptian which originated from Ethiopia (Knott and Anderson, 1956). Both are club wheats but are genetically distinct. McMurachy was used as a rust resistant parent in North America and Red Egyptian was used in Kenya to produce a number of wheats that were subsequently widely used as parents, for example Kenya 58 in North America and Kenya W743 in Australia.

Pathogenic Variability

When first exploited *Sr6* was very effective in most geographical areas. Now virulence occurs in most areas (Luig, 1983; Huerta-Espino, 1992). Progressive increases in virulence for *Sr6* were documented by Watson and Luig (1968). Some cultures with intermediate pathogenicities would be classified as virulent in regular surveys.

Reference Stocks

i: ISr6-Ra C.I.14163 (Loegering and Harmon, 1969); Kenya 58/6*Marquis (Green *et al.*, 1960); Kenya 58/10*Marquis (Knott, 1965); Sr6/9*LMPG (Knott, 1990).

s: Chinese Spring*5/Red Egyptian 2D (Sears *et al.*, 1957).

v: McMurachy (Knott and Anderson, 1956; Watson and Luig, 1963). Red Egyptian *Sr8a Sr9a* (Sears *et al.*, 1957).

Source Stocks

Many wheats carry *Sr6* (see Luig, 1983; McIntosh, 1988a).

Africa: Bonza 63 *Sr8a Sr9b*. Many Kenyan lines (Knott, 1962a; Watson and Luig, 1963).

Australia: Eureka; Wongoondy. Songlen *Sr2 Sr5 Sr8a Sr36*. Gamut *Sr5 Sr8a Sr12*; Oxley *Sr5 Sr8a Sr12*. Bayonet *Sr8a*. See Luig (1983).

North America: Milam; Shield; Trapper; Travis; Twin. Selkirk *Sr2 Sr7b Sr9d Sr17 Sr23*. Manitou *Sr5 Sr7a Sr9g Sr12*. Centurk *Sr5 Sr8a Sr9a Sr17*. Butte *Sr8a Sr9g*. Colt *Sr8a Sr9a Sr17 Sr24*. Onas *Sr10*; White Federation 54 *Sr10*. Homestead *Sr17*.

Use in Agriculture

Sr6 was a very effective source of resistance when first used in Australia (Eureka) and Canada (Selkirk). *Sr6* continues to be present in a range of modern cultivars; however, virulence is now widely distributed in most geographic areas (Luig, 1983).

Knott and Anderson (1956) found that *Sr6* was dominant when segregating populations were tested with race 56 but was recessive when they were tested with race 15B.

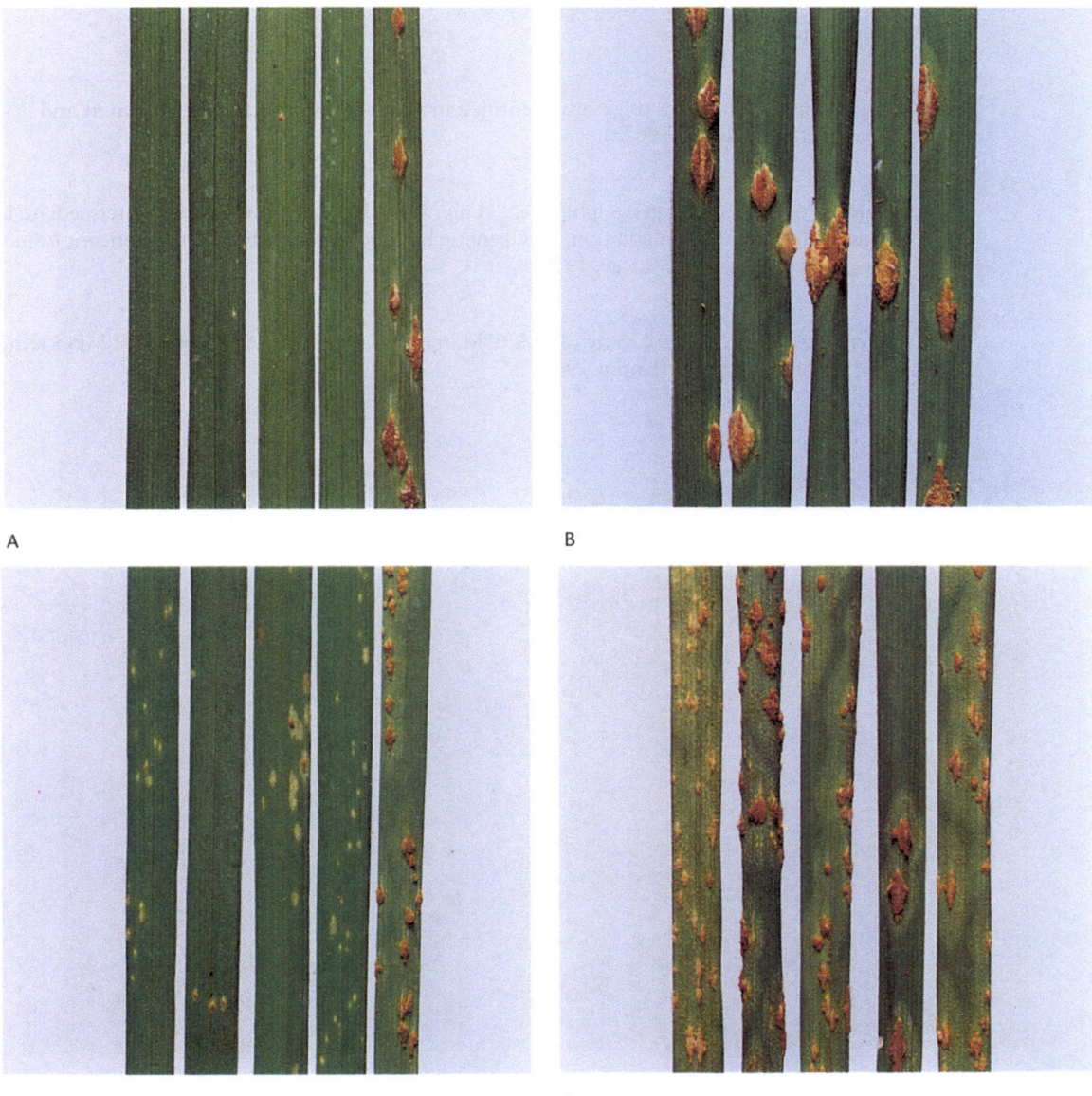

A   B

C   D

PLATE 3-3. *Sr6*

Seedling leaves of (L to R): McMurachy, Eureka, I*Sr6*-Ra, Kenya 58/6*Marquis and Chinese Spring. Upper sets were infected with A. pt. 126-5,6,7,11 [*P6*] and B. pt. 126-1,5,6,7,11 [*p6*] and incubated at 21°C. Lower sets were inoculated with pt. 34-2 [*P6*] and incubated at C. 21°C and D. 28°C. The upper sets demonstrate specificity between culture and host gene, whereas the lower sets demonstrate temperature sensitivity.

Sr7 (Knott and Anderson, 1956)

Chromosome Location

4A (Sears *et al.*, 1957; Knott, 1959; Loegering and Sears, 1966); 4AL. *Sr7* is genetically independent of the centromere (RA McIntosh, unpublished 1973).

Sr7a (Loegering and Sears, 1966)   (Plate 3-4)

Synonym

Sr7 (Knott and Anderson, 1956).

Low Infection Type

1 to 3C (Knott, 1989); yellow chlorosis/necrosis surrounding uredia is common and characteristic.

Environmental Variability

Presumed to be low.

Origin

Common wheat. *Sr7a* was originally identified in a number of Kenyan wheats (Knott and Anderson, 1956; Knott, 1962a).

Pathogenic Variability

Virulence occurs in most geographic areas (Luig, 1983; Huerta-Espino, 1992). Intermediate low responses are common. In addition, host genetic background probably has a significant influence on response (Roelfs and McVey, 1979).

Reference Stocks

i: Egypt Na 101/6*Marquis; Kenya 117A/6*Marquis (Green *et al.*, 1960); Sr7a/9*LMPG (Knott, 1990); Sr7a/10*Marquis (Knott, 1965).

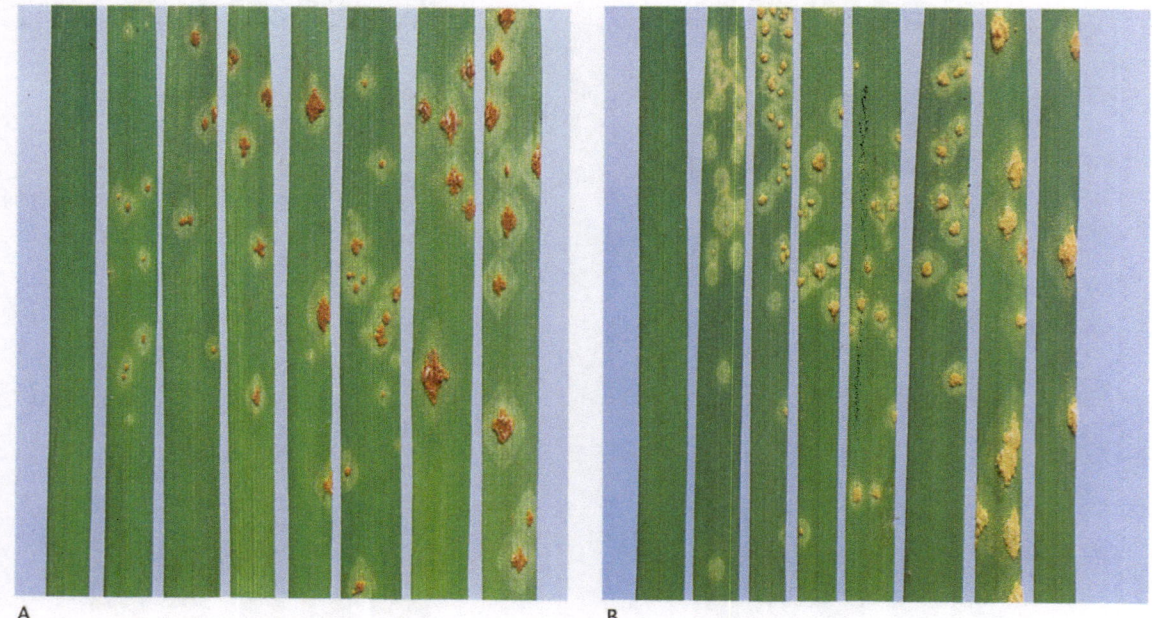

A   B

PLATE 3-4. *Sr7a*
Seedling leaves of (L to R): Chris, Chris Derivative W3747, Mendos Derivative, Marquis + *Sr7a*, CS/Kenya Farmer 4A, Sr7a/9*LMPG, Chinese Spring and Marquis; infected with A. pt. 116-2,3,7 and B. yellow culture 80-E-2. The first six wheats carry *Sr7a*. However, Chris and Chris Derivative show lower responses to both cultures due to avirulence for *Sr5* in the former, and avirulence for additional genes in the latter. Note the lower pathogenicity of culture 80-E-2 on seedlings with *Sr7a* and the lower pathogenicity of pt. 116-2,3,7 on Marquis relative to Chinese Spring.

s: Chinese Spring*7/Kenya Farmer 4A; Chinese Spring*8/Sapporo 4A (Loegering and Sears, 1966).

v: Kenya 117A C.I.13140 *Sr9b* (Knott, 1957b). Egypt Na 101 P.I.139599 *Sr9b Sr10* (Knott, 1957a).

Source Stocks

See lists in Luig (1983) and McIntosh (1988a).

Africa: Kentana 52 *Sr6*. Kenya wheats (Knott, 1957a, 1957b, 1959, 1962a).

Australia: Khapstein *Sr2 Sr13 Sr14* (Knott 1962b). Mendos *Sr11 Sr17 Sr36*. Chris derivative W3747 *Sr12*.

Canada: Manitou *Sr5 Sr6 Sr9g Sr12*.

Japan: Sapporo P.I.81790-2 (Knott and Shen, 1961).

USA: Chris *Sr5 Sr8a Sr9g Sr12* (Singh and McIntosh, 1987).

Use in Agriculture

Sr7a has not been added purposefully as a means of achieving resistance, and has been found in wheats studied after commercial release. Singh and McIntosh (1986b) reported interactions of *Sr7a* and *Sr12* resulting in significantly higher levels of resistance than that conferred by either gene acting alone.

Sr7b (Loegering and Sears, 1966)   (Plate 3-5)

Low Infection Type

2 to 3⁻.

Environmental Variability

Low.

Origin

Common wheat; present in the Stakman *et al.* (1962) differentials Marquis C.I.3641 and Kota C.I.5878.

Pathogenic Variability

Virulence occurs at high frequency in most geographic areas (Luig, 1983). Moderate to high levels of avirulence may be found in parts of southern Africa (Huerta-Espino, 1992), the USA and Australia.

Reference Stocks

i: ISr7b-Ra C.I.14165 (Loegering and Harmon, 1969).

s: Chinese Spring*6/Hope 4A (Sears *et al.*, 1957).

v: Marquis *Sr18 Sr19 Sr20* (Knott, 1965).

Source Stocks

Sr7b is very common and is not fully catalogued in many wheat cultivars [see list in Luig (1983)].

Australia: Spica *Sr17*.

Canada: Red Bobs *Sr10* (Dyck and Green, 1970).

CIMMYT: Kiric 66 *Sr6* (McVey and Roelfs, 1975).

Europe: Fertödi 293 (Luig, 1983); Halle 9H39 (Luig, 1983). Roussalka *Sr8a* (McVey and Roelfs, 1975).

USA: Geneva; Hart; TAM102. Hope *Sr2 Sr9d Sr17* (Sears *et al.*, 1957). Caldwell *Sr9d Sr10*. Marfed *Sr10*. Nell *Sr17* (Wells *et al.*, 1983). Ceres *Sr28*; Kota *Sr28* (McIntosh, 1978). C.I.12632 *Sr36*.

Use in Agriculture

Sr7b has not been consciously selected as a source of stem rust resistance. Avirulence on plants with *Sr7b* was probably present when rust research began in Australia in the 1920s. It disappeared with the extinction of pathotypes present at that time, only to reappear in 1968 with putative introductions from southern Africa (Watson and de Sousa, 1983). The current predominant Australian pathotype is avirulent (Park and Wellings, 1992) and *Sr7b* confers an effective degree of resistance.

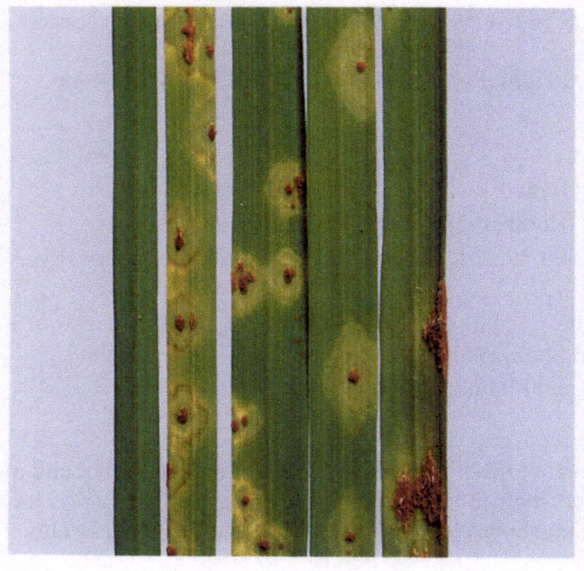

A

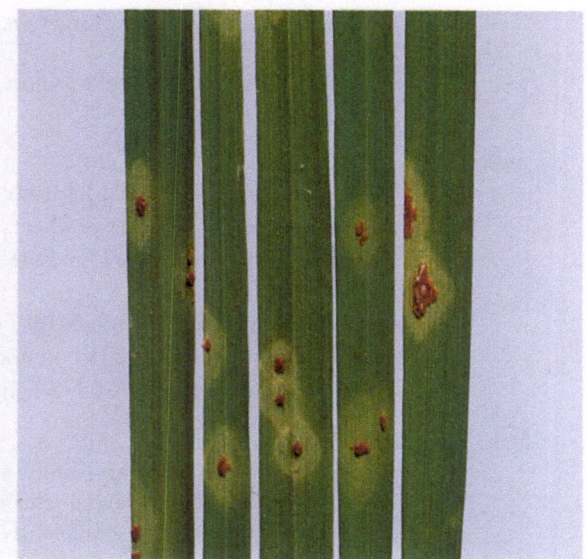

B

C

PLATE 3-5. *Sr7b*
Seedling leaves of (L to R): C.I.12632 = W1656, Marquis, Spica, CS/Hope 4A and Chinese Spring; infected with A. pt. 98-1,2,3,5,6 [*P7bp17P36*], B. pt. 343-1,2,3,4,5,6 [*P7bp17p36*] and C. pt. 21-1,2,3,7 [*p7bP17P36*]. Sets A. and B. were incubated at 20°C whereas set C. was incubated at 18°C to ensure that resistance would be conferred by *Sr17*. The resistance of C.I.12632 (leaf 1) in A. and C. is conferred by *Sr36*; the resistance of Spica (leaf 3) in C. is conferred by *Sr17*.

Sr8 (Knott and Anderson, 1956)

Chromosome Location

6A (Sears, 1954; Sears *et al.*, 1957); 6AS (McIntosh, 1972). Sr8 is genetically independent of the centromere (McIntosh, 1972).

Sr8a (Singh and McIntosh, 1986*a*)   (Plate 3-6)

Synonym

Sr8 (Knott and Anderson, 1956).

Low Infection Type

2⁻ to 3.

Environmental Variability

Low (Roelfs and McVey, 1979).

Origin

Common wheat. This gene was first described in Red Egyptian (Knott and Anderson, 1956) but subsequent work demonstrated its presence in many European and Mexican lines. It probably became widespread in modern spring wheats through the use of Italian lines and their derivatives in South America and elsewhere.

Pathogenic Variability

P. graminis populations in most geographic regions are polymorphic for pathogenicity on seedlings with *Sr8a* (Luig, 1983; Huerta-Espino 1992).

Reference Stocks

i: I*Sr*8-Ra C.I.14176 (Loegering and Harmon, 1969); Red Egyptian/10*Marquis (Knott,1965); *Sr*8a/9*LMPG (Knott, 1990).
s: Chinese Spring*5/Red Egyptian 6A (Sears *et al.*, 1957).
v: Mentana (Luig and Watson, 1965). Red Egyptian *Sr6 Sr9a* (Knott and Anderson, 1956).

Source Stocks

Sr8a is a very common gene. Examples from lists in Luig (1983), McIntosh (1988a) and elsewhere include the following.
Australia: Jacup. Songlen *Sr2 Sr5 Sr6 Sr36*. Hartog = Pavon 'S' *Sr2 Sr12 Sr30* (Brennan, 1983). Warigal *Sr5 Sr9b Sr15*. Condor *Sr5 Sr12*. Egret *Sr9b Sr12*.
China: An-Hewi II *Sr5*.
CIMMYT: Nuri 70 *Sr2 Sr5 Sr6*. Penjamo 62 *Sr5 Sr6 Sr9b*. Inia 66 *Sr9a Sr11*.
Europe: Victor 1 *Sr5 Sr6*. Golden Valley *Sr7b*; Roussalka *Sr7b*.
South America: Frontana *Sr9b*.
USA: Geneva. Arthur *Sr2 Sr5 Sr36*. Centurk *Sr5 Sr6 Sr9a Sr17*. Era *Sr5 Sr6 Sr17*. Chris *Sr5 Sr7a Sr8a Sr9g Sr12*. Butte *Sr6 Sr9g*. Olaf *Sr9b Sr12*. Benhur *Sr10*.

Use in Agriculture

Despite its high frequency in wheat cultivars, *Sr8a* probably has limited effects on field responses to stem rust. Wheats with *Sr8a* may become susceptible even with avirulent pathotypes.

A   B

PLATE 3-6. *Sr8a*
Seedling leaves of (L to R): Mentana, CS/Red Egyptian 6A, Sr8a/9*LMPG, Chinese Spring and LMPG; infected with A. pt. 21-0 [*P8a*] and B. pt. 126-5,6,7,11 [*p8a*]. The first three wheats carry *Sr8a*.

Sr8b (Singh and McIntosh, 1986*a*)   (Plate 3-7)

Synonym
SrBB (Luig, 1983).

Low Infection Type
X$^=$ to X.

Environmental Variability
Moderate, with lower infection types at cooler temperatures (Luig, 1983).

Origin

Common wheat. *Sr8b* appears to be derived from Barleta which was probably introduced to Argentina from Spain (Luig, 1983).

Pathogenic Variability

Australian populations are differentiated by their pathogenicity on wheats with *Sr8b*. Virulence was common in North America and South America, Europe and Africa in the early 1970s (Luig, 1983). In the survey of Huerta-Espino (1992) high levels of virulence were found in western Asia and eastern Europe, northern Africa and western Europe and South America. Populations in southern Africa showed a wide range of variability and those in China were generally avirulent.

Reference Stocks
v: Barleta Benvenuto C.I.14196 (Singh and McIntosh, 1986*a*).

Source Stocks

Klein Titan (Singh and McIntosh, 1986*a*). Bezostaya *Sr5* (RA McIntosh and AK Khanzada, unpublished 1989). Klein Cometa *Sr30* (Singh and McIntosh, 1986*a*).

Use in Agriculture

Sr8b is a rare gene. Bezostaya is the only non-South American wheat reported to carry this gene. However, the pedigree of Bezostaya includes Klein 33, which was of Argentinian origin. Resistance is effective to avirulent pathotypes in field plots (Singh and McIntosh, 1986*a*).

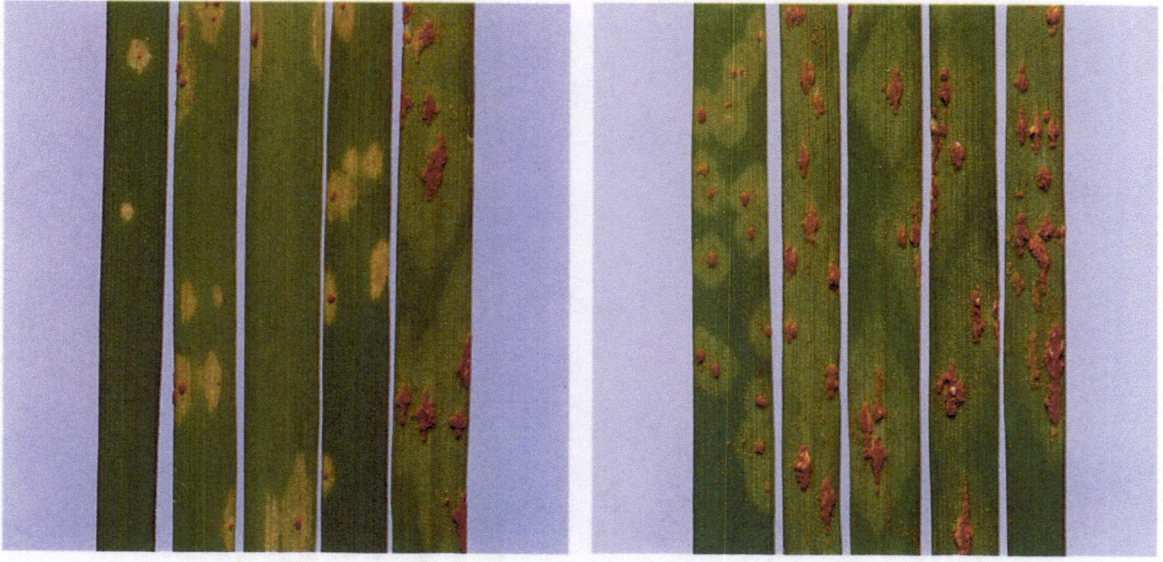

A   B

PLATE 3-7. *Sr8b*
Seedling leaves of (L to R): Klein Cometa, Barleta Benvenuto, Klein Titan, Bezostaya and Chinese Spring; infected with A. pt. 34-(1),2,3,7 [*p5P8bP30*] and B. pt. 126-5,6,7,11 [*p5p8bP30*]. *Sr8b* and *Sr30* in Klein Cometa interact to produce a lower infection type than conferred by *Sr8b* alone. The expression of *Sr30* in Klein Cometa is shown when infection occurs with the *Sr8b*-virulent pathotype (B.).

Sr9 (Knott and Anderson, 1956)

Chromosome Location

2B (Sears *et al.*, 1957; Knott, 1959; Knott, 1968; Loegering and Harmon, 1969). 2BL (Sears and Loegering, 1968; McIntosh and Baker, 1970b; McIntosh and Luig, 1973a; Williams and Maan, 1973). *Sr9* was mapped at 11 (Sears and Loegering, 1968) and 18 (McIntosh and Baker, 1969) map units from the centromere. It is closely linked with *Yr7* (McIntosh *et al.*, 1981) and presumably *Yr5*, and shows recombination of between 0.15 and 0.3 with *Lr13* and *Lr23* (McIntosh, 1988a).

Sr9a (Green *et al.*, 1960)   (Plate 3-8)

Synonym

Sr9 (Knott and Anderson, 1956).

Low Infection Type

1^- to 2, 23.

Environmental Variability

Low (Green *et al.*, 1960; Loegering and Harmon, 1969).

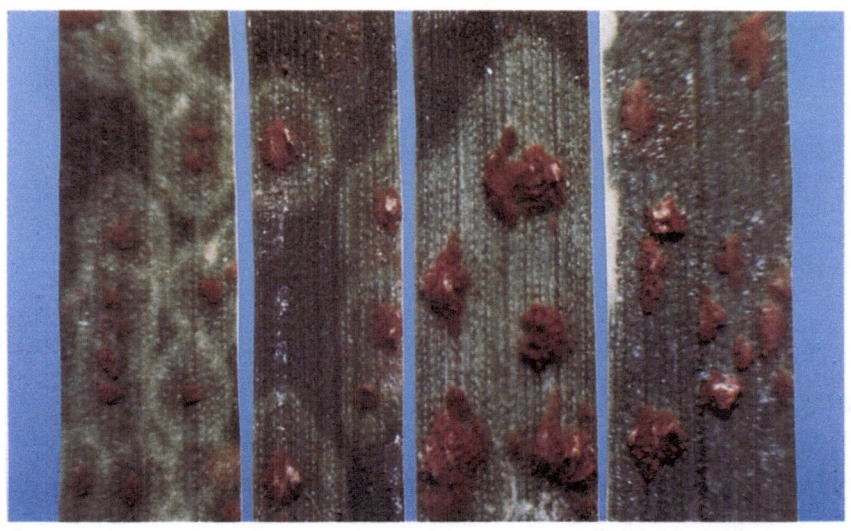

A

B

PLATE 3-8. *Sr9a*

A. Seedling leaves of (L to R): I*Sr9a*-Ra, Spica, Renown and Bowie; infected with University of Missouri culture 17-53B. Spica and Renown do not carry *Sr9a* but have a degree of resistance (IT 3C) relative to Bowie (3^+). No illustration of the low infection type was possible with cultures available in Australia.

B. Seedling leaves of I*Sr9a*-Ra infected with USDA Cereal Rust Laboratory pt. BDCN. Courtesy AP Roelfs.

Origin

Common wheat. *Sr9a* was first reported in Red Egyptian.

Pathogenic Variability

Wheats with *Sr9a* produce differential seedling responses in North America (Green *et al.*, 1960), central America, Brazil and possibly Kenya (Luig, 1983). *Sr9a* is ineffective in tests with Australian pathotypes (Green *et al.*, 1960) and on the Indian subcontinent.

Reference Stocks

i: ISr9a-Ra C.I.14169 (Loegering and Harmon, 1969); Red Egyptian/6*Marquis (Green *et al.*, 1960); Red Egyptian/10*Marquis (Knott, 1965); Sr9a/9*LMPG (Knott, 1990).
s: Chinese Spring*4/Red Egyptian 2B (Sears *et al.*, 1957).
v: Red Egyptian C.I.12345 *Sr6 Sr8a* (Knott and Anderson, 1956).

Source Stocks

CIMMYT: Lerma Rojo 64 *Sr2 Sr6 Sr7b*. Inia 66 *Sr2 Sr8a Sr11*.
Europe: French Peace *Sr7a Sr13* (Knott, 1983).
USA: TAM107. Centurk *Sr5 Sr6 Sr8a Sr17*. Sentinel *Sr6 Sr8a Sr17*. Colt *Sr6 Sr8a Sr17 Sr24*.

Use in Agriculture

Not widely used.

Sr9b (Green *et al.*, 1960) (Plate 3-9)

Synonym

SrKb1 (Athwal and Watson, 1954).

Low Infection Type

1^+ to 3. 2 (DR Knott, *pers. comm.* 1993).

Environmental Variability

Low (Roelfs and McVey, 1979).

Origin

Common wheat. *Sr9b* was first found in certain Kenyan wheats (Knott and Anderson, 1956). Its presence in the Brazilian cultivar Frontana must derive from Fronteira whose pedigree is Polysu/Alfredo Chaves 6 (Zeven and Zeven-Hissink, 1976).

Pathogenic Variability

Wheats with *Sr9b* produce differential responses to *P. graminis* in most geographic areas (Luig, 1983). In the survey of Huerta-Espino (1992) frequencies of virulence for *Sr9b* were generally high.

Reference Stocks

i: Kenya 117A/6*Marquis, W2402 (Green *et al.*, 1960; Watson and Luig, 1963); Kenya 117A/10*Marquis (Knott, 1965): Sr9b/10*LMPG (Knott, 1990).
s: Chinese Spring*7/Kenya Farmer 2B (McIntosh and Luig, 1973a).
v: Gamenya (Luig, 1983).

Source Stocks

Sr9b occurs in many wheats, especially those of Kenyan origin (see Luig, 1983; McIntosh, 1988a).
Africa: Romany *Sr5 Sr6 Sr7a Sr30*. Bonza 63 *Sr6 Sr8a*. Kenya 117A *Sr7a Sr10*. Kenya Farmer *Sr7a Sr10 Sr11*. Kenya W744 *Sr15*.
Australia: Gamenya. Aroona *Sr5 Sr8a Sr15*; Warigal *Sr5 Sr8a Sr15*. Egret *Sr8a Sr12*.
CIMMYT: Penjamo 62 *Sr5 Sr6 Sr8a*. Pitic 62 *Sr8a*. Nainari 60 *Sr11*.
South America: Frontana *Sr8a*; Rio Negro *Sr8a*.
USA: Atlas 66 *Sr10*. Lancota *Sr10 Sr17*.

Use in Agriculture

Sr9b is relatively common in spring wheats because the gene has been effective in many countries. In Frontana and many of its relatives, *Sr9b* is linked with *Lr13*. The presence of *Sr9b* in some wheats probably resulted from selection for leaf rust resistance.

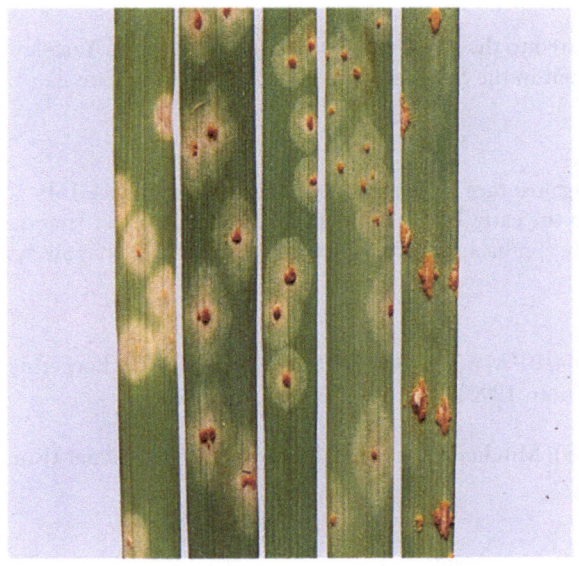

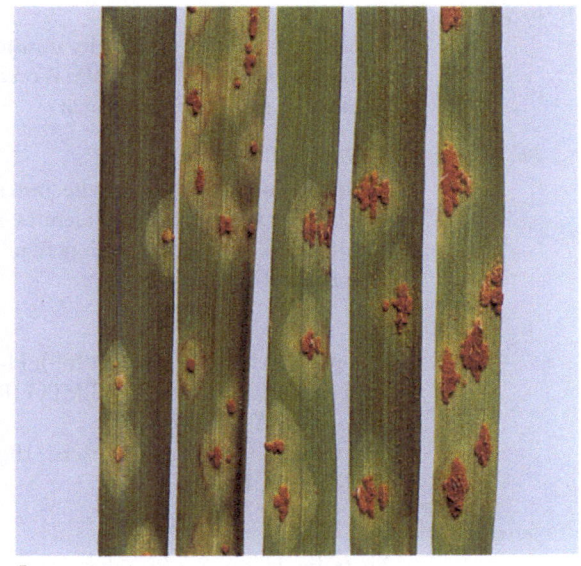

A

B

C

PLATE 3-9. *Sr9b*

Seedling leaves of (L to R): Kenya 117A, Gamenya, Kenya 117A/6*Marquis, CS*7/Kenya Farmer 2B and Chinese Spring; infected with A. pt. 126-5,6,7,11 [*P9b*], B. pt. 116-4,5 [*P9b*] and C. pt. 116-2,3,7 [*p9b*]. The low infection type with pt. 126-5,6,7,11 is usually characterised by smaller pustules and necrosis compared with that produced with pt. 116-4,5. The chlorotic response of Kenya 117A with pt. 116-2,3,7 can be attributed to *Sr7a*. It is difficult to attribute the lower response of Kenya 117A with pt. 126-5,6,7,11 to interaction between *Sr9b* and *Sr7a*, because this pathotype is considered virulent on seedlings with *Sr7a* alone.

Sr9c

This symbol was originally reserved for a gene derived from *T. timopheevii* and subsequently designated *Sr36*.

Sr9d (Knott, 1966)   (Plate 3-10)

Synonym

Sr1 (Ausemus *et al.*, 1946; Knott, 1966, 1968, 1971).

Low Infection Type

1^- to 2^+.

Environmental Variability

Probably low.

Origin

T. turgidum. Sr9d was originally transferred to the bread wheats Hope and H-44 from Yaroslav emmer (McFadden, 1930). *Sr9d* is present in the Stakman *et al.* (1962) durum differentials Mindum, Arnautka and Spelmar.

Pathogenic Variability

In North America, *Sr9d* was effective against race 56 but was not effective against race 15B which caused catastrophic epidemics in the early 1950s. In the international survey of Huerta-Espino (1992) virulence levels generally approached 100%, except for Turkey. *Sr9d* has always been ineffective in Australasia.

Reference Stocks

i: Hope/10*Marquis (Knott, 1968); H-44/10*Marquis (Knott, 1968); IHope 2B-Ra (Loegering and Harmon, 1969); Sr9d/8*LMPG (Knott, 1990).
v: Hope *Sr2 Sr7b Sr17*.
tv: Arnautka (Roelfs and Martens, 1988); Mindum (Roelfs and Martens, 1988); Spelmar (Roelfs and Martens, 1988).

Source Stocks

Australia: Hopps *Sr2* (Sunderwirth and Roelfs, 1980). Lawrence *Sr2 Sr7b Sr17*. Spica *Sr7b Sr17*.
Canada: Selkirk *Sr2 Sr6 Sr7b Sr17 Sr23*. Redman *Sr2 Sr7b Sr17*; Renown *Sr2 Sr7b Sr17*.
USA: Shawnee; Sturdy. Scout *Sr2 Sr17*.
tv: Nugget.

Use in Agriculture

Sr9d was of some use in both common wheats and durums in North America until the outbreaks of 15B in the early 1950s which affected both common and tetraploid wheats.

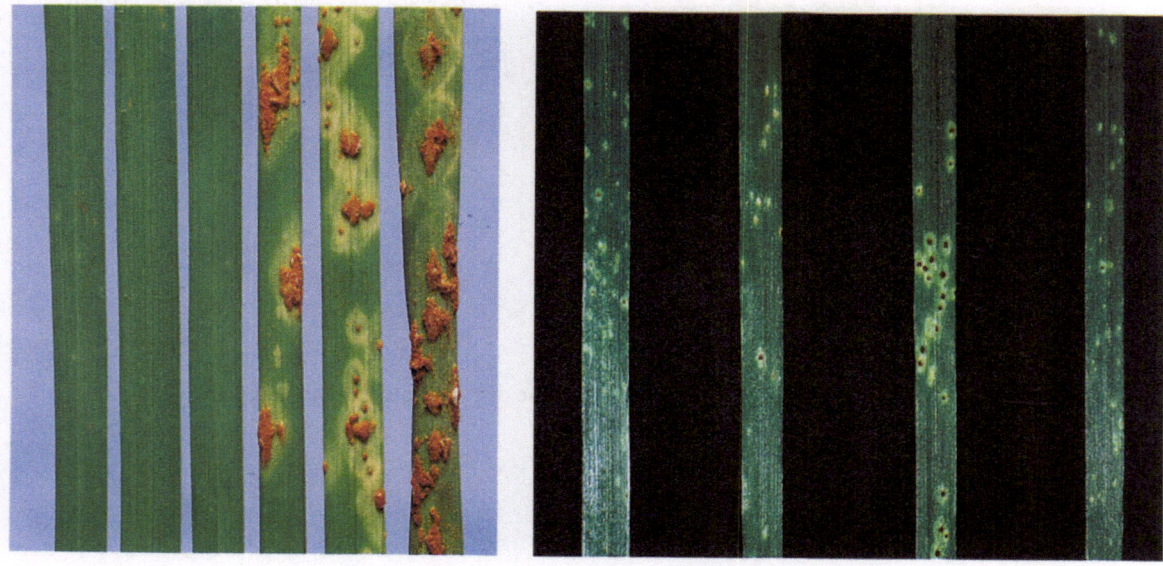

A   B

PLATE 3-10. *Sr9d*

A. Seedling leaves of (L to R): durums Mindum, Arnautka, Spelmar and common wheats IHope 2B-Ra (*Sr9d*), LMPG + *Sr9d* and LMPG; infected with an impure University of Missouri culture of 59-51A. The IT 2 pustules on the fourth and fifth leaves are typical of plants with *Sr9d* infected with avirulent cultures. Although the three durums carry *Sr9d*, a second more effective (non-designated) gene confers resistance to both culture 59-51A and the contaminating culture.

B. Seedling leaves of I*Sr9d*-Ra infected with USDA Cereal Rust Laboratory culture BDCN. Courtesy AP Roelfs.

Sr9e (McIntosh and Luig, 1973*a*)   (Plate 3-11)

Synonyms

Srv (Smith, 1957), Srd1v (Kenaschuk *et al.*, 1959).

Low Infection Type

1^- to 2^+.

Environmental Variability

Low.

Origin

T. turgidum. Sr9e occurs in Vernal emmer selected as a *P. graminis tritici* differential by Stakman *et al.* (1962). It is also present in some North American durum wheats. McVey (1990) postulated the presence of *Sr9e* in 174 of 578 spelt wheat accessions.

Pathogenic Variability

Virulence occurs at a high frequency in North America and at relatively low frequencies in other geographic areas (Luig, 1983; Huerta-Espino, 1992).

Reference Stocks

i: Sr9e/7*LMPG (Knott, 1990).
v: Vernstein (Luig and Watson, 1967).
tv: Vernal emmer (Smith, 1957).

Source Stocks

Australia: Sunstar *Sr8a Sr12*. Combination III *Sr36*.
South Africa: SST 3R (Sharma and Gill, 1983); SST-16 (Sharma and Gill, 1983); SST 33 (Le Roux and Rijkenberg, 1987*b*); SST-66 (Le Roux and Rijkenberg, 1987*b*).
tv: C.I.7778 (Luig and Watson, 1967). *Sr9e* is present in a range of North American durums (RA McIntosh, unpublished 1973). The Mexican cultivar Yavaros 79 *Sr12* and various durums with additional genes were shown to carry *Sr9e* (Singh *et al.*, 1992).

A   B   C

PLATE 3-11. *Sr9e*

Seedling leaves of (L to R): Vernal emmer, Sunstar, Vernstein and Chinese Spring; infected with A. pt. 126-5,6,7,11 [*P9e*] B. pt. 98-1,2,3,5,6 [*P9e*] and C. a mutant of pt. 98-1,2,3,5,6 virulent for *Sr9e* [*p9e*]. Tests were performed at 20°C. Note the lower sporulation with pt. 126-5,6,7,11, but the similar area of necrosis/chlorosis to pt. 98-1,2,3,5,6. The lower IT ;1 associated with Sunstar in A. and B. and IT 23 in C. is attributed to *Sr12*.

Use in Agriculture

The use of *Sr9e* as a source of resistance in South African bread wheats was followed by an increased pathogen virulence frequency. Although virulence for *Sr9e* was occasionally detected in Australian surveys, no isolate from wheat-growing areas was virulent on seedlings of Sunstar. Many North American durums appear to carry *Sr9e* as a component of their oligogenic resistances. However, *Sr9e* is not effective against race 15B which has been important since the 1950s.

Sr9f (Loegering, 1975)

Synonym

Srpl (Loegering, 1975).

Low Infection Type

;2⁻ (Roelfs and McVey, 1979).

Environmental Variability

Unknown.

Origin

Common wheat cv. Chinese Spring.

Pathogenic Variability

Most *P. graminis* f. sp. *tritici* isolates are virulent.

Reference Stocks

v: Chinese Spring; the contrasting line lacking *Sr9f* is ISr9a-Ra C.I.14169 (Loegering, 1975).

Use in Agriculture

Wheat possesses many resistance genes such as *Sr9f* that can be detected with laboratory cultures possessing unusual genes for avirulence. Most of these genes are not catalogued and are of little agricultural importance.

Plate

None available.

Sr9g (McIntosh and Luig, 1973*a*) (Plate 3-12)

Low Infection Type

2⁼ to 3.

Environmental Variability

Low.

Origin

T. turgidum var. durum, including the Stakman *et al.* (1962) differentials Acme and Kubanka. *Sr9g* was transferred from durum cv. Iumillo to Marquillo and Thatcher from which it was introduced into other common wheats.

Pathogenic Variability

Virulence is common. Apart from southern Africa and Australia, avirulence for *Sr9g* is uncommon (Luig, 1983; Huerta-Espino, 1992).

Reference Stocks

s: Chinese Spring*7/Marquis 2B *Sr16* (McIntosh *et al.*, 1981); Chinese Spring*4/Thatcher 2B *Sr16* (McIntosh *et al.*, 1981). The 'Marquis' accession used by Sheen and Snyder (1964) was probably Thatcher or a closely related line.
v: Thatcher *Sr5 Sr12 Sr16* (McIntosh *et al.*, 1981).
tv: Acme (McIntosh *et al.*, 1981); Kubanka (McIntosh *et al.*, 1981). Both Acme and Kubanka have resistance genes additional to *Sr9g*.

Source Stocks

Australia: Hartog *Sr2 Sr8a Sr30*. Corella *Sr5* (heterogeneous) *Sr8a Sr12*. Celebration *Sr12 Sr16*. Eagle *Sr26* (Luig, 1983).
Europe: Hochzücht *Sr5 Sr12*.
North America: Lee *Sr11 Sr16*. Many Thatcher derivatives carry *Sr9g* (McIntosh *et al.*, 1981).
tv: Iumillo which possesses additional genes including, presumably, *Sr12* (McIntosh *et al.*, 1981).

Use in Agriculture

Sr9g is one of the genes transferred to Marquillo, and subsequently Thatcher, from the durum cv. Iumillo. Although not commonly effective in North America, avirulence for *Sr9g* is relatively frequent in Australia, southern Africa and India. Because resistance to avirulent cultures is effective, *Sr9g* is a useful resistance gene in these areas when used in combinations to ensure protection against a wide range of pathotypes. Because *Sr9g* is closely linked with *Yr7* its presence in some instances may have resulted from selection for resistance to stripe rust.

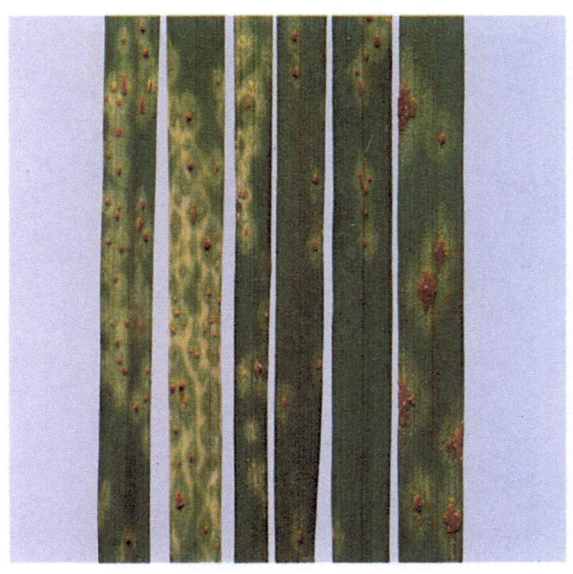

A

B

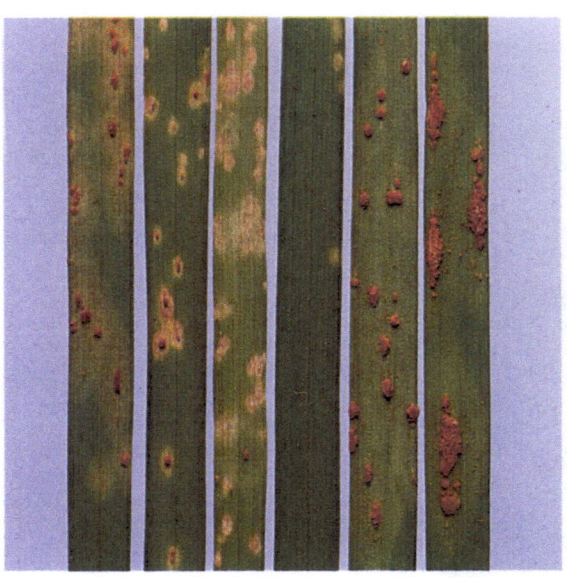

C

PLATE 3-12. *Sr9g*
Seedling leaves of (L to R): Acme durum, Kubanka durum, Celebration, Lee, CS/Marquis 2B and Chinese Spring; infected with A. pt. 222-1,2,3,5,6 [*P9g*], B. pt. 34-1,2,3,4,5,6,7 [*p9g*] and C. pt. 126-5,6,7,11 [*p9g*] and incubated at 21°C. The first five wheats carry *Sr9g* which is usually slightly more effective in durums. Each of these wheats express a different gene, or interactions, when infected with pt. 126-5,6,7,11. Note how Acme (IT 3CN) and Kubanka (IT 11+) are distinguished with the last culture. Also with the third culture, Celebration shows evidence of interaction of *Sr12* and *Sr9g*, Lee possesses *Sr11* and the slightly retarded growth on CS/Marquis 2B relative to Chinese Spring is probably caused by *Sr16*.

Sr10 (Knott and Anderson, 1956)   (Plate 3-13)

Chromosome Location

2B (DR Knott, *pers. comm.* 1993). Federation possesses a resistance gene, presumably *Sr10*, in chromosome 2B (RA McIntosh, unpublished 1980).

Low Infection Type

0;1N to 3C.

Environmental Variability

Temperature sensitive; sensitivity appears to vary between pathogen isolates (Green *et al.*, 1960; Roelfs and McVey, 1979). More effective at lower temperatures.

Origin

Common wheat; first documented in a Kenyan source and Egypt Na95, a derivative of Kenyan parents. However, more recent work indicated this gene is present in Australian wheats developed prior to 1900.

Pathogenic Variability

Virulence for *Sr10* was frequent in North America (Roelfs and McVey, 1979). Luig (1983) reported that *Sr10* differentiated between pathogen isolates in South America, Israel and South

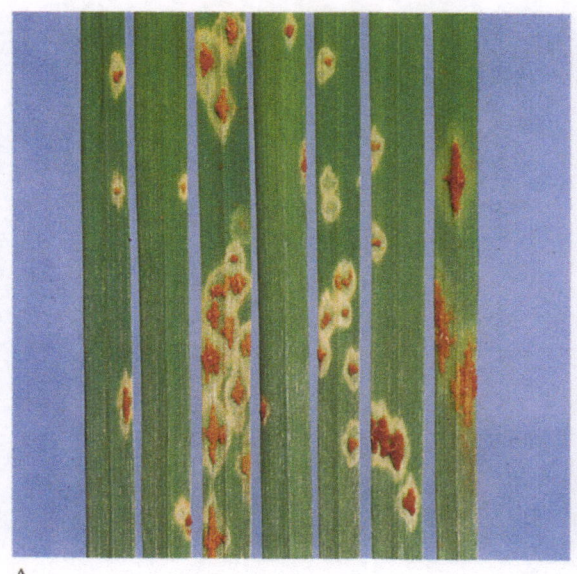

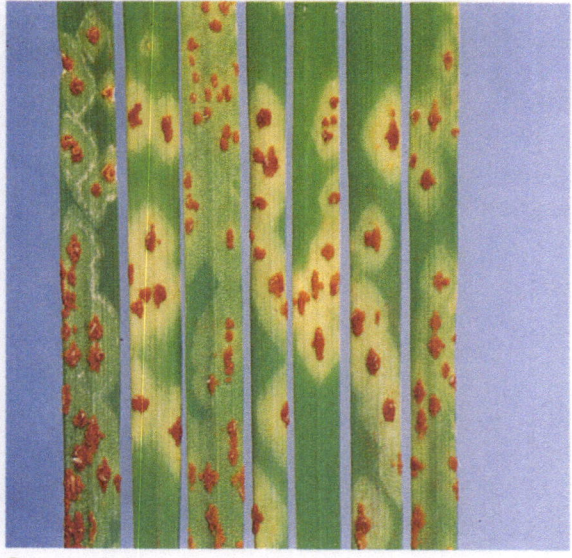

A

B

C

PLATE 3-13. *Sr10*

A. and B. Seedling leaves of (L to R): Line F (W2691 + *Sr10*), W2404 (Marquis + *Sr10*), Federation, Red Bobs, Marquis, Thew and Norka; infected with A. pt. 34-(4),(7),10 [*P10*] and B. pt. 21-0 [*p10*] and incubated at 23/25°C. Although not distinguishable from the other five resistant lines, the response of Marquis with the first pathotype cannot be due to *Sr10* and could be due to *Sr19*. Both Thew and Norka carry *Sr15* but the temperatures imposed were sufficiently high to overcome its effect against these *Sr15*-avirulent pathotypes. The resistance of Thew to pt. 34-(4),(7),10 may be due to *Sr10*. Marquis, Red Bobs and possibly Thew carry a gene(s) conferring IT 3C to pt. 21-0. Both Marquis and Red Bobs carry *Sr7b* but both cultures are virulent for this gene.

C. Seedling leaves of Line F (W2691 + *Sr10*) infected with USDA Cereal Rust Laboratory culture LCBB. Courtesy AP Roelfs.

Africa. Huerta-Espino (1992) found moderate levels of avirulence among cultures from Ethiopia and Turkey and high levels among those from Pakistan, Nepal and China.

Reference Stocks

i: Egypt Na95/4*Marquis (Green *et al.*, 1960) = W2404 (Luig, 1983); Line F = W2691 + *Sr10* (Luig, 1983).

v: Egypt Na95 *Sr7a Sr9b* (Knott and Anderson, 1956); Kenya 117A *Sr7a Sr9b* (Knott and Anderson, 1956).

Source Stocks

Africa: Kenya Farmer *Sr7a Sr9b Sr11* (Green *et al.*, 1960). Other Kenyan wheats (Knott, 1957a, 1957b, 1962a).

Australia: Federation (AP Roelfs, *pers. comm.* 1993).

North America: Geneva (Sorrells and Jensen, 1987); Lemhi; McNair 1003; Saluda; Springfield. Red Bobs *Sr7b* (Dyck and Green, 1970). Caldwell *Sr7b Sr9d*. Benhur *Sr8a*. Atlas 66 *Sr9b*.

Use in Agriculture

Green and Knott (1962) reported that *Sr10* conferred adult plant resistance but there are few data to indicate its real value. Roelfs and McVey (1979) noted that this gene was common in spring wheats in western USA. *Sr10* was probably transferred by chance through the use of Kenyan material in the CIMMYT program (Knott, 1990).

Sr11 (Knott and Anderson, 1956) (Plate 3-14)

Synonym

Kc2 (Watson and Stewart, 1956). Knott and Anderson (1956) originally asssigned symbols *Sr11* and *Sr12* assuming that linked complementary genes were involved. It was later shown that abnormal gametic transmission rates were responsible for disturbed genetic ratios obtained in crosses involving Chinese Spring as the susceptible parent (Luig, 1960, 1968; Sears and Loegering, 1961) and *Sr11* became the accepted symbol.

Chromosome Location

6B (Plessers, 1954; Sears, 1954; Knott, 1959); 6BL (Sears, 1966). *Sr11* was mapped more than 60cM from the 6B centromere and from the awn inhibitor *B2* which is near the centromere (Sears, 1966). Genetically, it shows close repulsion linkage with *Lr9* derived from *T. umbellulatum*, but ER Sears (*pers. comm.* 1966) obtained a rare recombinant with both genes (available as Sydney University accession C66.10). Heyne and Johnston (1954) reported linkage of 23 cM between *Sr11* and *Lr3*, but workers at The University of Sydney failed to obtain recombination between these genes (Luig, 1964). We do not know of any wheat accession that carries *Sr11* and an *Lr3* resistance allele.

Low Infection Type

; to 2⁻. Roelfs and McVey (1979) reported infection types of 2 to 2^+3^- with certain cultures.

Environmental Variability

Low.

Origin

T. turgidum var. durum cv. Gaza. It is assumed that all sources of *Sr11* derive from Bobin W39*2/Gaza material originally produced in Australia, but subsequently widely distributed. Watson and Stewart (1956) concluded that Timstein was derived from this material rather than a *T. timopheevii* cross.

Pathogenic Variability

Variability occurs in all geographic areas. Virulence frequencies are extremely high in Australia (Zwer *et al.*, 1992), South Africa (Le Roux and Rijkenberg, 1987a), Canada (Harder and Dunsmore, 1990) and the USA (Roelfs *et al.*, 1991), but relatively low on the Indian subcontinent and Europe (Luig, 1983). These results were supported by Huerta-Espino (1992) who also reported low to moderate frequencies of virulence in certain regions of South America and North Africa and no virulence in China.

PLATE 3-14. *Sr11*
Seedling leaves of (L to R):
Gabo, Lee, CS*7/Kenya
Farmer 6B and Chinese
Spring; infected with
A. pt. 116-4,5 [*P11*] and
B. pt. 116-2,3,7 [*p11*].
Lower infection types (; to
;1⁻) may be obtained with
some cultures.

A   B

Reference Stocks

i: I*Sr11*-Ra C.I.14171 (Loegering and Harmon, 1969); Lee/10*Marquis (Knott, 1965).
s: Chinese Spring*7/Kenya Farmer 6B (Loegering and Sears, 1966); Chinese Spring*9/Timstein 6B (Sears *et al.*, 1957).
v: Charter W1371 (Luig and Watson, 1965); Gabo W1422 C.I.12795; Timstein C.I.12347 (Knott and Anderson, 1956; Sears *et al.*, 1957); Yalta W1373 (Luig and Watson, 1965). Lee *Sr9g Sr16* C.I.12488 (Knott and Anderson, 1956).

Source Stocks

Sr11 is present in a large number of Australian and Kenyan wheats (see McIntosh, 1988*a*) and CIMMYT cultivars (Roelfs and McVey, 1979).
China: Qing-Chung 5 *Sr5 Sr6*.
Europe: Flevina.
India: N.P.790 *Sr5*.

Use in Agriculture

When first exploited in Australia in the mid 1940s *Sr11* was widely effective but its widespread use was followed by increased virulence frequencies. This increase was so spectacular that the pathogen population became genetically fixed for the corresponding pathogen gene for virulence. The consequence is that *Sr11* can be detected only with avirulent cultures held in the laboratory and its presence or absence is no longer of relevence to Australian wheat breeders. Charter has been used in India to differentiate pathotypes virulent for *Sr11*; that is, Charter carries a second gene which cannot be detected using Australian isolates (see Luig, 1983).

Sr12 (Sheen and Snyder, 1964)   (Plate 3-15)

Sr12 was first used in conjunction with *Sr11* to designate complementary genes thought to be present in Gabo, Lee and Timstein (Knott and Anderson, 1956). Luig (1960) and Sears and Loegering (1961) showed that a single gene (*Sr11*) was involved and that abnormal gametic transmission rates were responsible for the disturbed genetic ratios originally observed by Knott and Anderson (1956). Sheen and Snyder (1964) then used *Sr12* to designate a gene in a Thatcher derivative.

Chromosome Location

3B (Sheen and Snyder, 1964; Knott, 1984); 3BS (McIntosh *et al.*, 1980).

Low Infection Type

;, 2, X to 3.

Environmental Variability

Most effective at temperatures below 20°C.

Origin

Sr12 is derived from *T. turgidum* var. durum cv. Iumillo and was transferred to Marquillo and, eventually, Thatcher as part of the transference of rust resistance from Iumillo to bread wheat (Hayes *et al.*, 1920).

Pathogenic Variability

Virulence is reported to be common. However, RA McIntosh believes this gene is more effective than reported in the literature. *Sr12* also shows interactions with other genes, especially the *Sr9* alleles, *Sr9b* and *Sr9g*.

Reference Stocks

s: Chinese Spring*7/Marquis 3B (Sheen and Snyder, 1964). Chinese Spring*5/Thatcher 3B *Sr16* (Sheen and Snyder, 1964).
v: Thatcher *Sr5 Sr9g Sr16* (Luig, 1983). Marquillo *Sr9g* (Knott, 1984).

Source Stocks

Africa: Kenya Plume *Sr2 Sr5 Sr6 Sr7a Sr8a Sr9b Sr17* (Singh and McIntosh, 1986*b*).

A

B

C

PLATE 3-15. *Sr12*

Seedling leaves of (L to R): Thatcher, Celebration, Lee, Condor, CS/Marquis 3B and Chinese Spring; infected with A. pt. 34-2,4,5,7,11 [*P12P8a*] and B. and C. pt. 34-1,2,3,5,6,7 [*P12p8a*]. A. and B. were incubated at 18°C and C. at 23/28°C. *Sr12* confers IT X but there is no 'typical' response because of the temperature effects, genetic background effects and interactions. Repeated testing over many years has shown that pt. 34-2,4,5,7,11 is more avirulent on seedlings with *Sr12* than pt. 34-1,2,3,5,6,7. The low responses of Thatcher and Celebration are typical of these genotypes. Lee, which shares *Sr9g* and probably *Sr16* with them, is susceptible with both cultures. The low response of Condor with pt. 34-2,4,5,7,11 is due to both *Sr8a* and *Sr12*. In this comparison CS/Marquis 3B appears to be resistant in A. (IT X) but susceptible in B. (IT 3⁺) reflecting increased virulence of the second pathotype relative to the first. All of these wheats, except perhaps Condor, were susceptible when incubated at 23/28°C.

Australia: Tincurrin; Windebri. Celebration *Sr9g Sr16* (RA McIntosh, unpublished 1980). Tentative results based on multi-pathotype surveys in Australia indicate that *Sr12* may be relatively common. Unfortunately, *Sr12* is often difficult to distinguish from *Sr6* with Australian *P. graminis* pathotypes and appropriate genetic studies have not been pursued.

North America: Chris *Sr5 Sr8a Sr9g* (Singh and McIntosh, 1987). Olaf *Sr8a Sr9b*. *Sr12* is likely to occur in Thatcher derivatives.

tv: Yavaros 79 *Sr9e* (Singh *et al.*, 1992). Other durums (Singh *et al.*, 1992).

Use in Agriculture

Sr12 was highly effective against pre-1950 Australian pathotypes. After 1950 the effectiveness of this gene in wheats such as Windebri and Celebration was reduced but not completely overcome. In wheats combining *Sr12* with *Sr9b* or *Sr9g* the presence of *Sr12* can be recognised through its effects on the expression of the *Sr9* alleles in seedlings inoculated with avirulent pathotypes. These interactions also appear to occur in adult plants. One pathotype in the University of Sydney collection is known to be unusually virulent on seedlings of Iumillo durum. This culture (pt. 21-7,9 [71178]) is potentially virulent on wheats known to carry *Sr12*. Both the seedling and adult plant resistances of Marquillo are more effective than those of Thatcher and its derivatives. The reasons for this are not fully understood but *Sr12* appears to be involved (RA McIntosh, unpublished 1980; see also Knott, 1984).

In North America *Sr12* was very effective in conferring resistance to race 56. Race 15B is virulent.

Sr13 (Knott, 1962*b*) (Plate 3-16)

Chromosome Location
Distally located in 6AL (McIntosh, 1972).

Low Infection Type
1, 11$^+$, 2$^-$ to 23$^-$. 2$^-$ at 30°C and 2$^+$3 at 18°C (Roelfs and McVey, 1979).

Environmental Variability
Lowest infection types are obtained at temperatures of 20–28°C, that is, the higher temperature range used for seedling tests.

Origin
T. turgidum var. dicoccum cv. Khapli C.I.4013 in the differential set of Stakman *et al.* (1962). *Sr13* was transferred to the common wheat line, Khapstein by WL Waterhouse.

Pathogenic Variability
Virulence for *Sr13* appears to be extremely rare except in India and Pakistan (Luig, 1983) where Khapli emmer is cultivated. Occasional cultures from Europe and Africa produced high infection types on Khapstein (Luig, 1983). Knott (1990) reported a virulent culture of North American race 11. Huerta-Espino (1992) detected one isolate with virulence in Ethiopian samples, one in Turkish and two in Spanish collections.

Reference Stocks
i: Khapstein/10*Marquis (Knott, 1965); Sr13/9*LMPG (Knott, 1990).
v: Khapstein *Sr2 Sr7a Sr14* (Knott, 1962*b*).
tv: Khapli emmer *Sr7a Sr14*.

Source Stocks
Australia: Wialki (Luig, 1983). Madden *Sr2 Sr9b Sr11* (Luig, 1983).
Europe: French Peace *Sr7a Sr9a* (Knott, 1983).
tv: St464. Probably present in some durums especially those derived from St464.

Use in Agriculture
Because of its widespread effectiveness, *Sr13* is potentially useful as a source of resistance to stem rust and has been exploited in some Australian wheats. However, RA McIntosh, RG Rees and GJ Platz (unpublished, 1985) showed that wheats with *Sr13* expressed comparatively high adult plant reactions to avirulent cultures and experienced grain weight losses of about 40% relative to chemically protected controls. Related lines lacking the *Sr13* allele gave much higher losses. The origin of *Sr2* (RA McIntosh, unpublished 1980) and *Sr7a* (Knott, 1962*b*) in Khapstein must be

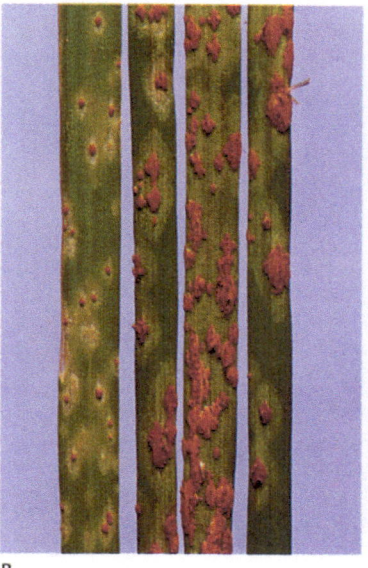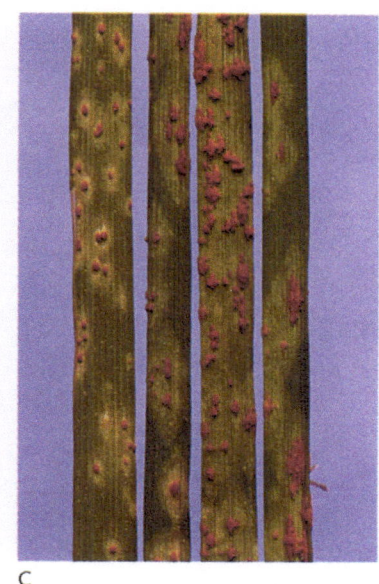

A   B   C

PLATE 3-16. *Sr13*

Seedling leaves of (L to R): Khapli emmer, Khapstein, Marquis + *Sr13* (= Khapstein/10*Marquis) and Marquis; infected with pt. 34-1,2,3,6,7,8,9 and incubated at A. 23/28°C, and B. and C. at 18°C. B. and C. show the lower and upper leaf surfaces, respectively. *Sr13* was ineffective at the lower temperature. Both Khapli and Khapstein possess genes additional to *Sr13* and some interactions involving *Sr13* are evident, especially necrosis associated with *Sr14*.

questioned because neither gene has been reported in Khapli. In his early studies on hexaploid/tetraploid wheat crosses Waterhouse (1933) reported that cv. Steinwedel and its derivatives were the only hexaploids not producing chlorotic (hybrid chlorosis) hybrids with Khapli. Sterility in pentaploids may have led to outcrossing and *Sr2* and *Sr7a* may be derived from other sources. RA McIntosh (unpublished, 1985) showed that a gene (*Srdp2*) in a Golden Ball-derived hexaploid wheat, accession W3504, was allelic with *Sr13*.

Sr14 (Knott, 1962*b*) (Plate 3-17)

Chromosome Location
1BL (Baker and McIntosh, 1973; McIntosh, 1980). *Sr14* is located very close to the centromere.

Low Infection Type
1N to 3C.

Environmental Variability
Lowest responses associated with *Sr14* are produced under high temperature and high light conditions (Luig, 1983; Gousseau *et al.*, 1985). These appear to enhance the distinct necrosis which is very characteristic of this gene (Knott, 1989).

Origin
T. turgidum var. dicoccum cv. Khapli. *Sr14* and other genes were transferred to the hexaploid cv. Steinwedel, resulting in cv. Khapstein (Waterhouse, 1933).

Pathogenic Variability
Maximum levels of avirulence are produced by relatively few isolates in most geographic areas. Under high temperature and high light conditions slight reductions in infection type and some necrosis can be detected with many pathogen cultures normally considered virulent (RA McIntosh, unpublished 1975). Luig (1983) reported possible instances of avirulence in European, Israeli and South African cultures, whereas Huerta-Espino (1992) observed avirulence in occasional cultures from Ethiopia and Ecuador. In North America avirulence is limited to race 56.

Reference Stocks
i: Khapstein/10*Marquis (Kao and Knott, 1969); Line A = W2691*2/Khapstein (McIntosh, 1980).

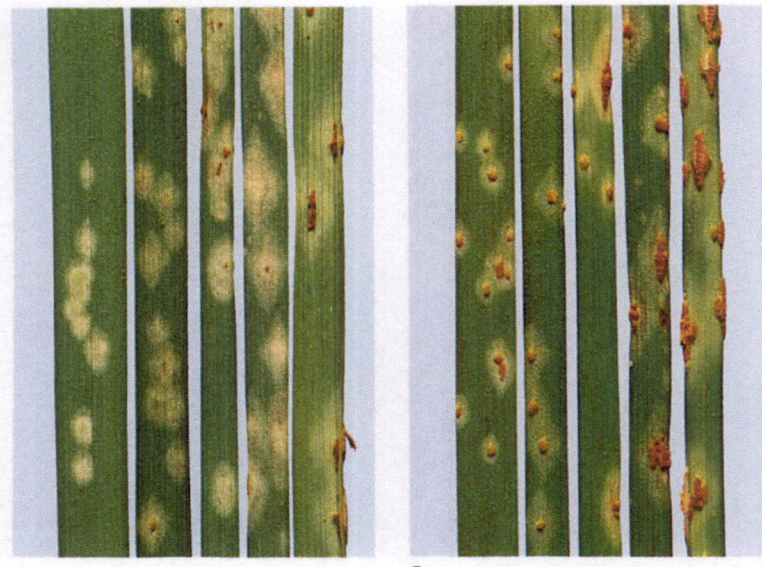

A   B

PLATE 3-17. *Sr14*
Seedling leaves of (L to R): Khapli emmer, Khapstein, Marquis + *Sr14* (= Khapstein/10*Marquis), Line A and W2691; infected with pt. 126-5,6,7,11 [*P14*] and incubated at 23/28°C. A. and B. show symptoms on the upper and lower leaf surfaces, respectively. Necrosis is more evident on the upper surfaces. Both Khapli emmer and Khapstein carry *Sr13* which is also effective against this culture. Necrosis associated with *Sr14* is influenced by temperature and light and can be induced to a limited extent with most Australian pathotypes at high temperatures.

v: Khapstein *Sr2 Sr7a Sr13* (Knott, 1962*b*).

tv: Khapli emmer *Sr13*.

Source Stocks

Possibly present in some durum wheats where it is combined with other genes such as *Sr9e*, *Sr9d*, *Sr13* or *Srdp2*. *Sr14* is probably present in the USA durum cv. Yuma (AP Roelfs, *pers. comm.* 1993).

Use in Agriculture

Sr14 has rarely been deployed in commercial wheats.

Sr15 (Watson and Luig, 1966)   (Plate 3-18)

Sr15 is completely associated with genes *Lr20* and *Pm1* (Watson and Luig, 1966; McIntosh, 1977), hence the genetics of these factors are relevent to those for *Sr15*. On the basis of mutation studies, McIntosh (1977) concluded that *Sr15* and *Lr20* were the same gene.

Chromosome Location

7A (Sears, 1954); 7AL (Sears and Briggle, 1969). Distally located to *Sr22* (The and McIntosh, 1975).

Low Infection Type

0; to X^{++}.

Environmental Variability

Temperature sensitive; lowest reactions develop at greenhouse temperatures of 15–18°C. Ineffective at temperatures above 26°C (Gousseau *et al.*, 1985). Roelfs and McVey (1979) list IT ;1^{+}N at 18°C, 3N at 21°C and 44C at 23°C.

Origin

Common wheat. Although *Sr15* was first described in cv. Norka, it was known to be present in some unrelated cultivars.

Pathogenic Variability

Pathogenic polymorphisms occur for *Sr15* in most geographical areas. However, virulence levels tend to be very high except in Australia, parts of Africa and North America (Luig, 1983; Huerta-Espino, 1992). Pathotypes can be grouped according to the infection types produced on selected genotypes under controlled conditions. At least five levels of pathogen responses can be

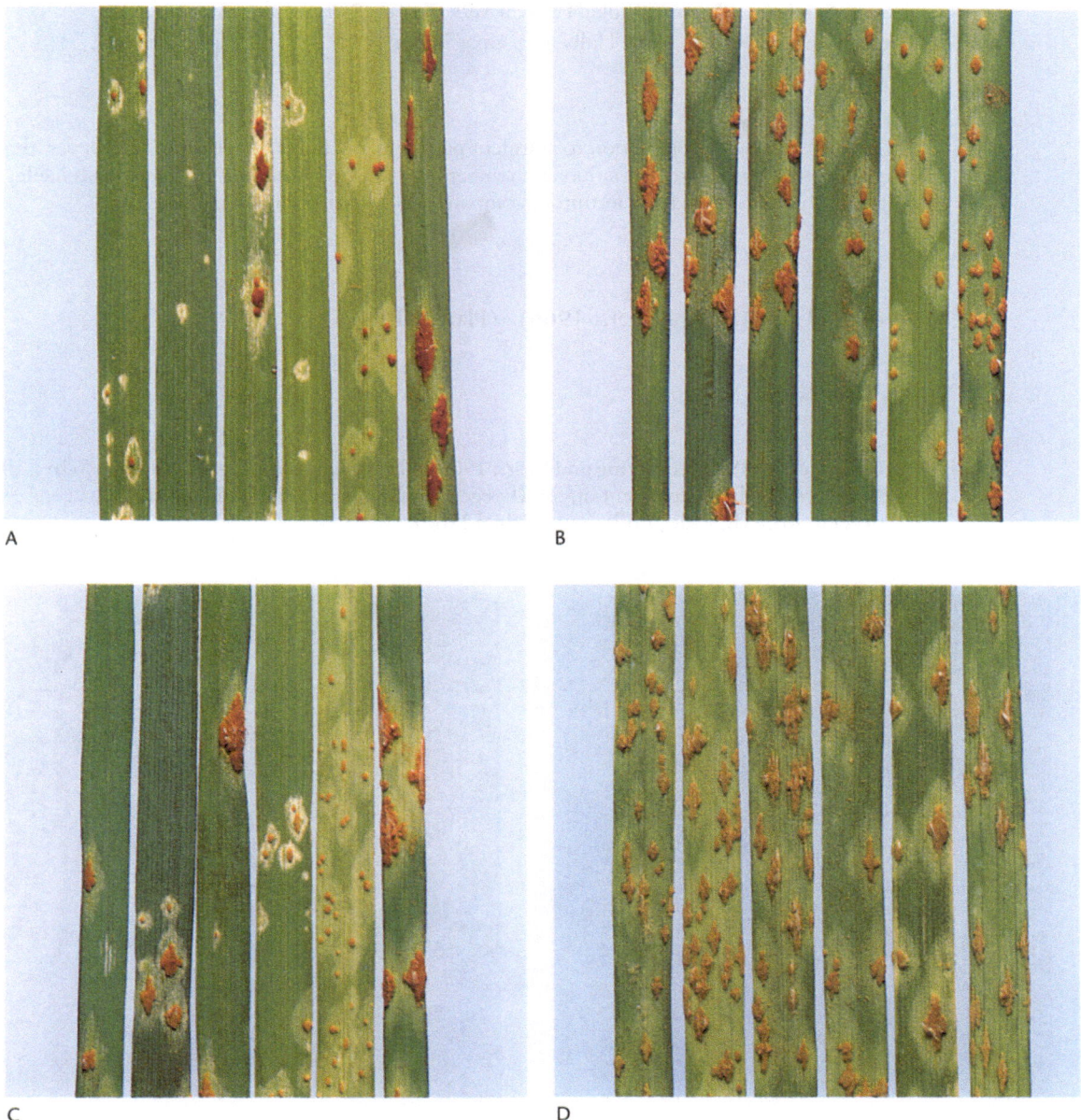

PLATE 3-18. *Sr15*

Seedling leaves of (L to R): Norka, Thew, CS*5/Axminster 7A, 77.5056, 77.5058 and Chinese Spring; infected with pt. 343-1,2,3,5,6 [*P15*] at A. 18°C and B. 23/28°C. Set C. is infected with pt. 194-2 [*P15*] and D. with pt. 21-1,2,3,7 [*p15*] and incubated at 20°C. Comparisons of the first three genotypes show the effects of genetic background on infection type. Lines 77.5056 and 77.5058 are non-mutant and mutant lines, respectively, extracted from a mutation experiment aimed at the *Sr15,/Lr20* gene. In addition to the small difference in incubation temperature, pt. 194-2 is less avirulent than 343-1,2,3,5,6 under comparable conditions. Pt. 21-1,2,3,7 is fully virulent at all temperatures on seedlings with *Sr15*. Lines 77.5056 and 77.5058 display a lower IT indicating a second gene in A., B. and C.

distinguished among the Australian pathotype collection (Luig, 1983). Pathotypes giving the higher intermediate responses would be classified as virulent in many laboratories, or if temperatures exceeded 18–20°C.

Reference Stocks

> **s**: Chinese Spring*5/Axminster 7A.
> **v**: Line AB = W2691*2/Norka (Luig, 1983); Norka (Watson and Luig, 1966).

Source Stocks

> All wheats with *Lr20* and *Pm1*. However, some mutants lacking *Pm1* but possessing *Lr20* and *Sr15* were generated in the mutagenesis study of McIntosh (1977).
> **Africa**: Kenya W744 *Sr9b*.

Australia: Angas; Aroona; Fedka. Festival *Sr9b*. Tatiara *Sr9b Sr12*.
Europe: Anfield; As II; Maris Halberd; Sappo; Timmo.
USA: Wared.

Use in Agriculture

Sr15 gives satisfactory protection to avirulent pathotypes at lower temperatures. However, the high degree of pathogenic variation and temperature sensitivity make it extremely unreliable. At best, it has been used as a fortuitous component of oligogenic resistances.

Sr16 (Loegering and Sears, 1966)   (Plate 3-19)

Synonym

Srrl2 (Rondon *et al.*, 1966).

Chromosome Location

2B (Sears *et al.*, 1957; Loegering and Sears, 1966); 2BL (Sears and Loegering, 1968). *Sr16* is distal to, and genetically independent of, *Sr9* (Loegering and Sears, 1966). It is allelic with a gene (*SrKt2*) in Kota (RA McIntosh, unpublished 1980).

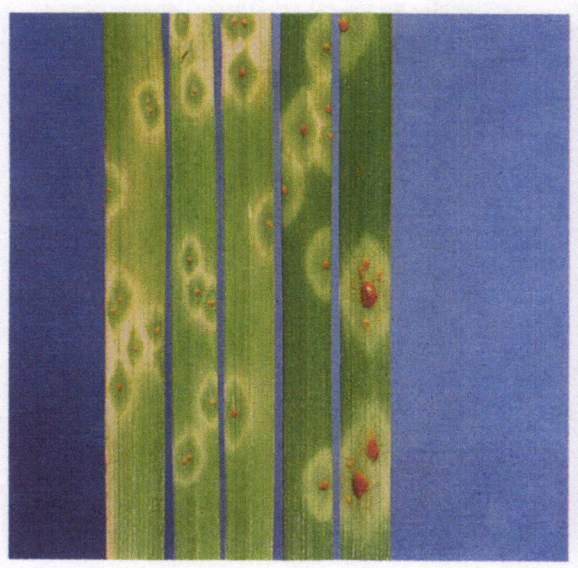

A

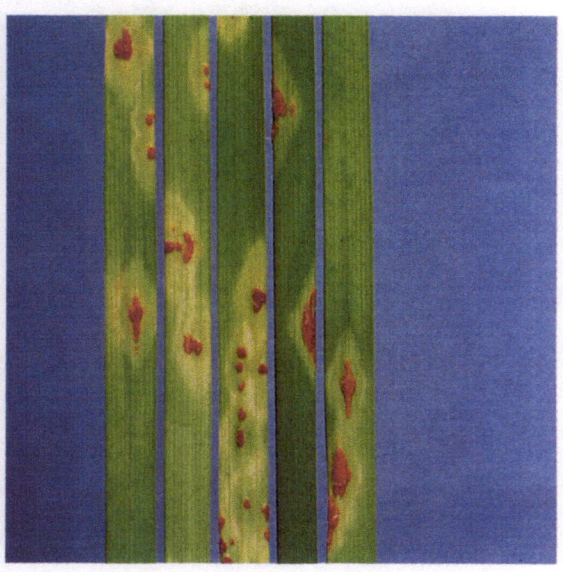

B

C

PLATE 3-19. *Sr16*

A. and B. Seedling leaves of (L to R): I*Sr16*-Ra, CS/Thatcher 2B, CS/Marquis 2B, Line AE and Chinese Spring; infected with A. culture 80-E-2 [*p9gP16PKt2*] and B. pt. 126-5,6,7,11 [*p9gp16pKt2*]. The first three lines with *Sr16* are usually not as susceptible as Chinese Spring to Australian pathotypes. On the other hand, Line AE which possesses gene *SrKt2* in a W2691 background, is fully susceptible to the second culture.

C. Four seedling leaves of I*Sr16*-Ra; infected with USDA Cereal Rust Laboratory pathotype BDCN. Courtesy AP Roelfs.

Low Infection Type
> 2 to 3.

Environmental Variability
> No variation was detected by Roelfs and McVey (1979).

Origin
> Common wheat cv. Reliance. *Sr16* is probably present in Kanred, a Crimean wheat used as one of the parents leading to Reliance and Thatcher.

Pathogenic Variability
> Avirulence is uncommon. Huerta-Espino (1992) recorded avirulent cultures among collections from Ethiopia, Kenya, Nepal, Chile and Paraguay. Most isolates of the wheat stem rust are normally considered to be virulent. However, comparison of rust infections on seedlings of ISr16-Ra and Chinese Spring often indicate very slight reductions in uredinial development on the line with *Sr16* (RA McIntosh, unpublished 1976).

Reference Stocks
> i: ISr16-Ra (Loegering and Harmon, 1969); ITha 2B-Ra (Loegering and Sears, 1973).
> s: Chinese Spring*7/Marquis 2B *Sr9g* (Williams and Kaveh, 1976); Chinese Spring*5/Thatcher 3B *Sr12* (Loegering and Sears, 1973); Chinese Spring*4/Thatcher 2B *Sr16* (Sears *et al.*, 1957).
> v: Reliance *Sr5* (Rondon *et al.*, 1966). Thatcher *Sr5 Sr9g Sr12* (Loegering and Sears, 1966).

Source Stocks
> Because of the high frequencies of virulence in the pathogen, searches for *Sr16* are rarely undertaken. Consequently, *Sr16* probably occurs in more wheat cultivars than has been reported. It is common in Thatcher derivatives (Luig, 1983), for example Manitou *Sr5 Sr6 Sr7a Sr9g Sr12*, Neepawa *Sr5 Sr7a Sr9g Sr12*, Lee *Sr9g Sr11*, Celebration *Sr9g Sr12*.

Use in Agriculture
> Not intentionally deployed in agriculture.

Sr17 (McIntosh, 1988*a*)   (Plate 3-20)

Synonym
> *sr17* (McIntosh *et al.*, 1967). Because this gene is recessive, it was originally designated with a lower case letter. However, adoption of the Rules of Genetic Nomenclature for Wheat (McIntosh, 1988*a*) required that all genes involved in disease response should be designated with upper case letters irrespective of dominance. This decision was taken in order to provide for less ambiguity in verbal communication of genetic information.

Chromosome Location
> 7B (Law and Wolfe, 1966); 7BL (McIntosh *et al.*, 1967). *Sr17* is genetically linked with *Pm5* and *Lr14a* (see McIntosh, 1988*a*).

Low Infection Type
> ;, X, X$^+$.

Environmental Variability
> Temperature-sensitive (Roelfs and McVey, 1979). *Sr17* is more effective at low temperatures, becoming ineffective above 25°C.

Origin
> Assumed to be *T. turgidum* var. dicoccum cv. Yaroslav emmer used in the development of Hope and H-44 by McFadden (1930). *Sr17* has not been detected in a tetraploid wheat by formal genetic studies. Indeed an accession of Yaroslav that possesses the relevant genes of Hope wheat, that is, *Sr2, Sr17, Lr14a* and *Pm5*, is not available.

Pathogenic Variability
> *P. graminis* populations in North America and Australia are polymorphic (Roelfs and McVey, 1979; Luig, 1983). Huerta-Espino (1992) found low frequencies of avirulence in North Africa and Spain, Kenya and the Malagasy Republic, Turkey and South America.

Reference Stocks

s: Chinese Spring*6/Hope 7B (McIntosh *et al.*, 1967).
v: Hope *Sr2 Sr7b Sr9d* (McIntosh *et al.*, 1967; Knott, 1971); H-44 *Sr2 Sr7b Sr9d* (McIntosh *et al.*, 1967; Knott, 1971). Spica *Sr7b* (McIntosh *et al.*, 1967).

Source Stocks

Sr17 is present in many USA and Mexican wheats (Roelfs and McVey, 1979), particularly those with *Lr14a* and *Pm5* as all three genes are linked. European workers (Heun and Fischbeck, 1987a, 1987b; Hovmøller, 1989; Lutz *et al.*, 1992) reported that the gene *Mli* for resistance to powdery mildew is identical to *Pm5*. It would be of interest to determine if wheats reported to possess *Mli* also carried *Sr17* or *Lr14a*.

Africa: Giza 144 *Sr11*.

Australia: Glenwari; Hofed. Gala *Sr2*. Warigo *Sr2 Sr7b*. Lawrence *Sr2 Sr7b Sr9d*. Mendos *Sr7a Sr11 Sr36*. Hopps *Sr9d*.

Canada: Selkirk *Sr2 Sr6 Sr7b Sr9d Sr23*.

CIMMYT: Nadadores *Sr11*.

Europe: Sava *Sr5*. Adam *Sr5 Sr8a*. Dunav-1 *Sr9b*.

India: Kalyansona.

New Zealand: Aotea.

USA: Auburn; Brule; Gage; Larned; Riley 67; Scoutland; Winalta. Scout *Sr2*. Newthatch *Sr2 Sr5 Sr7b Sr12*. Redman *Sr2 Sr7b Sr9d*; Renown *Sr2 Sr7b Sr9d*. Lancer *Sr2 Sr9d*. Era *Sr5 Sr6 Sr8a*. Centurk *Sr5 Sr6 Sr8a Sr9a*. Homestead *Sr6*. Colt *Sr6 Sr8a Sr9a Sr24*. Lancota *Sr9b Sr10*. Osage *Sr24*.

Use in Agriculture

From the time of their release in the 1920s, Hope and H-44 have displayed durable resistance to stem rust. A vast literature on these resistances indicates involvement of genes conferring both seedling and adult plant resistances. At various times, and using a limited range of pathogenic variability, researchers have obtained different results when determining the resistance genotype [see Hare and McIntosh (1979) for review]. *Sr17* is a significant component of this resistance and can be found in a wide range of Australian, Mexican, USA, Canadian and Indian cultivars. However, the durability of resistance in Hope and H-44 and their derivatives is associated with *Sr2*.

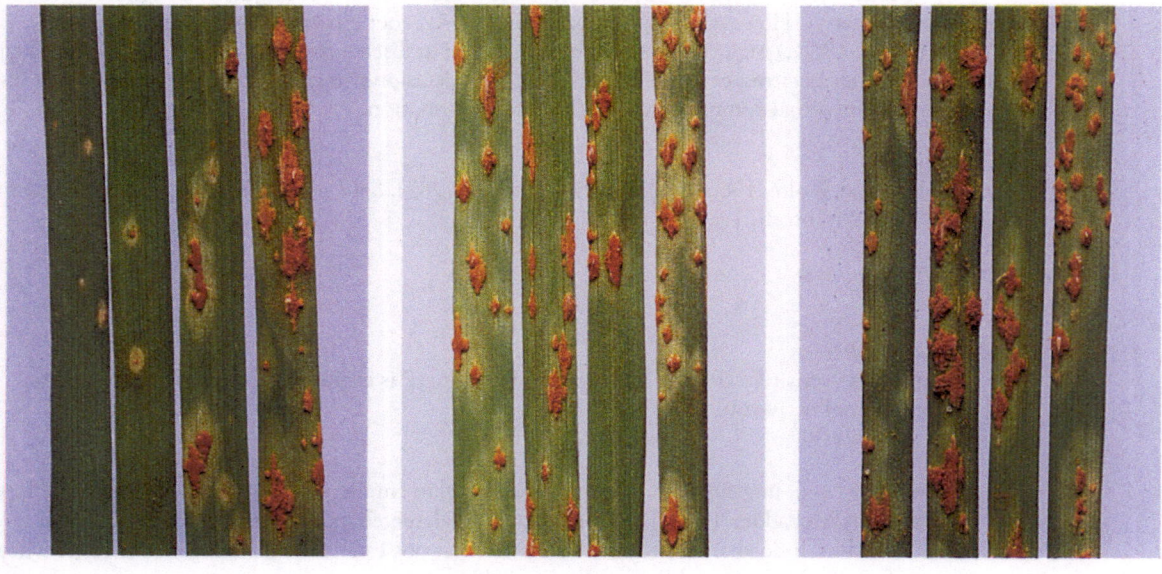

A   B   C

PLATE 3-20. *Sr17*
Seedling leaves of (L to R): Renown, Spica, CS*6/Hope 7B and Chinese Spring; infected with A. pt. 194-2 [*P17*], B. pt. 21-2,4,5 [*p17*] and C. pt. 21-(1),2 [*P17*]. A. and B. were incubated at 18°C, and C. at 23/28°C. Set B. shows the effect of virulence for *Sr17* whereas C. shows the effect of higher temperatures.

Sr18 (Baker *et al*, 1970)   (Plate 3-21)

Synonyms

SrG2 (Luig and Watson, 1965); Srrl1 (Rondon *et al.*, 1966); Srmq1 (Berg *et al.*, 1963); SrPs1 and SrMn1 (Sanghi and Baker, 1972); R1 (Loegering and Powers, 1962).

Chromosome Location

1D (Sears *et al.*, 1957; Baker *et al.*, 1970; Anderson *et al.*, 1971); 1DL (Williams and Maan, 1973).

Low Infection Type

; to ;2$^=$.

Environmental Variability

Low.

Origin

Sr18 is present in a very high proportion of common wheat lines (Baker *et al.*, 1970).

Pathogenic Variability

Avirulence in *P. graminis* f. sp. *tritici* is rare (Roelfs and McVey, 1979; Luig, 1983). However, avirulence for this gene is widespread in collections of *P. graminis* f. sp. *secalis* and in hybrids between *P. g. tritici* and *P. g. secalis* (Luig and Watson, 1972).

Reference Stocks

i: I Hope 1D-Ra (Loegering and Harmon, 1969); Sr18/8*LMPG (Knott, 1990); Mq-A and Rl-A (Anderson *et al.*, 1971).
s: Chinese Spring*6/Hope 1D (Sears *et al.*, 1957).
v: Mona (Sanghi and Baker, 1972); Pusa (Sanghi and Baker, 1972). Hope Sr2 Sr7b Sr9d Sr17 (Luig, 1983). Reliance Sr5 Sr16 Sr20 (Anderson *et al.*, 1971). Marquis Sr7b Sr19 Sr20 (Anderson *et al.*, 1971). Gabo Sr11 (Luig and Watson, 1965).

Source Stocks

This gene is very difficult to identify with certainty because pathogen isolates avirulent for *Sr18* usually possess additional genes for avirulence which render them unable to attack most wheat cultivars. Nevertheless, workers at The University of Sydney demonstrated its presence in most wheats (Baker *et al.*, 1970). Indeed McIntosh (1988a) considered it more convenient to list wheats not possessing *Sr18*. These included Chinese Spring (Loegering and Harmon, 1969); Eureka (Baker *et al.*, 1970); Federation (Baker *et al.*, 1970); Little Club (Loegering and Harmon, 1969); Morocco (Baker *et al.*, 1970); Prelude (Loegering and Harmon, 1969) and Yalta (Baker *et al.*, 1970) as well as W2691 and Line E, both of which were especially bred at The University of Sydney for susceptibility to pathogen cultures with unusual genes for avirulence, including *P18*. The presence of *p18* in *P. graminis* f. sp. *tritici* would be essential because *Sr18* occurs in most common wheats. In contrast, *P18* is unnecessary in *P. graminis* f. sp. *secalis* due to the absence of *Sr18* in cereal rye. In this way genes such as *Sr18* play a role in host specialisation.

Use in Agriculture
None.

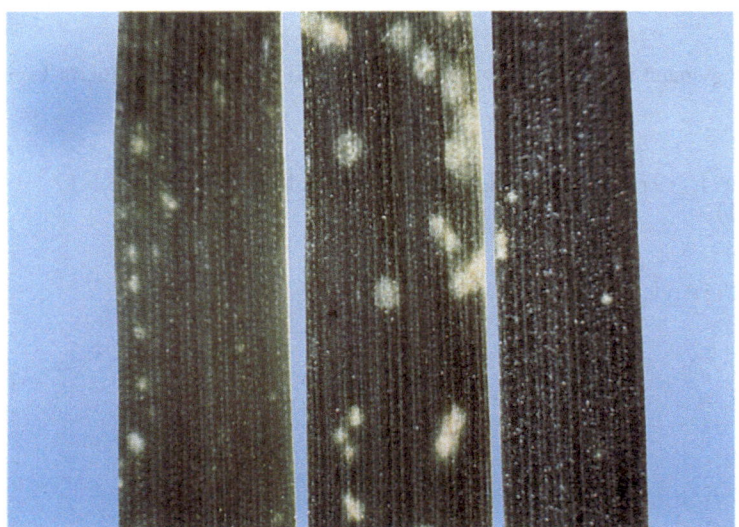

PLATE 3-21. *Sr18*
Seedling leaves of (L to R): IHope 1D-Ra, W2691/Purple Straw Selection, and Line AD = W2691*5/Kota; infected with University of Missouri culture 111 x 36.#97. The first two wheats show the reaction typical of plants with *Sr18* whereas the high resistance of the third plant is conferred by *Sr28*.

Sr19 (Anderson *et al.*, 1971)

Synonym
Srmq2 (Berg *et al.*, 1963).

Chromosome Location
2B (Anderson *et al.*, 1971); 2BS (Williams and Maan, 1973).

Low Infection Type
1^- to 1^+ (Roelfs and McVey, 1979), 2C (Huerta-Espino, 1992).

Environmental Variability
Not known.

Origin
Common wheat cv. Marquis. Other sources of *Sr19* have not been identified.

Pathogenic Variability
Most collections of *P. graminis* f. sp *tritici* are virulent (Roelfs and McVey, 1979; Luig, 1983). Huerta-Espino (1992) recorded two cultures producing IT 2C on lines with *Sr19*. One of these was collected in the Malagasy Republic and the other in Chile.

Reference Stocks
v: Mq-B (Anderson *et al.*, 1971). Marquis *Sr7b Sr18 Sr20* (Anderson *et al.*, 1971).

Source Stocks
None known.

Use in Agriculture
None.

Plate
Not available.

Sr20 (Anderson *et al.*, 1971)

Synonyms
Srmq3 (Berg *et al.*, 1963); Srrl3 (Rondon *et al.*, 1966); R2 (Loegering and Powers, 1962).

Chromosome Location
2B (Anderson *et al.*, 1971).

Low Infection Type
$2^=$ to 3 (Anderson *et al.*, 1971; Roelfs and McVey, 1979; Roelfs *et al.*, 1992).

Environmental Variability
Not known.

Origin
Common wheat cv. Marquis.

Pathogenic Variability
Most collections of *P. graminis* f. sp *tritici* are virulent (Roelfs and McVey, 1979; Huerta-Espino, 1992).

Reference Stocks
v: Mq-C (Anderson *et al.*, 1971); RI-C (Anderson *et al.*, 1971). Marquis *Sr7b Sr18 Sr19* (Anderson *et al.*, 1971).

Source Stock
Reliance *Sr5 Sr16 Sr18* (Anderson *et al.*, 1971).

Use in Agriculture
None.

Plate
Not available.

Sr21 (The, 1973*a*) (Plate 3-22)

Chromosome Location

2A (The, 1973*a*); 2AL (The *et al.*, 1979). *Sr21* mapped 2 cM from the centromere and 48 cM from *Pm4*.

Low Infection Type

; to 23⁻. The low infection type becomes higher with increasing levels of ploidy (The, 1973*a*).

Environmental Variability

Low (Roelfs and McVey, 1979).

Origin

T. monococcum accessions, including Einkorn C.I.2433 which was adopted by Stakman *et al.* (1962) as a pathotype differential (The, 1973*a*). A gene with identical specificity in chromosome 1D of *T. tauschii* (McIntosh, 1981) is sometimes referred to as *SrX* (D The, unpublished 1992).

Pathogenic Variability

Polymorphism occurs in most geographic areas. The frequencies of virulence in North America (Roelfs and McVey, 1979) and South America (Huerta-Espino, 1992) are very high compared to other regions. Virulent mutants have been detected in Australian field surveys but none has become established as a common field pathotype (Zwer *et al.*, 1992).

Reference Stocks

i: Sr21/8*LMPG (Knott, 1990).
v: W3586 = Glossy Huguenot/Einkorn//unknown hexaploid (The, 1973*a*); R.L.5406 = Tetra Canthatch/*T. tauschii* R.L.5289 (McIntosh, 1981).
dv: Einkorn C.I.2433.

Source Stocks

i: Sr21/5*Aroona; Sr21/5*Condor; Sr21/5*Egret; Sr21/5*Halberd; Sr21/5*Lance; Sr21/5*Oxley; Sr21/5*Teal (The *et al.*, 1988).
v: Hexaploid derivatives of *T. monococcum* produced by The (1973*b*). Hexaploid derivatives of *T. tauschii* possessing *SrX* (see Origin).
tv: Tetraploid derivatives of *T. monococcum* (The, 1973*b*).
dv: *T. monococcum* accessions (The, 1973*a*, 1973*b*). *T. tauschii* accessions including R.L.5289 (McIntosh, 1981; RA McIntosh, unpublished 1991).

Use in Agriculture

This gene has potential for limited use in areas with low frequencies of virulence if deployed together with other genes. If used alone, rapid increases in the frequency of virulent pathotypes

A   B

PLATE 3-22. Sr21
Seedling leaves of (L to R): Einkorn (*T. monococcum*), W3586 (Glossy Huguenot/Einkorn//unknown hexaploid), R.L.5406 (Tetra Canthatch/*T. tauschii* R.L.5289) and Steinwedel; infected with A. pt. 194-2 [*P21*] and B. pt. 17-1,2,3,7 [*p21*] and incubated at 20°C. Note the similarity in response of Einkorn and its derivative with the *T. tauschii* derivative.

can be anticipated. In addition, current lines with *Sr21* are not adequately resistant to avirulent pathotypes as adult plants. The *et al.* (1988) showed that, in the absence of disease, a number of backcross-derived lines possessing *Sr21* gave significantly lower yields than near-isogenic counterparts that did not carry the gene.

Sr22 (The, 1973*a*)   (Plate 3-23)

Chromosome Location

7A (Kerber and Dyck, 1973); 7AL (The, 1973*a*). *Sr22* is located 30 cM from the centromere (The, 1973*a*), 2 cM from *cn-A1* (chlorina phenotype) and more than 50 cM proximal to *Pm1/Lr20/Sr15* (The and McIntosh, 1975).

Low Infection Type

; (in diploid wheat) to 2⁻ (in hexaploid wheat) (Kerber and Dyck, 1973); 1⁻ to 2 (Huerta-Espino, 1992).

Environmental Variability

Temperature sensitive; more effective at lower temperatures.

Origin

T. monococcum R.L.5244 (Kerber and Dyck, 1973; The, 1973*a*). The (1973*b*) found that *Sr22* was present in a range of wild einkorn wheats.

Pathogenic Variability

Virulent cultures were not detected in surveys in Mexico (Singh, 1991), USA (Roelfs and McVey, 1979), South Africa (Le Roux and Rijkenberg, 1987*a*) and Australia (Zwer *et al.*, 1992). Huerta-Espino (1992) also failed to detect virulence in his international survey. However, Gerechter-Amitai *et al.* (1971) reported virulence in ten of 12 pathotypes used in Israel.

Reference Stocks

i: *Sr22*/9*LMPG (Knott, 1990).
v: Chinese Spring/3/Steinwedel*2//Spelmar/*T. boeoticum* G-21 (The, 1973*a*); Marquis*5//Stewart*3/*T. monococcum* R.L.5244 (Kerber and Dyck, 1973; The, 1973*a*); Steinwedel*2//Spelmar2*/*T. boeoticum* G-21 (The, 1973*a*).
tv: Spelmar*2/*T. boeoticum* G-21 (Gerechter-Amitai *et al.*, 1971; The, 1973*a*); Stewart*6/*T. monococcum* R.L.5244 (Kerber and Dyck, 1973).
dv: Various *T. monococcum* accessions including G-21 and R.L.5244 (Kerber and Dyck, 1973; The, 1973*a*, 1973*b*).

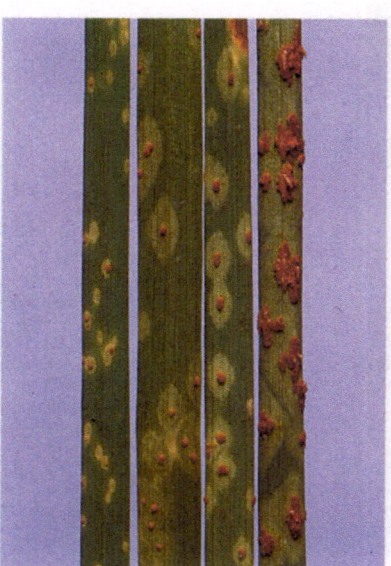 

PLATE 3-23. *Sr22*
Seedling leaves of (L to R): *T. monococcum* C68.114 (2n = 14), Glossy Huguenot*2/ *T. monococcum* C85.105 (2n = 28), Marquis*5//Stewart*3/ *T. monococcum* R.L.5244 (2n = 42) and Marquis; infected with pt. 21-1,2 and incubated at A. 18°C and B. 23/28°C. Note the distinctly higher response of the third (hexaploid) line at the higher temperature. On the other hand, the second (tetraploid) line appears to be more resistant at the higher temperature.

A   B

Australia: BT-Schomburgk; Schomburgk.

Use in Agriculture

Cultivar Schomburgk was released in Australia in 1986. This and a boron-tolerant derivative, BT-Schomburgk are the only commercial cultivars with this gene. Lines carrying *Sr22* were reported to be moderately susceptible in the field by Roelfs and McVey (1979).

Sr23 (McIntosh and Luig, 1973*b*) (Plate 3-24)

Chromosome Location

2BS (RA McIntosh, unpublished 1978). This gene is completely associated with *Lr16*. It was initially believed to be located in chromosome 4B, but a Rescue monosomic series was used to establish the location and Rescue carries a 2B-4B reciprocal translocation relative to Chinese Spring (RA McIntosh, unpublished 1980; EDP Whelan, *pers. comm.* 1982). *Sr23/Lr16* are genetically independent of, and presumably distal to, *Sr36* (RA McIntosh, unpublished 1980).

Low Infection Type

1^{++}N to 3^{+}CN.

Environmental Variability

Sr23 is expressed only under conditions of high temperature and high light intensity (Luig, 1983). The low infection type is often characterised only by necrosis surrounding some of the otherwise fully compatible uredinia. Roelfs and McVey (1979) observed slight increases in sporulating area with increasing temperature.

Origin

Sr23 was first noted and characterised in common wheat cultivars Selkirk, Exchange and Warden.

Pathogenic Variability

A necrotic response and low infection type with the old Australian pathotype 126-5,6,7,11 is characteristic of plants with this gene. With other pathotypes some necrosis is expressed in otherwise large 'compatible' uredinia under high light and temperature conditions. Huerta-Espino (1992) used Exchange as the host tester for *Sr23* but it apparently also carries the uncatalogued gene *SrMcN*. Despite the statement that "all isolates were virulent to ...*Sr23* and *SrMcN*" (Huerta-Espino, 1992), three cultures, one each from Burundi, Turkey and Nepal, were recorded as producing ITs 2C to 23C on seedlings of Exchange. These responses seemed to be too high to be conferred by *SrMcN* (2^{-}; Roelfs and McVey, 1979), but could be associated with *Sr23*.

PLATE 3-24. Sr23
Lower and upper surfaces, respectively, of seedling leaves of (L to R) Warden, Exchange and Chinese Spring; infected with pt. 126-5,6,7,11 [*P23*] and incubated at 27/38°C. The resistant response conferred by *Sr23* is usually that of necrosis, which is enhanced by high light intensity and high temperatures, superimposed on large pustules. Necrosis of some pustules can be detected under these conditions even with 'virulent' cultures. Necrosis is reduced under lower light and temperature.

Reference Stocks

v: Exchange (McIntosh and Luig, 1973*b*). Selkirk *Sr2 Sr7b Sr9d Sr17* (McIntosh and Luig, 1973*b*).

Source Stocks

Warden (McIntosh and Luig, 1973*b*). Etoile de Choisy *Sr29* (McIntosh and Luig, 1973*b*). All stocks carrying *Lr16* are assumed to carry *Sr23* (see *Lr16*).

Use in Agriculture

None. *Sr23* is not expected to provide significant levels of protection when deployed in susceptible backgrounds (McIntosh and Luig, 1973*b*).

Sr24 (McIntosh *et al.*, 1976)   (Plate 3-25)

Chromosome Location

3D (McIntosh *et al.*, 1976; Smith *et al.*, 1968; Sears, 1973); 3DL (Hart *et al.*, 1976); 3Ag (Sears, 1977). Two other translocation lines produced by Sears involve chromosome 3BL (Sears, 1977). A further translocation is present in cultivar Amigo (see *Lr24*). *Sr24* is completely associated with *Lr24* (McIntosh *et al.*, 1976).

Low Infection Type

1^- to 22^+.

Environmental Variability

Low (Roelfs and McVey, 1979).

Origin

Th. ponticum.

Pathogenic Variability

Virulence for *Sr24* has been reported in South Africa (Le Roux and Rijkenberg, 1987*b*) and India (Bhardwaj *et al.*, 1990). Sydney University culture 57096 [pt. 34-(4),7] produces IT 3 on seedlings with *Sr24*. This Australian culture is assumed to have arisen from somatic hybridisation between *P. graminis* f. sp. *tritici* and *P. graminis* f. sp. *secalis* (Luig, 1983). Huerta-Espino (1992) did not find virulence for *Sr24* among a wide range of international collections.

Reference Stocks

i: *Sr24*/9*LMPG (Knott, 1990); various 3D/Ag translocation lines produced by Sears (1972*b*, 1973, 1977) can be considered near-isogenic to Chinese Spring.
s: Chinese Spring 3Ag(3D) (Sears, 1973).
v: Agent (McIntosh *et al.*, 1976).

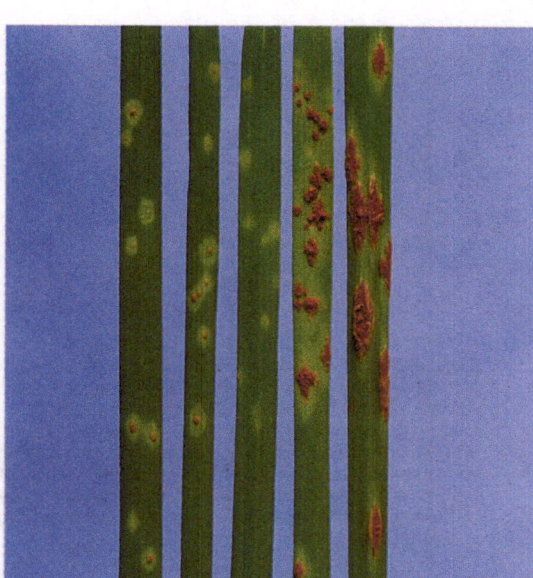

PLATE 3-25. Sr24
Seedling leaves of (L to R): Agent, Amigo, Sunco (a Cook derivative with *Sr24*), Cook and Chinese Spring; infected with pt. 34-1,2,3,4,5,6,7.

Australia: Janz *Sr5*; Torres *Sr5* (Brennan *et al.*, 1983); Vasco *Sr5*. Sunco *Sr5 Sr6 Sr8a Sr36*. Sunbird *Sr5 Sr8a*. Sunelg *Sr26*.

North America: Abilene; Amigo (The *et al.*, 1992); Arapahoe; Arkan; Blueboy II (McIntosh *et al.*, 1976); Centura; Century; Cimmaron; Cloud; Cody; Coker 9733; Collin; Fox; Mesa; Norkan; Parker 76; Payne; Rio Blanco; TAM200; Terral 101; Thunderbird; Timpaw; Trailblazer; Twain; Wanken. Karl *Sr2 Sr9d*. Siouxland *Sr5 Sr31* (Schmidt *et al.*, 1985). Butte 86 *Sr6* (Modawi *et al.*, 1985). Colt *Sr6 Sr8a Sr9a Sr17*. Osage *Sr17*; Sage *Sr17* (Livers, 1978). Jasper *Sr31*; Longhorn *Sr31*.

South Africa: Gamka; Karee; Kinko; Palmiet; SST25; SST44 = T4R; SST102; Wilge (Le Roux and Rijkenberg, 1987*b*); SST23 (Sharma and Gill, 1983).

Use in Agriculture

Derivatives of Agent were widely used in the USA and South Africa as a source of leaf rust resistance. *P. graminis* pathotypes with virulence for *Sr24* were found in South Africa in 1984 (Le Roux, 1985) and in India in 1989 (Bhardwaj *et al.*, 1990). All backcross derivatives with *Lr24/Sr24* from Agent added to white-seeded Australian wheats were found to be red seeded (RA McIntosh, unpublished 1973). Several of Sears's 3D/Ag translocation stocks were then used, and of these, transfers #3 and #14 gave white-seeded derivatives. These lines were used to develop Australian cultivars, the first of which, Torres, was released in 1983. Australian lines were subsequently used in Indian backcrossing programs but virulence for *Sr24* was detected before cultivars could be commercialised.

The *et al.* (1992) found that Amigo carried *Sr24/Lr24*. Amigo carries an independent translocation with stem rust resistance from rye and, although the chromosome location of *Sr24* in this red-seeded stock was unknown, white-seeded derivatives were easily selected in Australian backcrossing programs.

Sr25 (McIntosh *et al.*, 1976)   (Plate 3-26)

Chromosome Location

7D (Sharma and Knott, 1966). 7DL (McIntosh *et al.*, 1976; Dvorak and Knott, 1977); 7AL (Eizenga, 1987); 7Ag (Sears, 1973). In most stocks *Sr25* is associated with *Lr19*.

Low Infection Type

1⁻ to 23. Low infection types recorded by Knott (1980, 1990) were generally lower than those obtained in Australia.

Environmental Variability

Low (Roelfs and McVey, 1979), but probably inadequately researched (Luig, 1983). Gough and Merkle (1971) suggested that lines with *Sr25* may become more susceptible at high temperatures.

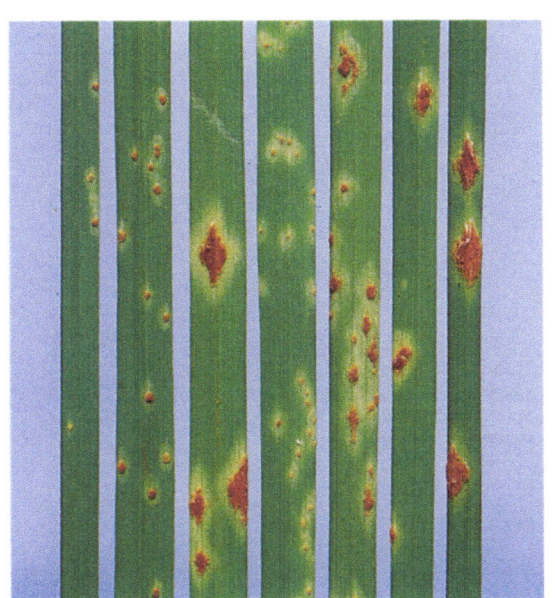

PLATE 3-26. Sr25
Seedling leaves of (L to R): Agatha, Sears' 7D/Ag#3, Sears' 7D/Ag#11 (C75.39), Knott's Agatha Mutant-1 (C80.1), Knott's Agatha Mutant-2 (C80.2), Thatcher and Chinese Spring; infected with pt. 34-1,2,3,4,5,6,7. Sears' 7D/Ag#11 carries *Lr29* which is not associated with *Sr25*. Knott's Agatha Mutant-2 lost *Sr25*, but retained *Lr19*. The slightly lower response of this line and Thatcher compared to 7D/Ag#11 and Chinese Spring is attributed to the presence of additional genes in Thatcher.

Origin

Th. ponticum.

Pathogenic Variability

Luig (1983) mentioned an Israeli isolate with putative virulence. Huerta-Espino (1992) identified one virulent culture from Ethiopia and two virulent cultures from Nepal.

Reference Stocks

i: Sr25/9*LMPG (Knott, 1990). Sears' independently derived 7D/Ag translocation stocks (Sears, 1973) can be considered near-isogenic to Chinese Spring. Agatha *Sr5 Sr9g Sr12 Sr16* = T4 (Sharma and Knott, 1966) is near-isogenic to Thatcher.

su: Chinese Spring 7Ag(7D) (Sears, 1973). Agrus, a 7Ag(7D) stock (Sharma and Knott, 1966).

Source Stocks

Because of its close association with *Lr19* (McIntosh *et al.*, 1976), *Sr25* should be present in Oasis F86, Indis and Sunnan (see *Lr19*).

Use in Agriculture

Lines of Chinese Spring with *Lr19/Sr25* can become moderately rusted with avirulent cultures in breeding nurseries and losses may occur (McIntosh *et al.*, 1976; Roelfs and McVey, 1979). The use of this gene in breeding can be justified by its linkage with *Lr19*. However, lines with *Sr25/Lr19* from Agatha and Sears' translocations are characterised by high levels of yellow pigment in the endosperm. Knott (1980) obtained two mutants of Agatha with reduced levels of yellow pigment in the flour. One of these mutants lacked *Sr25*. Marais (1992a) reported that a gene very similar to *Sr25*, and designated *Sr25d*, was present in the Inia 66 x *Th. distichum* derivative, Indis. Marais (1992a, 1992b) also obtained mutants with reduced yellow pigment in Indis derivatives and some of these lacked *Sr25d*.

Sr26 (McIntosh *et al.*, 1976)   (Plate 3-27)

Chromosome Location

6A (6A/6Ag translocation) (Knott, 1961); 6AL (J Fisher, *pers. comm.* 1975).

Low Infection Type

0; to 2⁻.

Environmental Variability

Low (Roelfs and McVey, 1979).

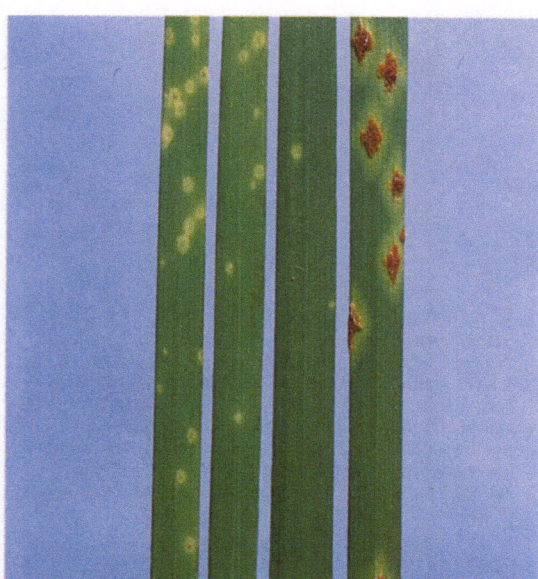

PLATE 3-27. *Sr26*
Seedling leaves of (L to R): Eagle, Kite, Harrier and Chinese Spring; infected with pt. 34-1,2,3,4,5,6,7. The resistant response is very uniform.

Origin

Th. ponticum

Pathogenic Variability

Virulence has not been confirmed in field isolates (Luig, 1983; Huerta-Espino, 1992).

Reference Stocks

i: Sr26/9*LMPG (Knott, 1990).
s: Chinese Spring 6Ag and 6AgL substitutions for 6A, 6B and 6D, respectively (D The, unpublished 1990).

Source Stocks

Australia: Avocet (J Fisher, *pers. comm.* 1979); Flinders (Syme, 1983); Harrier; Kite (Luig, 1983); Takari (Fletcher, 1983). Eagle *Sr9g* (McIntosh *et al.*, 1976). Sunelg *Sr24*. Bass *Sr36* (Syme *et al.*, 1983).

Use in Agriculture

Sr26 has been used as a source of resistance only in Australia where the first cultivar, Eagle, was released in 1971. The *et al.* (1988) showed that backcross-derived lines with *Sr26* yielded 9% less than *sr26* sibs. However, cultivars Flinders, Harrier, Kite, Takari, Eagle and Sunelg were widely grown and competed satisfactorily with contemporary cultivars.

Sr27 (McIntosh, 1988a)   (Plate 3-28)

Chromosome Location

3A (3A.3R translocation) (Acosta, 1962).

Low Infection Type

0; to 12$^=$.

Environmental Variability

Low (Roelfs and McVey, 1979).

Origin

S. *cereale* cv. Imperial.

Pathogenic Variability

Virulence for *Sr27* is rare. Harder *et al.* (1972) isolated an east African culture virulent on a Pembina line with *Sr27*. Luig (1983) reviewed Australian work showing that cultures of

PLATE 3-28. *Sr27*
Seedling leaves of (L to R): CS WRT 238.5, Coorong triticale, Satu triticale and Chinese Spring; infected with A. pt. 34-2 [*P27*] and B. pt. 34-2,12 [*p27*]. The resistant response of wheat with *Sr27* is lower than that of Coorong. Both CS WRT 238.5 and Coorong are susceptible with pt. 34-2,12 which was originally collected from Coorong. Satu triticale appears to be more resistant with pt. 34-2,12 but this effect is due to a significantly lower intensity of infection.

A   B

P. graminis f. sp. *secalis* and certain hybrids of *P. graminis* f. sp. *tritici* and *P. graminis* f. sp. *secalis* were virulent. McIntosh *et al.* (1983) showed that isolates of *P. graminis* f. sp. *tritici* from triticale cv. Coorong were virulent on wheat seedlings with *Sr27*. A greenhouse mutant with virulence on Coorong was also virulent on seedlings with *Sr27*. The results were accepted as evidence that the resistance gene in Coorong and many other triticale lines developed in Mexico was *Sr27*. Virulence on triticale cultivars with *Sr27* was found in South Africa in 1988 (Smith and Le Roux, 1992).

Reference Stocks

i: Chinese Spring WRT 238.5 (Acosta, 1962). Justin, Selkirk and Pembina derivatives of WRT 238.5 (Stewart *et al.*, 1968). Sr27/9*LMPG (Knott, 1990).
ad: Chinese Spring + Imperial 3R (2n = 44) (ER Sears, *pers. comm.* 1969).

Source Stocks

Widespread in triticales, for example Coorong, Towan, Dua, Tyalla, Arabian, Bura S (McIntosh *et al.*, 1983). Some sources of Setter carry *Sr27*.

Use in Agriculture

Wheats with *Sr27* have not been released in agriculture. Pathogen samples collected from Coorong triticale in eastern Australia were shown to be virulent for *Sr27* (McIntosh *et al.*, 1983). Coorong and many other triticale lines were extremely susceptible to this pathotype. McIntosh *et al.* (1983) further showed that *Sr27* occurred at high frequency in lines present in nurseries distributed from CIMMYT and gave warning of genetic vulnerability in triticale. Cultivar Satu was recommended in Australia as a replacement for Coorong, but a further mutant of the 'Coorong' pathotype quickly developed. Genetic studies indicated that a single gene in Satu was allelic with the gene (*Sr27*) in Coorong.

Sr28 (McIntosh, 1978)   (Plate 3-29)

Synonym

SrKta1 (Berg *et al.*, 1963).

Chromosome Location

2BL (McIntosh, 1978). *Sr28* was located 18 cM distal to *Sr9* and 33 cM proximal to *Sr16*.

Low Infection Type

0; to $2^=$.

Environmental Variability

Low (Roelfs and McVey, 1979).

Origin

Common wheat cv. Kota. Kota was included in the differential set of Stakman *et al.* (1962).

Pathogenic Variability

Virulence is common in all geographic areas (Luig, 1983). Huerta-Espino (1992) found significant levels of avirulence among cultures from Ethiopia and Nepal; otherwise virulence levels were high.

Reference Stocks

i: Line AD = W2691*5/Kota (McIntosh, 1978).
s: Ceres *Sr7b* (McIntosh, 1978); Kota *Sr7b* (McIntosh, 1978).

Source Stocks

None.

Use in Agriculture

Sr28 is of limited use in current wheat breeding. However, it was the basis for resistance in cultivar Ceres released in the USA and rendered susceptible by race 56 in 1935.

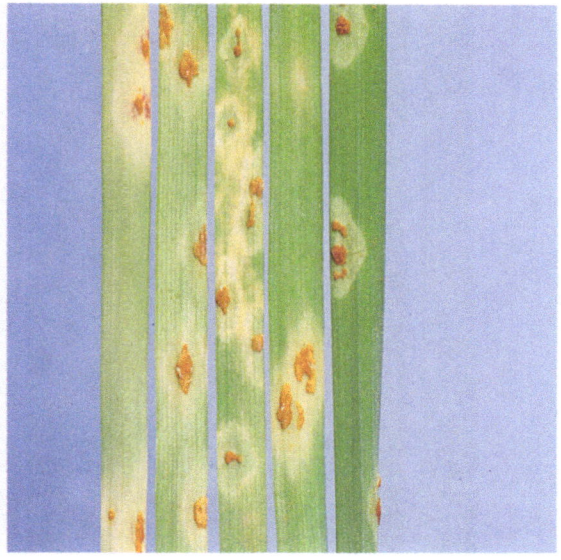

PLATE 3-29. *Sr28*

A. and B. Seedling leaves of (L to R): Line AD (W2691*5/Kota), Kota, Ceres, Line AE (W2691*5/Kota) and Chinese Spring; infected with A. culture 80-E-2 [*P28*)] and B. pt. 126-5,6,7,11 [*p28*]. Line AD carries *Sr28* whereas Line AE has a second gene conferring IT 2 (see *SrKt2*). Pt. 126-5,6,7,11 is virulent for both of these genes as well as *Sr7b* present in Kota.

C. Seedling leaves of a selection of Line AD infected with USDA Cereal Rust Laboratory pathotype LBBL. Courtesy AP Roelfs.

Sr29 (Dyck and Kerber, 1977*b*) (Plate 3-30)

Synonym

SrEC (McIntosh *et al.*, 1974).

Chromosome Location

6DL (Dyck and Kerber, 1977*b*); 6DS (Zeller and Oppitz, 1977); *Sr29* appeared to be genetically independent of the centromere (Dyck and Kerber, 1977*b*).

Low Infection Type

1 to 3.

Environmental Variability

Low (Roelfs and McVey, 1979).

Origin

Common wheat. *Sr29* appears to be a gene of European origin.

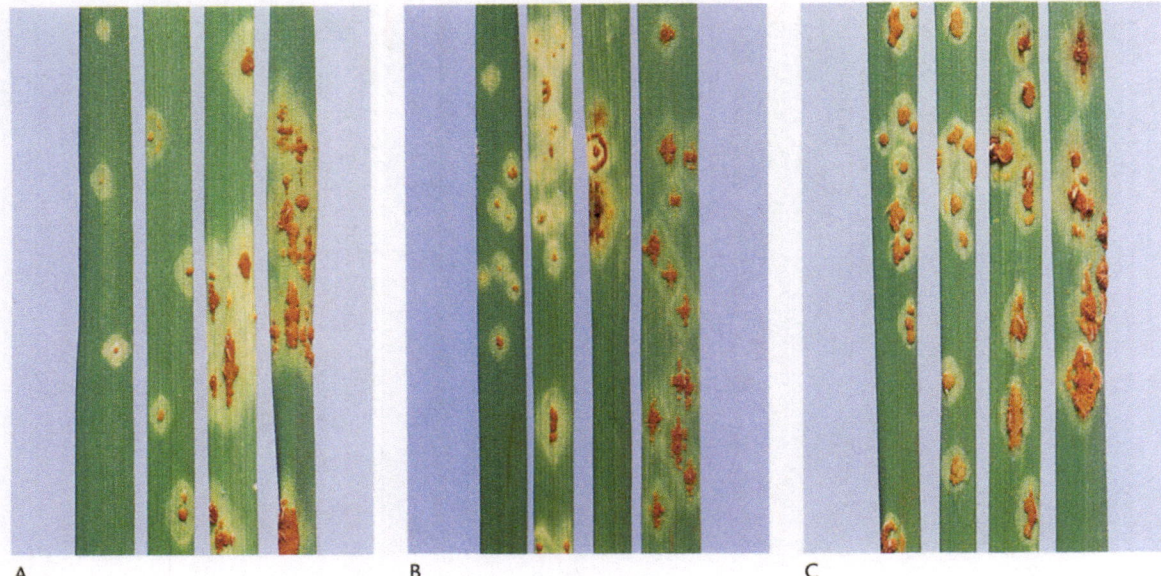

A   B   C

PLATE 3-30. *Sr29*

Seedling leaves of (L to R): Etoile de Choisy, Prelude/8*Marquis//Etoile de Choisy, Marquis and Chinese Spring; infected with A. and B. pt. 21-1,2,3,7 and C. pt. 126-5,6,7,11. A. and C. were incubated at 23/20°C and B. at 27/23°C. The brown necrosis often present in pustules on Etoile de Choisy is due to *Sr23* which is also present in this cultivar. With Australian pathotypes, Marquis is usually less susceptible than Chinese Spring. The uppermost uredium on the second leaf in A. is a leaf rust contaminant.

Pathogenic Variability

Roelfs and McVey (1979) reported that a few avirulent cultures gave low infection types that were slightly higher than usual. Huerta-Espiño (1992) reported virulence only among cultures from western Asia, eastern Europe, Egypt, Ethiopia and Turkey.

Reference Stocks

i: Prelude/8*Marquis//Etoile de Choisy (Dyck and Kerber, 1977b).
v: Etoile de Choisy W3550 *Sr23* (McIntosh *et al.*, 1974).

Source Stocks

Hela; Mara; Slavia; Vala (Bartos and Stuchlikova, 1986). Moisson *Sr23* (Luig, 1983).

Use in Agriculture

Adult plants with this gene are moderately susceptible (Roelfs and McVey, 1979). The gene has been used for stem rust control in European wheats, but has not been incorporated into North American winter wheats or spring wheats grown outside of Europe.

Sr30 (Knott and McIntosh, 1978)   (Plate 3-31)

Chromosome Location

5DL (Knott and McIntosh, 1978). *Sr30* is genetically independent of *Pm2* (in 5DS) and *Lr1* (Knott and McIntosh, 1978).

Low Infection Type

1^+ to 3. Some lines with *Sr30* produce lower responses than others (Roelfs and McVey, 1979; Knott, 1990).

Environmental Variability

Low (Roelfs and McVey, 1979).

Origin

Common wheat cv. Webster which was introduced to the USA from Russia.

Pathogenic Variability

Luig (1983) reported that *Sr30* was generally effective in North America and Europe, although Roelfs and McVey (1979) mentioned that race 11-RHR was virulent. In Australia, several virulent pathotypes became prevalent after 1968 (Luig, 1983). Virulence was reported as frequent in South Africa (Le Roux and Rijkenberg, 1987*a*). Huerta-Espino (1992) found virulence in a number of countries with moderate to high levels among samples from Spain, Ethiopia, Turkey, Pakistan and a number of South American countries.

Reference Stocks

i: Sr30/7*LMPG-1; Sr30/7*LMPG-2; Sr30/7*LMPG-3 (Knott, 1990).
v: Festiguay W2706 (Knott and McIntosh, 1978); Webster W973 (Knott and McIntosh, 1978); Mediterranean W1728 (Singh and McIntosh, 1985). Klein Cometa *Sr8b* (Singh and McIntosh, 1986*a*).

Source Stocks

Work at The University of Sydney Plant Breeding Institute has demonstrated the presence of *Sr30* in several Mexican wheats and Australian derivatives.
Australia: Dollarbird *Sr2 Sr8a Sr9g*. Hartog (= Pavon 'S') *Sr2 Sr8a Sr9g Sr12*. Batavia *Sr2* (heterogeneous) *Sr8a Sr12*. Houtman *Sr2 Sr9g Sr17*. Rosella *Sr5 Sr7b Sr8a Sr12*. Lark *Sr5 Sr8a*.

A

B

C

PLATE 3-31. Sr30

A. and B. Seedling leaves of (L to R): Festiguay, Webster, Banks, Condor and Federation; infected with A. pt. 34-1,2,3,4,5,6,7 [*P30*] and B. pt. 34-1,2,3,6,7,8,9 [*p30*]. The small degree of resistance of Condor relative to Federation in A. may be due to *Sr12*.

C. Seedling leaves of (L to R): Lerma Rojo 64A, Mediterranean W2706 and Webster; infected with (L) pt. 34-1,2,3,4,5,6,7 [*P30*] and (R) 34-1,2,3,6,7,8,9 [*p30*]. *Sr30* is relatively common in Mexican wheats. Courtesy D The.

Sunfield *Sr5 Sr8a Sr9b Sr12*. Osprey *Sr5 Sr8a Sr12*. Katunga *Sr8a*. Banks *Sr8a Sr9b Sr12*; Vulcan *Sr8a Sr9b Sr12*. Sunstar *Sr8a Sr9e Sr12*. Lilimur *Sr8a Sr17* (heterogeneous). Cranbrook (= Flicker 'S') *Sr8a Sr9g Sr12 Sr17*.

CIMMYT: Lerma Rojo 64A.

Use in Agriculture

Because of the absence of virulence, Webster was once considered to be almost universally resistant (Hart, 1931). However, when the resistance was deployed in the Australian cultivar Festiguay, virulent pathotypes increased on this cultivar. These pathotypes declined after Festiguay was withdrawn from cultivation. More recently, a distinctive virulent pathotype was isolated in eastern Australia (Park and Wellings, 1992). Although this pathotype can overcome the resistance of some current wheats with *Sr30*, it has remained at extremely low levels.

Sr31 (McIntosh, 1988*a*)   (Plate 3-32)

Chromosome Location

1BS (1BL.1RS translocations) or 1R(1B) substitutions (Mettin *et al.*, 1973; Zeller, 1973). Some wheat cultivars comprise both substitution and translocation biotypes (Zeller, 1973).

Low Infection Type

1^- to 2.

Environmental Variability

None reported.

Origin

S. cereale cv. Petkus. Most wheats with *Sr31* were derived from wheat x rye hybrid derivatives produced in Germany in the 1930s (see Mettin *et al.*, 1973; Zeller, 1973).

Pathogenic Variability

Huerta-Espino (1992) recorded a virulent culture in a collection from Turkey.

Reference Stocks

i: Federation*4/Kavkaz (RA McIntosh and CR Wellings, unpublished 1992); Thatcher*6/ ST-1.25, R.L.6078 (PL Dyck, *pers. comm.* 1986).
v: Aurora (Zeller, 1973); Kavkaz (Zeller, 1973).

Source Stocks

All wheats with *Lr26* and *Yr9* (see *Lr26, Yr9*). *Sr31* is present in many European wheats and some Chinese and USA wheats as well as being widely used in wheats distributed by the CIMMYT program (e.g. Bobwhite and Veery selections) and continues to occur at high frequencies in CIMMYT breeding populations.

Australia: Grebe; Warbler.

China: Feng-Kang 2; Feng-Kang 8; Jan 7770-4; Jin-Dan 106; Lu-Mai 1; Yi 78-4078. Dong Xie 3 *Sr5*; Dong Xie 4 *Sr5*. See Hu and Roelfs (1986).

CIMMYT: Alondra; Angostura 88; Bacanora 81; Bobwhite S; Cumpas 88; Curinda 87; Genaro 81 (=Veery #3); Glennson 81 (=Veery #1); Guasave 81; Mochis 88; Seri 82 (Veery #5); Ures 81 (=Veery #2). See Singh and Rajaram (1991) and Singh (1993). Many of these wheats or sibs are grown in other countries (Villareal and Rajaram, 1988).

Europe: Aurora; Benno; Bezostaya 2; Burgas 2; Clement; Kavkaz; Lovrin 10; Lovrin 13; Mildress; Neuzücht; Skorospelka 35; Weique; Zorba.

Indian Subcontinent: CPAN 1922; HUW 206; Pakistan 81; Sarhad 82. See Singh and Gupta (1991).

South Africa: Gamtoos (= Veery #3).

USA: Excel; Freedom; Salmon (USA). Siouxland *Sr5 Sr24*. Longhorn *Sr24*.

Use in Agriculture

The value of *Sr31* as a source of protection against stem rust is difficult to determine. The widespread international distribution of wheats with *Sr31* may reflect the broad agronomic adaptability of these materials rather than the unique contribution of stem rust resistance.

A   B

PLATE 3-32. *Sr31*

Seedling leaves of (L to R): Kavkaz, Skorospelka 35, Mildress, Amigo (1RS ex *S. cereale* cv. Insave), CS 1DL.1RS (1RS ex *S. cereale* cv. Imperial) and Chinese Spring; infected with A. pt. 34-1,2,3,4,5,6,7 and B. pt. 126-5,6,7,11 and incubated at 18°C. The first three lines with 1RS from Petkus rye gave responses that were lower than those of Amigo and CS 1DL.1RS. Amigo also carries *Sr24* which could influence the response. CS 1DL.1RS produced slightly higher responses with both cultures. Note the distinctly lighter uredial colour of the second pathotype.

Despite successful use in many areas, wheats with *Sr31* have not been widely grown in Australia due to potential problems in bread-making. The only cultivars registered are Grebe, an Egret derivative used for biscuit (cookie) quality flour, and Warbler, a feed wheat. Other wheat derivatives with 1RS have stem rust resistance characterised by low infection types similar to those produced by lines with *Sr31*. KW Shepherd and coworkers produced 1D.1RS (Koebner and Shepherd, 1986) and 1B.1RS derivatives of Imperial rye, whereas The *et al.* (1992) described a gene presumably associated with 1AL.1RS in Amigo. No cultivar with 1RS from Imperial has been produced but several derivatives of Amigo are grown in the USA. The 1AL.1RS chromosome carries the gene *Gb5* for resistance to greenbug (*Schizaphis graminum*). It is not known if the gene from Imperial rye and the gene in Amigo are the same as *Sr31*. The presence of *Sr31* in wheat is readily confirmed by the concurrent presence of *Lr26* and *Yr9* (see respective sections) as well as by cytological and biochemical methods. These were reviewed or described by Javornik *et al.* (1991), May and Wray (1991), Gupta and Shepherd (1992).

Sr32 (McIntosh, 1988*a*) (Plate 3-33)

Chromosome Location
Independent translocations located in chromosomes 2A (RA McIntosh, unpublished 1974), 2B (ER Sears, *pers. comm.* 1982) and 2D (ER Sears, *pers. comm.* 1982).

Low Infection Type
1$^+$ to 2C.

Environmental Variability
None reported.

Origin
T. speltoides.

Pathogenic Variability
The international survey of Huerta-Espino (1992) as well as pathogenicity surveys in the USA (Roelfs *et al.*, 1991), Canada (Harder and Dunsmore, 1990), Mexico (Singh, 1991), South Africa (Le Roux and Rijkenberg, 1987*a*) and Australia (RF Park, unpublished 1992) failed to find virulence.

A   B   C

PLATE 3-33. Sr32

A. and B. Seedling leaves of: C77.19 (L) and Chinese Spring (R); infected with A. pt. 34-1,2,3,4,5,6,7 and B. pt. 126-5,6,7,11 and incubated at 18°C. No Australian pathotype is virulent for *Sr32*.

C. Seedling leaves of (L to R): C82.2 (Sears' 2D translocation), C90.1 = R.L.5711 (with *Sr39*) and Chinese Spring; infected with pt. 34-1,2,3,6,7,8,9 (L), pt. 126-5,6,7,11 (C) and pt. 34-(4),(7),10 (R). Note the similar responses of the two *T. speltoides* derivatives with the first two cultures, and the higher but resistant, response of C90.1 with the third culture indicating that *Sr32* and *Sr39* are not identical.

Reference Stocks

i: W3531, a Chinese Spring stock produced by ER Sears and involving a translocation to chromosome 2A (RA McIntosh, unpublished 1974). This stock is characterised by having adhering glume fragments, especially in the crease region of the grain. This primitive feature tends to be associated with the absence of at least a part of wheat chromosome 2A.

C77.19, a CS/*T. speltoides* derivative, is a cleaner threshing line with *Sr32* present in chromosme 2B. Sears later produced four further transfers with *Sr32* present in chromosomes 2B and 2D. These are accessioned in The University of Sydney cytogenetics collection as C82.1 (chromosome 2B), C82.2 (2D), C82.3 (2D) and C82.4 (2D). The lines designated CS *Sr32* (Le Roux and Rijkenberg, 1987*b*) and ER 5155 (Roelfs and Martens, 1988) probably correspond to W3531 or C77.19.

Source Stocks

Australian backcross lines produced at The University of Sydney Plant Breeding Institute.

Use in Agriculture

Early breeding studies at The University of Sydney using W3531 failed to separate resistance from the adherent glume phenotype. Backcross derivatives with *Sr32* derived from C77.19 were produced and distributed to wheat breeders but no line was commercialised or used in further breeding. The reasons for this are unknown. Sears (*pers. comm.* 1982) suggested that C82.2 was the most normal of the translocation lines.

Sr33 (McIntosh, 1988*a*)   (Plate 3-34)

Synonym

SrSQ (Kerber and Dyck, 1979).

Chromosome Location

1DL (Kerber and Dyck, 1979). Because of linkage with *Lr21*, *Rg2* and *Gli-D1* (Jones *et al.*, 1990), *Sr33* must be located in 1DS. Czarnecki and Lukow (1992) mapped it 9 cM from *Gli-D1* whereas Jones *et al.* (1991) obtained estimates of 5.6 and 7.6 cM for the same interval. *Sr33* was proximal to *Gli-D1*.

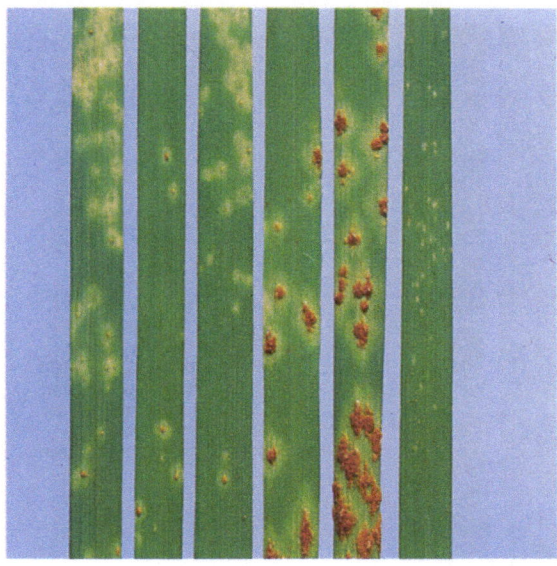

PLATE 3-34. Sr33

Seedling leaves of (L to R): R.L.5405, 66505 (*Sr33*/5*Oxley), 66604 (*Sr33*/5*Condor), Oxley, Chinese Spring and CS/R.L.5406 1D carrying a substituted chromosome 1D from R.L.5406; infected with pt. 34-1,2,3,4,5,6,7. Note the distinctive low infection type produced by the *Sr21*-like gene in CS/R.L.5406 1D compared with the first three lines with *Sr33* (see *Sr21*).

Low Infection Type

> ;1⁻ in diploid stocks, 2 in hexaploid derivatives (Kerber and Dyck, 1979). Huerta-Espino (1992) recorded ITs 1⁻C to 2.

Environmental Variability

> None reported.

Origin

> *T. tauschii* R.L.5288.

Pathogenic Variability

> Virulence has not been reported, but lines with *Sr33* have not been tested widely. Pathogenicity surveys in the USA (Roelfs *et al.*, 1991), Mexico (Singh, 1991) and Australia as well as the study of Huerta-Espino (1992) failed to detect virulence for *Sr33*.

Reference Stocks

> **v**: R.L.5405 = C78.15 = Tetra Canthatch/*T. tauschii* R.L.5288 (Kerber and Dyck, 1979).
> **dv**: *T. tauschii* R.L.5288 (Kerber and Dyck, 1979).

Source Stocks

> Several Australian backcross derivatives produced at The University of Sydney Plant Breeding Institute.

Use in Agriculture

> No wheat with *Sr33* has been commercialised. Kerber and Dyck (1979) reported that *Sr33* was closely linked in repulsion with *Lr21*. RA McIntosh (unpublished, 1981) tested F3 lines of the *Sr33*/*Lr21* cross produced by ER Kerber, and demonstrated the proximity of *Sr33* to both *Lr21* and a stem rust resistance gene with the same specificity as *Sr21* and present in the *Lr21* stock. Although he believed that recombination of *Sr33* and *Sr21* had occurred he was unable to confirm this because of possible meiotic instability in the amphiploid materials being examined. The study is being repeated using backcross derivatives. Preliminary results indicate a genetic distance of 4 cM between *Sr33* and the *Sr21*-like gene.

Sr34 (McIntosh *et al.*, 1982)   (Plate 3-35)

Chromosome Location

> 2A (2A/2M, translocation), 2D (2D/2M translocation), 2M (McIntosh *et al.*, 1982). *Sr34* is present in wheats with Yr8.

Low Infection Type

> 1N to 3. 2C to 23CN (Huerta-Espino, 1992).

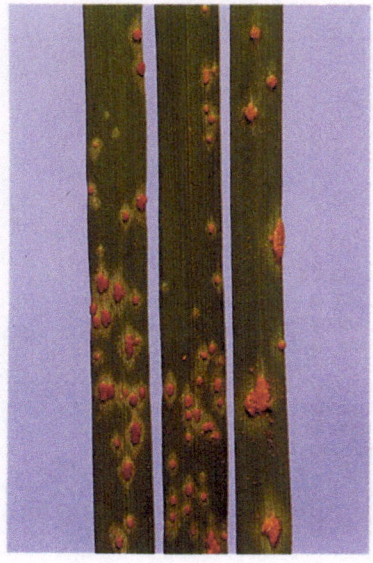

A   B

PLATE 3-35. Sr34
Seedling leaves of (L to R):
Compair, CS 2D/2M 3/8 =
C77.1 and Chinese Spring;
infected with pt. 126-5,6,7,11
and incubated at A. 18°C and
B. 28/23°C. This and a closely
related culture are the only
Australian pathotypes avirulent
for *Sr34*.

Environmental Variability

Most effective at low temperatures.

Origin

T. comosum.

Pathogenic Variability

Avirulence for *Sr34* is relatively rare in Australia with only two older related pathotypes
producing low ITs on tester stocks (McIntosh *et al.*, 1982). Knott (1990) reported avirulence in a
Canadian culture of race 111 which is known to be widely avirulent. In the survey of cultures
carried out by Huerta-Espino (1992) avirulence was common with levels approaching 50% in
south Asia (Pakistan and Nepal), Ethiopia, Kenya, the Malagasy Republic and South America.
He found no virulence among collections from China but he cited unpublished 1984 results from
CC Hu and AP Roelfs indicating a virulence level of 84%. Thus for international comparisons of
pathogenicity, a line with *Sr34* would be a useful differential.

Reference Stocks

i: Sr34/6*LMPG (Knott, 1990); Compair; Chinese Spring 2D/2M 3/8 (C77.1); Chinese Spring
2A/2M 4/2 (C77.2) (McIntosh *et al.*, 1982).
su: Chinese Spring 2M(2A) (McIntosh *et al.*, 1982).

Source Stocks

None.

Use in Agriculture

Translocations with *Sr34* have not been deployed in commercial cultivars.

Sr35 (McIntosh *et al.*, 1984)   (Plate 3-36)

Synonym

$SrTm_1$ (Valkoun *et al.*, 1986).

Chromosome Location

3AL (McIntosh *et al.*, 1984). *Sr35* was located 41.5 cM from the centromere and showed 1%
recombination with *R2* derived from the same source.

Low Infection Type

0; to ;.

Environmental Variability

None reported.

Origin

T. monococcum. Sr35 is present in a selection from The University of Sydney accession C69.69 (P.I.264935) and in the Canadian accession G2919 (McIntosh *et al.*, 1984). It was independently transferred from these accessions to tetraploid and hexaploid wheats in Canada and to hexaploid wheat in Australia.

Pathogenic Variability

Sr35 is effective against most Australian pathotypes except cultures of standard race 126 and a mutant culture (82-L-2) isolated during studies in the greenhouse. Many North American cultures are avirulent (McIntosh *et al.*, 1984). In the survey of Huerta-Espino (1992) virulence was found in Ethiopia, Kenya, the Malagasy Republic, Nepal, China, Argentina, Brazil and Chile.

Reference Stocks

i: C81.42 = R.L.6071*7/G2919 *Sr9e* (McIntosh *et al.*, 1984).

v: M80.3990, M80.4635, M80.4636 derived from C69.69 (McIntosh *et al.*, 1984). Backcross derivatives of cvv. Zlatka and Yubinlenaya (Valkoun *et al.*, 1986).

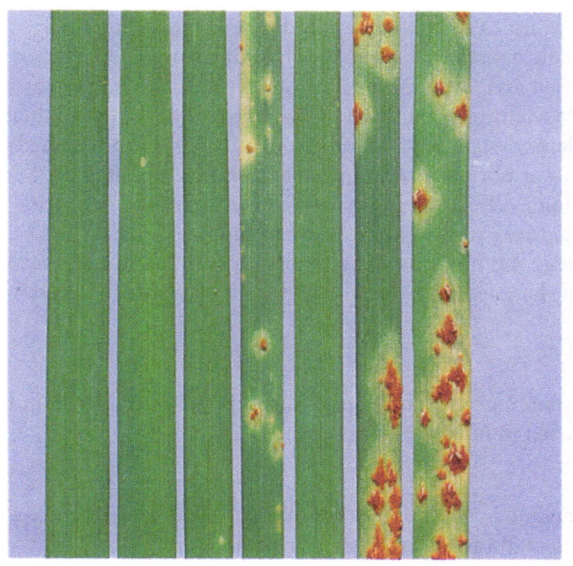

A

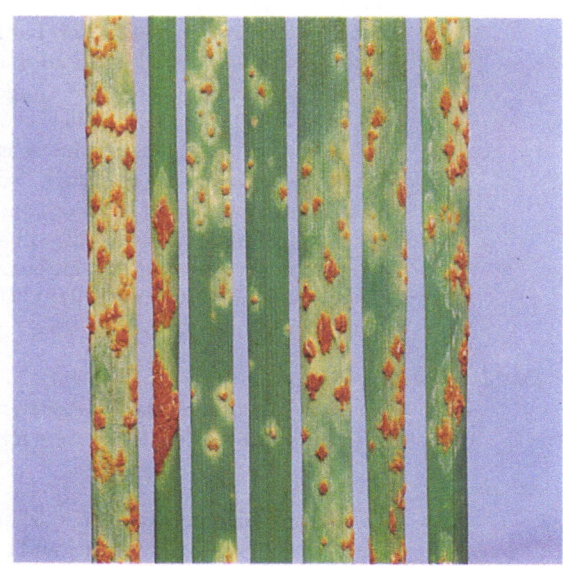

B

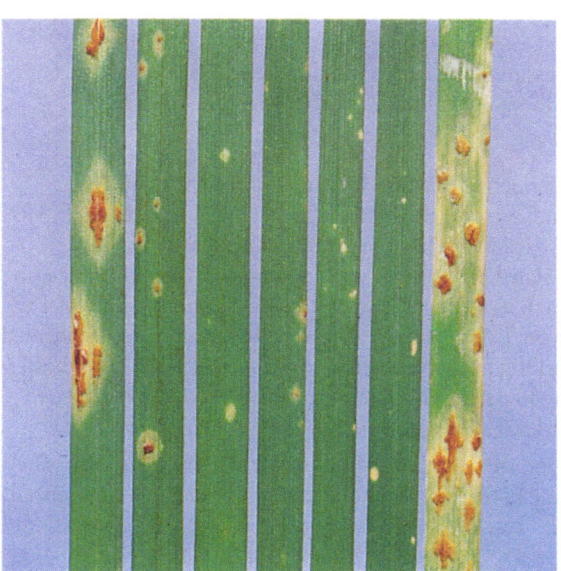

C

PLATE 3-36. *Sr35*

Seedling leaves of (L to R): M80.3990 (an Australian hexaploid derivative of C69.69), C81.42 (R.L.6071*5/G2919), 103335 (*Sr35*/3*77W549), 77W549, 67093 (*Sr35*/4*Canna), Canna and Chinese Spring; infected with A. pt. 34-1,2,3,4,5,6,7, B. culture 82-L-2, a mutant derivative of 34-1,2,3,4,5,6,7 virulent for *Sr35* and C. pt. 126-5,6,7,11 which is also virulent for *Sr35*. The first, second, third and fifth lines carry *Sr35*. The second and third cultures are virulent for *Sr35* but note the resistance of 77W549 to culture 82-L-2 and of various wheats to pt. 126-5,6,7,11. The isolate of 82-L-2 used for this photograph was contaminated with leaf rust, visible on some leaves.

tv: Stewart*8/G2919 *Sr9e* (McIntosh *et al.*, 1984).
dv: Components of C69.69 = P.I.264935; G2919 (McIntosh *et al.*, 1984).

Source Stocks

Backcross derivatives produced at The University of Sydney, for example
Sr35/3*77W549, *Sr35*/4*Canna.

Use in Agriculture

Although not yet exploited, *Sr35* should have a role if used in gene combinations. White-seeded
derivatives with *Sr35* were recovered in the genetic studies conducted by McIntosh *et al.* (1984).

Sr36 (McIntosh, 1988*a*)   (Plate 3-37)

Synonym

SrTt1 (McIntosh and Gyarfas, 1971).

Chromosome Location

2B (Nyquist, 1957); 2BS (Gyarfas, 1978). Stem rust and powdery mildew resistances in the
wheats C.I.12632 and C.I.12633 were originally thought to be determined by duplicate linked
factors (Allard and Shands, 1954) but Nyquist (1962) demonstrated that a single gene was
involved with differential fertilisation resulting in variable ratios. The mildew resistance factor
was subsequently designated *Pm6* (Jorgensen and Jensen, 1973). In early studies at The University
of Sydney, *Sr36* behaved as an allele of *Sr9*. McIntosh and Luig (1973*a*) reported a rare recombinant
combining *Sr36* and *Sr9e*. The line was later designated Combination III (Luig, 1983). In further
genetic studies involving this stock, *Sr36* was normally transmitted and showed about 20%
recombination with *Sr9* (McIntosh and Luig, 1973*a*). *Sr36* recombined with *Sr9b* but not with
Lr23, located in chromosome 2BS. In a separate study with progeny of a monosomic 2B line
derived from Combination III, chromosome misdivision products that separately carried *Sr9e* and
Sr36 were obtained (Gyarfas, 1978). This demonstrated that the genes were located in different
chromosome arms.

Low Infection Type

Usually $0;^=$, but sometimes X. Apparent heterogenous infection types of 0;3 (Knott, 1990) and
0;4 (Roelfs and McVey, 1979) were reported in North America.

Environmental Variability

Under Australian greenhouse conditions, *Sr36* appears to produce higher infection types in
winter when both light intensity and temperatures are lower than in summer. AP Roelfs (*pers.
comm.* 1988) observed mixed responses with some cultures in autumn and spring but not in
summer or winter.

Origin

T. timopheevii. Independent transfers to hexaploid wheat resulted in C.I.12632 and C.I.12633
(Allard and Shands, 1954), Timvera (Pridham, 1939) and C.I.13005 (Atkins, 1967).

Pathogenic Variability

Virulence has arisen on at least two independent occasions in Australia, resulting in the
susceptibility of Mengavi in 1961 (Luig and Watson, 1970) and Cook in 1984 (Zwer *et al.*, 1992).
A similar experience occurred in South Africa following the deployment of *Sr36* in wheat-
growing areas. In North America, early isolates of race 15B were characterised by virulence on
T. timopheevii and, by inference, wheats such as C.I.12632 and C.I.12633. Later isolates of 15B
were virulent on C.I.12632 and avirulent on *T. timopheevii*. These observations were confirmed
by RA McIntosh (unpublished, 1970) working at the University of Missouri. In later studies,
wheats with *Sr36* were described as having low receptivity to pathotypes normally considered
virulent (Rowell and McVey, 1979; Roelfs and McVey, 1979; Rowell 1981*a*, 1981*b*, 1982).

Huerta-Espino (1992) found pathogenic variation in most of the major regions from which he
obtained samples. This indicated that a line with *Sr36* would be a useful worldwide differential.

Reference Stocks

i: Sr36/8*LMPG (Knott, 1990); Line C, a W2691 backcross derivative with *Sr36* (Luig, 1983).
v: C.I.12632 (= W1656); C.I.12633 (= W1657); Idaed 59; Mengavi; Timvera (McIntosh and Gyarfas, 1971).

Source Stocks

Australia: Songlen *Sr2 Sr5 Sr6 Sr8a*. Timgalen *Sr5 Sr6*. Cook *Sr5 Sr6 Sr8a*. Mendos *Sr7a Sr11 Sr17*. Combination III W3486 *Sr9e* (Luig, 1983).
Mexico: Zaragosa 75 (Le Roux and Rijkenberg, 1987*b*).
South Africa: Dipka; Flamink; Gouritz; SST101; SST107. See Le Roux and Rijkenberg (1987*b*) and Sharma and Gill (1983).
USA: Hand; Kenosha; Purdue; Roughrider; Vernum; Wisconsin Supremo. Arthur *Sr2 Sr5 Sr8a*; Arthur 71 *Sr2 Sr5 Sr8a*.

Use in Agriculture

Sr36 has been a very valuable gene for wheat production in Australia. It was initially used in cv. Mengavi which was rendered very susceptible by a new pathotype detected very soon after registration. *Sr36* was then combined with *Sr7a*, *Sr11* and *Sr17* and released in cv. Mendos which in turn succumbed to stem rust after a few years. In 1967 Timgalen was released as a prime hard quality wheat in rust-prone areas. The source of *Sr36* in Timgalen is unknown but is presumed to be C.I.12632 or a derivative. Timgalen was followed with the derivatives, Timson, Songlen, Shortim and Cook over subsequent years. Stem rust was found on Cook wheat in 1984 (Zwer *et al.*, 1992). Because of the activities of the National Wheat Rust Control Program, Cook derivatives with additional resistance genes (such as Sunco with *Sr24*) were released in 1985 and Cook was rapidly withdrawn. The Cook-attacking pathotype then declined to negligible levels, thus providing an enhanced resistance to the Cook derivatives (Park and Wellings, 1992).

According to Roelfs and McVey (1979) and supported by the studies of Rowell (1982) some pathotypes considered virulent on lines with *Sr36* developed more slowly and gave lower numbers of pustules on genotypes with *Sr36*. This contrasted with the Australian experience with Mengavi and Mendos which were seriously affected by the relevant virulent pathotypes.

A   B

PLATE 3-37. Sr36

Seedling leaves of (L to R): C.I.12632 (W1656), Mengavi, Arthur, Combination III and Chinese Spring; infected with A. pt. 34-1,2,3,6,7,8,9 [*P36*] and B. pt. 34-1,2,3,4,5,6,7 [*p36*]. Arthur carries an unknown resistance to pt. 34-1,2,3,4,5,6,7. This resistance could be due to *Sr7a* whereas resistance in Combination III to this pathotype is conferred by *Sr9e*.

Sr37 (McIntosh, 1988*a*)   (Plate 3-38)

Synonym

　　　SrTt2 (McIntosh and Gyarfas, 1971).

Chromosome Location

　　　4B (Gyarfas, 1978).

Low Infection Type

　　　0;, ;1, ;3C, depending on pathogen culture (McIntosh and Gyarfas, 1971). Susceptible off-types are relatively common.

Environmental Variability

　　　Low (Roelfs and McVey, 1979).

Origin

　　　T. timopheevii.

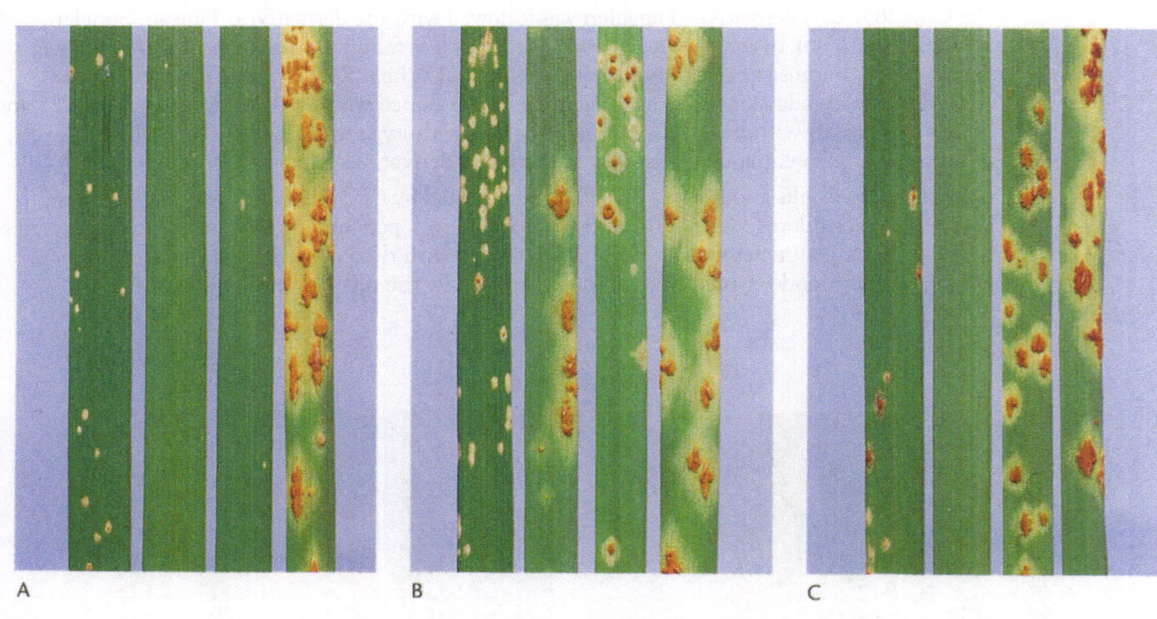

A   B   C

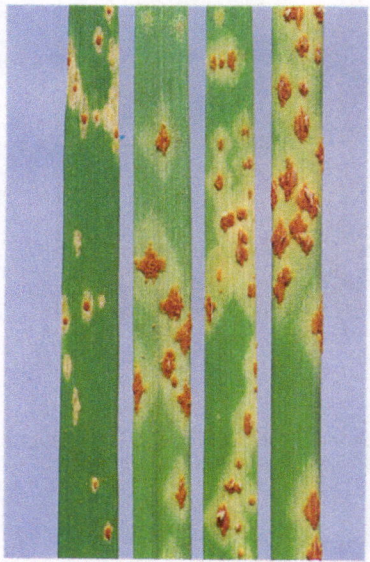

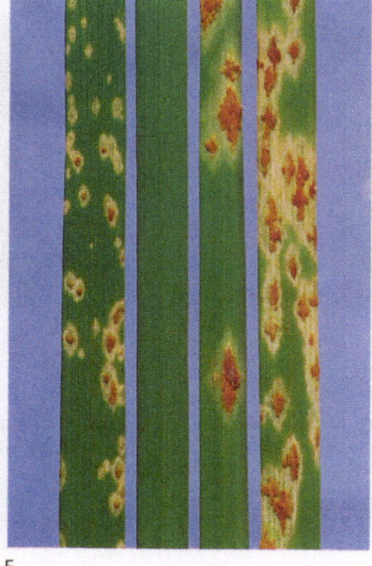

D   E

PLATE 3-38. Sr37

Seedling leaves of (L to R): Line W, Mengavi (*Sr36*), a Federation cross with *SrTt3* and Federation; infected with A. pt. 126-5,6,7,11 [*P37*], B. pt. 34-2,4,5,7,11 [*P37*], C. pt. 34-1,2,3,6,7,8,9 [*P37*], D. pt. 34-1,2,3,4,5,6,7, [*P37*] and E. pt. 34-(4),(7),10 [*P37*]. Line W is resistant to all cultures but there is variation in infection type. The second and fourth cultures are virulent for *Sr36* in Mengavi. The Federation line with *SrTt3* is highly resistant to pt. 126-5,6,7,11(A.) and gives an intermediate response in B., C. and D. Pt. 34-(4),(7),10 (E.) is virulent on the *SrTt3* stock but Federation is resistant probably due to the presence of *Sr10*. Earlier records indicated that lines with *Sr36* give intermediate responses with pt. 34-(4),(7),10. The high level of resistance to pt. 34-(4),(7),10 indicated that Mengavi may carry an additional gene for resistance to this atypical pathotype, which is suspected to be a somatic hybrid of *P. graminis* f. sp. *tritici* and *P. graminis* f. sp. *secalis*.

Pathogenic Variability

Luig (1983) discussed results of an international survey which indicated that virulence for *Sr37* occurred in the USA, Canada, Brazil, Pakistan, India, Italy and Israel. Virulence for *Sr37* has not been detected recently in the USA (Roelfs *et al.*, 1990, 1991), or in the regions surveyed by Huerta-Espino (1992).

Reference Stocks

v: Line W = W3563 (McIntosh and Gyarfas, 1971).
tv: Various accessions of *T. timopheevii* of which W1899 and C.I.11802 should be regarded as the standards (McIntosh and Gyarfas, 1971).

Source Stocks

Other sources of *T. timopheevii* are listed in McIntosh and Gyarfas (1971).

Use in Agriculture

Sr37 has not been exploited in cultivated durums or common wheat. Although potentially useful for gene combinations from a pathological viewpoint, the *T. timopheevii* chromosome segment (indeed most of the 4B chromosome) bearing *Sr37* has only limited homology with the equivalent chromosome in common wheat (Gyarfas, 1978; Dvorak *et al.*, 1990). Consequently, any common wheat line with *Sr37* would have to be derived from a translocation event leading to increased chromosomal homology with wheat chromosome 4B.

McIntosh and Gyarfas (1971) concluded there may be as few as three genes for resistance to stem rust in *T. timopheevii*, that is, *Sr36*, *Sr37* and possibly a third factor present in the common wheat derivative C.I.13005 (Atkins, 1967). Gyarfas (1978) later presented evidence for a fourth gene which she designated *SrTt3* in the common wheat derivative, Line AH (see *SrTt3*).

Sr38 (Bariana, 1991; Bariana and McIntosh, 1993) (Plate 3-39)

Chromosome Location

2AS. *Sr38* is completely linked with *Lr37* and *Yr17*, and is closely linked in repulsion with *Lr17* (Bariana and McIntosh, 1993).

Low Infection Type

X, often with very large pustules towards the leaf base.

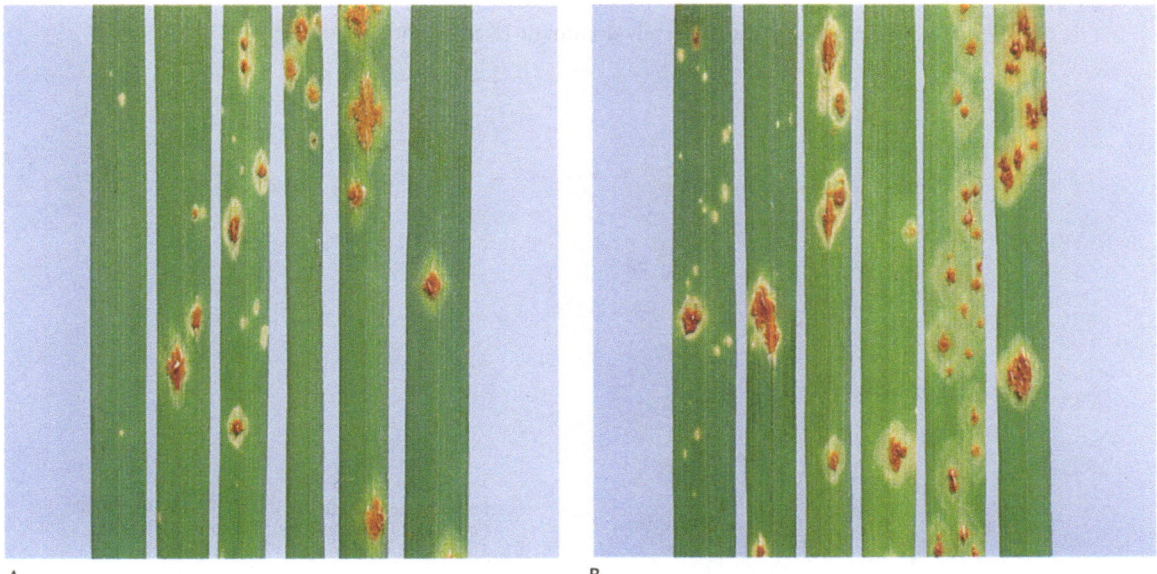

A   B

PLATE 3-39. *Sr38*

Seedling leaves of (L to R): VPM1 leaf tip, VPM1 mid leaf, Tc*8/VPM1, Trident, Chinese Spring and Marne; infected with pt. 34-1,2,3,4,5,6,7 and incubated at A. 20/23°C and B. 23/28°C. Resistance conferred by *Sr38* is more effective at lower temperatures. Uredia are often larger, even fully compatible, towards the leaf base.

Environmental Variability

More effective at lower temperatures.

Origin

T. ventricosum.

Pathogenic Variability

Virulence has not been reported, although this gene has not been widely tested.

Reference Stocks

i: Thatcher*8/VPM1, R.L.6081 (PL Dyck, *pers. comm.* 1991).
v: VPM1 (Doussinault *et al.*, 1981, 1988).

Source Stocks

Australia: Trident (= Spear*4/VPM1). Sunbri *Sr5 Sr6 Sr8a Sr36* (Brown *et al.*, 1991). Various backcross-derived lines generated at the PBI, Cobbitty.
UK: Rendezvous (Bariana, 1991).
USA: Hyak (RA McIntosh, unpublished 1990); Madsen (RA McIntosh, unpublished 1990).

Use in Agriculture

Because of the associated resistances to all three rust diseases, wheats with *Sr38*, *Lr37* and *Yr17* will be useful in most parts of the world. VPM1 was initially selected as a breeding parent in Australia because of high levels of resistance to all three rusts. The phenomenon of linkage was noted locally during the backcrossing, following tests on Rendezvous bred in the UK and on Hyak and Madsen bred in the Pacific Northwest of the USA, and was demonstrated formally by Bariana and McIntosh (1993). Rendezvous, Hyak and Madsen were bred primarily for resistance to eyespot or strawbreaker disease (caused by *Pseudocercosporella herpotrichoides*) controlled by gene *Pch* in chromosome 7DL. Selection for stripe rust resistance, also practiced by the breeders, presumably led to the leaf rust and stem rust resistances present in these cultivars. The Australian cultivar Sunbri was bred for rust resistance and is assumed not to carry *Pch*.

Sr39 (ER Kerber, *pers. comm.* 1991)   (Plate 3-40)

Chromosome Location

2B (Kerber and Dyck, 1990). *Sr39* showed 3% recombination with *Lr35* (Kerber and Dyck, 1990) and segregated independently of *Sr32* and *Lr13* (ER Kerber, *pers. comm.* 1991).

Low Infection Type

1 to 2. Resistance is incompletely dominant (Kerber and Dyck, 1990).

PLATE 3-40. S r39
Seedling leaves of (L to R): R.L.5711 (*Sr39*), C82.2 (*Sr32*) and Chinese Spring; infected with pt. 34-1,2,3,6,7,8,9 (L) and pt. 126-5,6,7,11 (R). The responses conferred by seedlings with *Sr39* and *Sr32* to these cultures were very similar.

Environmental Variability

None reported.

Origin

T. speltoides R.L.5344.

Pathogenic Variability

Virulence has not been reported, although wheats with this gene have not been tested widely. An intermediate response was produced with one University of Sydney PBI culture (see Plate 3-33 C).

Reference Stocks

i: Marquis-K*8/R.L.5344, R.L.5711 (Kerber and Dyck, 1990).
amphiploid: *T. speltoides* R.L.5344/*T. monococcum* R.L.5346, R.L.5347 (Kerber and Dyck, 1990).

Source Stocks

None.

Use in Agriculture

This gene has not been adequately assessed for use in agriculture.

Sr40 (Dyck, 1992)   (Plate 3-41)

Synonym

SrA (Dyck, 1992).

Chromosome Location

Probably 2BS (Dyck, 1992). The arm location was based on linkage with other genes in chromosome 2B, that is, 0.01 with *Lr13*, 0.05 with *Lr23*, 0.34 with *Lr16*, 0.22 with *Sr36* and 0.28 with *Sr9*.

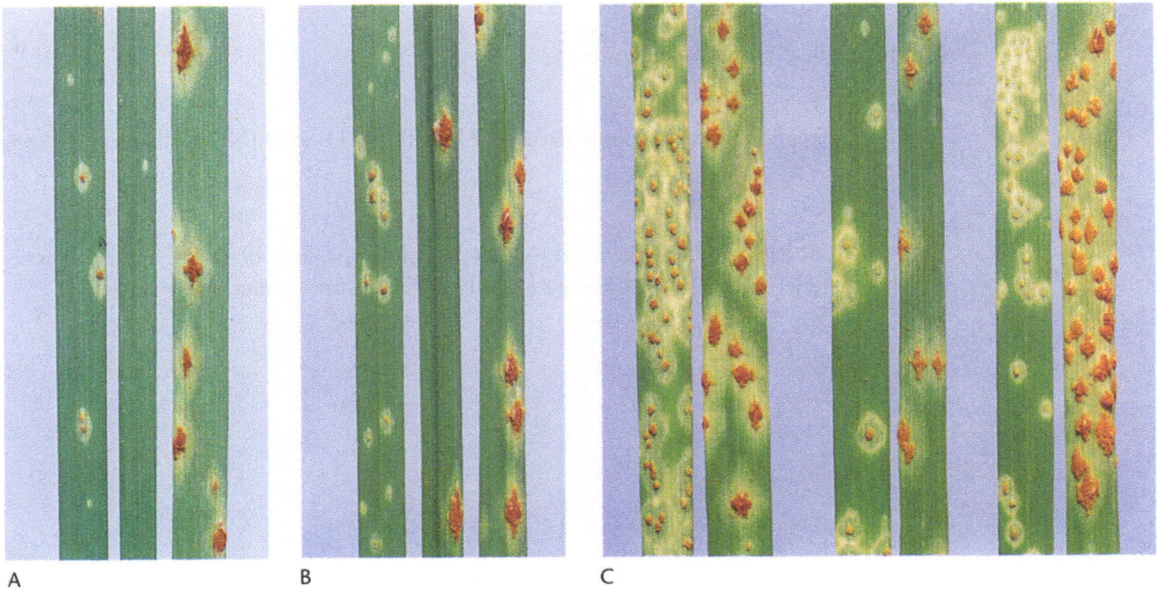

A   B   C

PLATE 3-41. *Sr40*

A. and B. Seedling leaves of (L to R): R.L.6087, C.I.12632 = W1656 (*Sr36*) and Chinese Spring; infected with A. pt. 17-1,2,3,7 [*P36P40*] and B. pt. 34-1,2,3,4,5,6,7 [*p36P40*]. *Sr40* can be distinguished from *Sr36* by low infection type (12⁻ v. 0;) and pathogenicity as shown by comparing A. and B.

C. Seedling leaves of: R.L.6087 and Chinese Spring; infected with pt. 34-2,4,5,7,11 (L), pt. 34-1,2,3,6,7,8,9 (C) and pt. 126-5,6,7,11 (R).

Low Infection Type

1^+ to 2^+.

Environmental Variability

None reported.

Origin

T. timopheevii (2n = 28, AAGG).

Pathogenic Variability

Sr40 confers resistance to many Canadian pathotypes (Dyck, 1992), but has not been tested elsewhere.

Reference Stocks

i: R.L.6071*7/PGR 6126, R.L.6087; R.L.6071*7/PGR 6195, R.L.6088 (Dyck, 1992).
tv: *T. timopheevii* accessions PGR 6126 and PGR 6195 (Dyck, 1992).

Source Stocks

None.

Use in Agriculture

It is too early to know if *Sr40* will have a role in agriculture.

Sr41 (ND Williams, *pers. comm.* 1993)

Chromosome Location

4D (ND Williams, *pers. comm.* 1993).

Low Infection Type

23C (ND Williams, *pers. comm.* 1993).

Environmental Variability

None reported.

Origin

Common wheat cv. Waldron.

Pathogenic Variability

This gene was effective against only pathotype LCBB (= 111-SS2) (ND Williams, *pers. comm.* 1993).

Reference Stocks

WDR-B1 (CR Riede, ND Williams, JD Miller and LR Joppa, unpublished 1993). Waldron *Sr5* (heterogeneous) *Sr11* (heterogeneous) (ND Williams, *pers. comm.* 1993).

Source Stocks

None.

Use in Agriculture

None.

Plate

Not available.

TEMPORARILY DESIGNATED AND MISCELLANEOUS STEM RUST RESISTANCE GENES

In this section we list genes that were allocated temporary designations, or are well documented from genetic studies. In some instances, adequate information is available for formal gene designations (e.g. *Srdp2, SrKt2* and Norin 40); in other cases, more genetic information is required such as chromosome location and tests of allelism with previously designated genes.

Temporary Designations

The following list of temporarily designated resistances is not complete. A number of additional genes, both in hexaploid wheats and tetraploid wheats have been described, but they were effective against rare laboratory cultures and were not tested by other laboratories.

SrCharter

Uppal and Gokhade (cited in Luig, 1983) reported a variant of *P. graminis* f. sp. *tritici* that was virulent on cv. Yalta but avirulent on cv. Charter, both of which were known to carry *Sr11*. The additional gene(s) in Charter has not been located to a particular chromosome. A monogenic stock developed by P Bahadur (*pers. comm.* 1986) has not been widely tested.

Low Infection Type

1, 2^-, X^- and 3 (Luig, 1983).

Pathogenic Variability

Charter was included in the International Gene Virulence Survey undertaken by Luig (1983). Avirulence on Charter among cultures virulent for *Sr11* was evident in tests conducted in Kenya, Israel and North America as well as India and Pakistan. Charter is used as a supplementary differential for stem rust work in India and is included with the B set of Nagarajan *et al.* (1986*b*) (Table 1-4).

Origin

Common wheat cv. Charter (Luig, 1983).

Reference Stocks

v: Charter W1371 *Sr11 Sr18* (Luig, 1983).

Srdp2 (Rondon *et al.*, 1966)   (Plate 3-42)

Srdp2 may be present in a range of durum wheats (Roelfs and McVey, 1979). McIntosh (unpublished, 1980) found that Golden Ball possessed two independent phenotypically indistinguishable genes. One of these factors was transferred to Golden Ball Derivative. In a separate study of hexaploid derivatives of Entrelago de Montijo W3560, a durum used as a differential tester in Australia (Table 1-4), two readily distinguishable genes were found, one of which was located in chromosome 6A and allelic with both *Srdp2* and *Sr13*. The results of Huerta-Espino (1992) indicate that a gene in Entrelago de Montijo cannot be *Srdp2*.

RA Hare, Agricultural Research Centre, Tamworth, NSW, Australia produced stem rust resistant hexaploid derivatives of durum cv. Guillemot. Genetic studies and phenotype comparisons indicated the presence of derivatives with three distinctive low infection types. One derivative produced IT 2 and the single gene involved was not only located in chromosome 6A but was also allelic with the gene in Golden Ball Derivative (RA McIntosh, unpublished 1992). A second derivative produced IT X controlled by a single gene and the third derivative with IT ; carried both genes.

Chromosome Location

6AS (RA McIntosh, unpublished 1980). McIntosh also showed that this gene was allelic with *Sr13* but was unable to distinguish it from *Sr13* using Australian pathotypes.

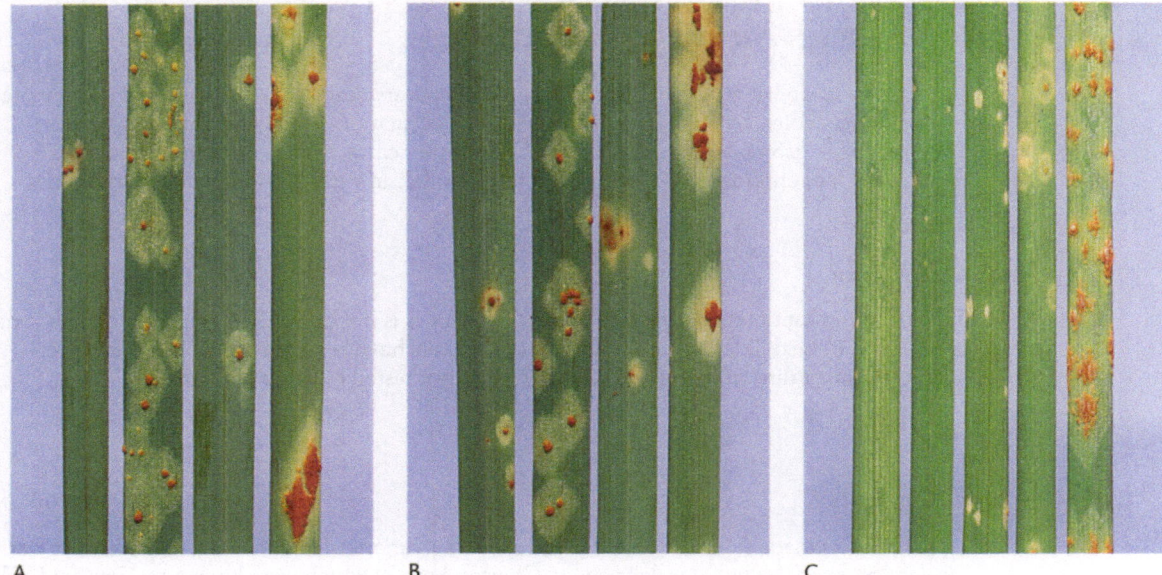

PLATE 3-42. *Srdp2*

A. and B. Seedling leaves of (L to R): Golden Ball (2n = 28), Golden Ball Derivative (2n = 42), Entrelago de Montijo (2n = 28) and Chinese Spring; infected with pt. 17-1,2,3,7 and incubated at A. 20/23°C and B. 23/28°C.

C. Seedling leaves of a plant of the Guillemot Derivative with IT; (L) and the array of distinctive infection types, that is, ;, X, 2 and 3[+] that appeared in the F2 progeny of a cross with Chinese Spring infected with pt. 34-1,2,3,4,5,6,7.

Low Infection Type

2[-] to 23.

Environmental Variability

Less effective at lower temperatures. In this respect plants with *Srdp2* responded similarly to those with *Sr13* (RA McIntosh, unpublished 1980). Roelfs and McVey (1979) found no significant temperature effects.

Origin

T. turgidum. This gene was transferred from Golden Ball to common wheat line, Golden Ball Derivative, by RA McIntosh.

Pathogenic Variability

In international surveys Luig (1983) and Huerta-Espino (1992) found pathogenic variation on Golden Ball Derivative, in most geographic areas. However, Luig (1983) indicated that genetic heterogeneity in the Golden Ball Derivative and the Golden Ball stock used in his survey introduced a degree of unreliability. Golden Ball was generally more resistant than the Derivative indicating the presence of additional genes. Golden Ball was used as a differential tester in North America (Stakman *et al.*, 1962). The pathogenic variation reported by Huerta-Espino (1992) confirmed that *Srdp2* and *Sr13* were different genes.

Reference Stocks

v: Golden Ball Derivative W3504.
tv: Golden Ball; Medea Ap9d (Roelfs and McVey, 1979).

Source Stocks

tv: Arabian P.I.145720 and Rojal de Almeria P.I.191194 (Kenaschuk *et al.*, 1959); C.I.3255, P.I.168906 and R.L.1714 (Heerman *et al.*, 1956); P.I.94701 (Rondon *et al.*, 1966).

SrGt

SrGt was identified as one of four resistance genes in the Australian cultivar Gamut (Rajaram *et al.*, 1971*b*). Subsequent research on this gene was not pursued in Australia because Gamut was considered to be susceptible to stem rust.

Low Infection Type

2 to 23.

Environmental Variability

Low (Roelfs and McVey, 1979).

Pathogenic Variability

Gamut was included as a test genotype in the international survey of pathogenicity undertaken by Luig (1983). He reported virulence in Australia, Europe, Canada, Israel, Pakistan and parts of Africa and South America. Virulence on seedlings of Gamut implied virulence for *SrGt*.

In North America, *SrGt* was transferred to a Baart background and the resulting selection (BtSrGtGt) was included as a tester in the international survey undertaken by Huerta-Espino (1992). He identified virulence in collections from Ethiopia, the Malagasy Republic, Turkey, Brazil and Paraguay.

Reference Stocks

v: BtSrGtGt (Roelfs and McVey, 1979; Huerta-Espino, 1992). Gamut *Sr6 Sr9b Sr11* (Luig, 1983).

Source Stocks

None.

Use in Agriculture

Gamut is the only cultivar known to carry this gene.

SrH (Green and Dyck, 1979)

Gene *SrH* was described by Green and Dyck (1979), who analysed differences in response between lines putatively near-isogenic for *Sr9d* and used as differentials at the Cereal Rust Laboratory, St Paul, Minnesota, and at Agriculture Canada Research Station, Winnipeg. The line in use at St Paul was derived from Hope, whereas the line used at Winnipeg was derived from H-44. The *Sr9d* (H-44) line differed from the *Sr9d* (Hope) line by a unique gene, *SrH*, which was isolated in Line 577.

Low Infection Type

Green and Dyck (1979) reported variable low infection types of 2, X and 23⁻, depending on pathogen culture and environment, to a limited number of pathotypes in the 11-32-113 race group of Stakman *et al.* (1962). Huerta-Espino (1992) reported low IT ranging from 0; to 23.

Pathogenic Variability

Lines with *SrH* (described as 'H44 derivative') were included in the pathogenicity survey of Huerta-Espino (1992). He found low to moderate frequencies of avirulence in south Asia, Turkey, north and central Africa and South America.

Reference Stocks

v: Line 577 from Prelude*8/Marquis//Sr9d (H-44) (Green and Dyck, 1979). H44 derivative (Huerta-Espino, 1992). It is not known if these two lines are the same. *Sr9d* (H-44) (Green and Dyck, 1979). H-44 *Sr2 Sr7b Sr9d Sr17*.

SrKt2 (Plate 3-29)

SrKt2 or *SrKt¢2¢* (Roelfs and McVey, 1979) was allocated by The University of Sydney rust research group to a gene present in Kota. RA McIntosh (unpublished, 1980) showed that this gene was allelic with *Sr16*, but could not distinguish the two genes with the pathogen cultures available. Data obtained by AP Roelfs (*pers. comm.* 1979) and by Huerta-Espino (1992) indicated that *SrKt2* should be designated as an allele of *Sr16*.

Chromosome Location

2BL. *SrKt2* was inherited independently of *Sr9* and of the centromere, but showed recombination of 0.29–0.38 with *Sr28* (McIntosh, 1978).

Low Infection Type

2 (McIntosh, 1978).

Environmental Variability

Low (Roelfs and McVey, 1979).

Origin

Common wheat cv. Kota.

Pathogenic Variability

Frequencies of virulence for *SrKt2* are generally high, as indicated in the response of Kota in many pathogenicity surveys (Luig, 1983). Huerta-Espino (1992) found avirulence among pathogen samples from Ethiopia, the Malagasy Republic, Nepal, Turkey, Chile and Parguay.

Reference Stocks

i: Line AE = W2691*5/Kota (Luig, 1983).

Source Stocks

Kota *Sr7b Sr18 Sr28* (McIntosh, 1978).

Plate

See *Sr28*.

SrLC

In listing the gene *SrLC*, Roelfs and McVey (1979) attributed its description to Luig *et al.* (1973). Huerta-Espino (1992) reported that the gene *SrLC* was also present in W2691 and Baart. However, *SrLC* cannot be the gene envisaged by Luig *et al.* (1973) because the gene to which they referred was not present in the Little Club derivative, W2691. *SrLC* is of no agricultural value and avirulence will only rarely be encountered in field collections of *P. graminis* f. sp. *tritici*.

Low Infection Type

2^- (Roelfs and McVey, 1979).

Origin

Common wheat cv. Little Club.

Pathogenic Variability

Roelfs and McVey (1979) reported avirulence only in isolates of aecial origin. Huerta-Espino (1992) identified a single isolate from Ethiopia that produced IT 2 or 2^+ on W2691 and Baart. Avirulence is more likely to occur in samples of *P. graminis* f. sp. *secalis* and in sexual and asexual hybrids between f. sp. *tritici* and f. sp. *secalis*.

Reference Stocks

v: Little Club (Roelfs and McVey, 1979).

Source Stocks

Baart; W2691 (Huerta-Espino, 1992).

SrMcN (Roelfs and McVey, 1979)

This temporary designation was given to a resistance gene originally identified in cv. McNair 701 which was being used as a susceptible host at the USDA Cereal Rust Laboratory (Roelfs and McVey, 1979).

Low Infection Type

2^-.

Pathogenic Variability

Avirulence was confined to a few cultures of aecial origin (Roelfs and McVey, 1979). Three wheats with *SrMcN*, that is, McNair 701, Maris Nimrod and Exchange *Sr23* were included in the international survey of pathogenicity undertaken by Huerta-Espino (1992). No culture was avirulent for *SrMcN*.

PLATE 3-43. *SrNorin 40*
Seedling leaf pairs of Norin 40 and Chinese Spring; infected with (L) pt. 21-1,2,3,7 (*PNorin 40*) and (R) pt. 126-5,6,7,11 (*pNorin 40*).

Reference Stocks

    v: McNair 701 (Roelfs and McVey, 1979); Maris Nimrod (Huerta-Espino, 1992).

Source Stocks

    Houser; Turbo. Auburn *Sr17*. Exchange *Sr23*.

SrNorin 40   (Plate 3-43)

Chromosome Location

    RA McIntosh (unpublished, 1983) located a single gene for resistance in the short arm of chromosome 6D. This gene is closely linked, but not allelic, with *Sr5*.

Low Infection Type

    2^-2. This gene is effective against most Australasian isolates of *P. graminis* f. sp. *tritici*.

Environmental Variability

    Low.

Pathogenic Variability

    Only Australasian pathotypes that were present 1926–1960 and occasional mutants selected in the greenhouse are virulent.

Reference Stocks

    Norin 40 (RA McIntosh, unpublished 1980).

SrTmp (Roelfs and McVey, 1975)

According to Roelfs and McVey (1979) this gene, present in cv. Triumph 64, has occurred in commercial cultivars in the USA since 1874 when cv. Turkey was introduced. *SrTmp* has provided resistance to a wide array of pathotypes except those described as standard races 56 and 15 on the Stakman *et al.* (1962) differential set. These pathotypes have been of major economic significance to wheat production in North America.

Low Infection Type

    $2^=$ to 23. Roelfs and McVey (1979) noted that the lowest infection types occur in the Triumph background, whereas other wheats such as Turkey show similar patterns of response but the low infection types are higher.

Environmental Variability

Roelfs and McVey (1979) reported slightly lower infection types at higher temperatures but differences were not consistent over cultures.

Pathogenic Variability

Only pathotypes classified earlier as standard races 15 and 56 are virulent on seedlings of Triumph in the USA. Huerta-Espino (1992) found pathogenic variation in most geographic areas except south Asia (Nepal and Pakistan) and China where all isolates were avirulent. Australian isolates are avirulent for *SrTmp*.

Reference Stocks

v: McN701*SrTmpTmp* (Huerta-Espino, 1992); Triumph 64 C.I. 13679 (Roelfs and McVey, 1979).

Source Stocks

Parker; Trison; Triumph; Triumph 64. Eagle *Sr2*. Karl *Sr2 Sr9d Sr24*. Winoka *Sr5*. Larned *Sr17*.

Use in Agriculture

Roelfs and McVey (1979) noted that major epidemics of stem rust in the USA after the barberry eradication campaign involved only pathotypes with virulence for *SrTmp*. This inferred that virulence on the hard red winter wheats possessing this gene was prerequisite to adequate winter survival and sufficient inoculum increase to cause an epidemic in the southern Great Plains and more northerly regions. *SrTmp* has not been transferred to the spring wheat cultivars of North America and elsewhere.

SrTt3 (Gyarfas, 1978)   (Plate 3-38)

Gyarfas (1978) produced a common wheat derivative of *T. timopheevii* that carried a resistance gene different from *Sr36* and *Sr37*. Although distinctive, *SrTt3* shows certain similarities with *Sr36*.

Chromosome Location

2B. *SrTt3* showed recombination of 0.16 with *Sr36* (Gyarfas, 1978).

Low Infection Type

;$12^=$ to 33C, commonly 23 (Gyarfas, 1978). 0 to 0;1^+C (Roelfs and McVey, 1979).

Origin

T. timopheevii.

Pathogenic Variability

Virulence was common in North American cultures of standard race 15 (Roelfs and McVey, 1979). Lines with *SrTt3* were included in the pathogenicity survey of Huerta-Espino (1992). In his tests, the predominant low infection type was 2C, virulence frequencies were relatively low worldwide and virulences for *Sr36* and *SrTt3* were independent.

Reference Stocks

Line AH = Steinwedel*2/4/W1906/*T. timopheevii*//W1906/3/Steinwedel; Federation/Line AH; Federation*2/Line AH (Roelfs and McVey, 1979).

Plate

See *Sr37*.

SrU (Loegering, 1968)

Gene *SrU* in the chromosome substitution line Chinese Spring (Red Egyptian 2D) was described by Loegering (1968).

Low Infection Type

21CN (Roelfs, 1985).

Pathogenic Variability

Lines with *SrU* were tested in the survey of Huerta-Espino (1992) and moderate levels of avirulence were found among cultures from most geographic areas except North America. Australasian pathogen populations are presumably virulent although no detailed survey has been undertaken.

Reference Stocks

v: Cns*SrURE* (sel.) (Huerta-Espino, 1992); *SrU* Ac (Huerta-Espino, 1992).
s: Chinese Spring*5/Red Egyptian 2D *Sr6* (Loegering, 1968).

SrWld1 (McVey and Roelfs, 1978)

SrWld1 was allocated to one of the genes present in wheat cv. Waldron (McVey and Roelfs, 1978), which carries at least six genes for resistance (ND Williams, *pers. comm.* 1993).

Low Infection Type

2.

Pathogenic Variability

This gene was effective against all North American isolates of the pathogen (McVey and Roelfs, 1978; Roelfs *et al.*, 1983). McVey and Roelfs (1978) suggested that the gene was widely effective throughout the world. However, both Luig (1983) and Huerta-Espino (1992) noted considerable variation in response in Waldron and a Baart/Waldron selection, respectively.

Reference Stocks

v: Baart/Waldron (Huerta-Espino, 1992); Ellar (Roelfs *et al.*, 1992). Waldron *Sr5* (heterogeneous) *Sr11* (heterogeneous) *Sr41* (ND Williams, *pers. comm.* 1993).

Source Stocks

Butte *Sr6 Sr8a Sr9g*. Olaf *Sr8a Sr9b Sr12*.

SrWld2 (McVey and Roelfs, 1978)

Low Infection Type

2.

Pathogenic Variability

SrWld2 conferred resistance to cultures of North American pt. 151-QFB and 17-HNL.

Reference Stocks

v: Waldron *Sr5* (heterogeneous) *Sr11* (heterogeneous) *Sr41* (ND Williams, *pers. comm.* 1993).

SrX (Roelfs and McVey, 1979)

Low Infection Type

23C to XC (Roelfs and NcVey, 1979). $2^=$ to 2 (Huerta-Espino, 1992).

Environmental Variability

Moderate. Roelfs and McVey (1979) were unable to separate the effects of environment and culture in explaining the infection type variability.

Origin

Common wheat cv. Marquis.

Pathogenic Variability

According to Roelfs and McVey (1979) certain races, such as 113-RTQ and 113-RKQ and 151-QCB and 151-QFB were avirulent on lines with this gene. Lines with *SrX* were included in the international survey conducted by Huerta-Espino (1992). Two cultures from Ethiopia, one from Kenya and one from Bolivia, were recorded as avirulent.

Reference Stocks

v: LCSrMqX = C.I.104087 (Huerta-Espino, 1992); Marquis *Sr7b Sr18 Sr19 Sr20*, C.I. 3641 (Roelfs and McVey, 1979); Prelude/8*Marquis *Sr19* (Roelfs and McVey, 1979).

Thinopyrum Derived Resistances

SrAgi   (Plate 3-44)

TAF2 (Vilmorin 27*3/*Th. intermedium*), an alien chromosome addition line (2n = 44) produced by Cauderon *et al.* (1973), possesses an added homoeologous group 7 chromosome from *Th. intermedium*. The line TAF2d carries the long arm telocentrics of the same chromosome. Stem rust resistance is located in the long arm. The and Baker (1970) produced alien substitutions of the full chromosome and the telocentric for the various group 7 wheat homoeologues in Chinese Spring. RA McIntosh (unpublished, 1978) produced wheat translocation stocks with the stem rust resistance gene. This gene is the only stem rust resistance gene present in the partial amphiploid (TAF 46, 2n = 56) generated by Cauderon *et al.* (1973).

Other partial amphiploids and addition lines involving genetic materials derived from *Th. intermedium* possess different genes for resistance (see below).

Low Infection Type

0 to X, frequently 1 to 2, often with necrosis.

Environmental Variability

Low.

Pathogenic Variability

TAF2, or a derivative, is used as a supplementary differential for Australian surveys of the stem rust pathogen. Luig (1983) reported that virulence for TAF2 was common in Australia; however, the recent frequency of virulence has been low (Park and Wellings, 1992). Pathogenic variation on TAF2 was found in east Africa, South Africa, Israel, Europe, China, North America and South America (Luig, 1983; Huerta-Espino, 1992) despite the absence of host lines with the resistance gene.

Reference Stocks

TAF2 (Cauderon *et al.*, 1973); TAF2 /Chinese Spring, W3592 (Luig, 1983).

PLATE 3-44. *SrAgi*
Seedling leaves of (L to R): TAF2, two derivatives of Banks with a translocated wheat 7Ai chromosome, Banks and Chinese Spring; infected with pt. 34-1,2,3,4,5,6,7. Banks carries *Sr30* which is effective against this pathotype.

Miscellaneous *Thinopyrum* Derived Resistances (Plate 3-45)

Certain stem rust resistant lines that possess genes derived from *Th. ponticum* and *Th. intermedium* and not discussed in earlier sections were developed in various laboratories. Most appear to involve homoeologous group 7 chromosomes and resistance in some lines is associated with genes producing unacceptable levels of flour yellowness caused by carotenoid pigments. The relevant lines include the following.

Thinopyrum ponticum (Kibirige-Sebunya and Knott, 1983)

7el2	(2n = 44)	Addition	LMq^6 -28	=	C83.2
7el2(7D)	(2n = 42)	Substitution	LMq^6 -28	=	C83.3
7D-7el2	(2n = 42)	Translocation	10-2	=	C83.4
	(2n = 42)	Translocation	23-9	=	C85.32
	(2n = 42)	Translocation	24-1	=	C85.33
	(2n = 42)	Translocation	24-2	=	C85.34

Thinopyrum intermedium (Hsam and Zeller, 1982)

Caribo/*Th. intermedium* C (2n = 44) Addition		=	C83.5

This addition chromosome belongs to homoeologous group 6 and is homologous with that in Cauderon's disomic addition line, L7.

W44	Substitution	7Ai#2(7D)	=	C91.47

The 7Ai chromosome in W44 also carries resistance to leaf rust (*Lr38*) and stripe rust. However, translocation lines with *Lr38* do not carry the stem rust and stripe rust resistances.

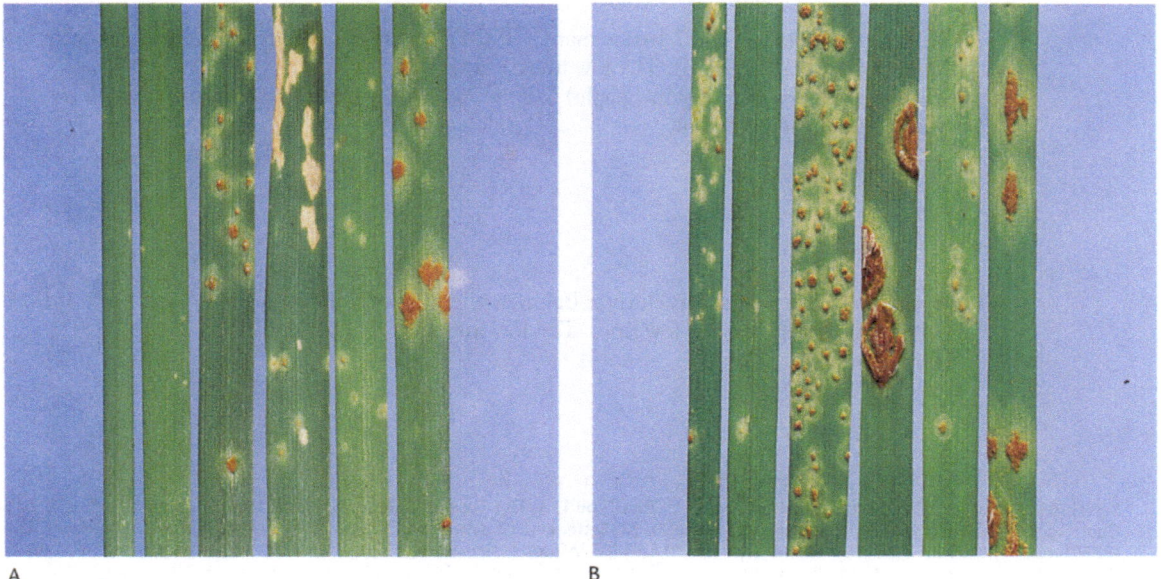

A   B

PLATE 3-45. Miscellaneous *Thinopyrum* Derived Resistances
Seedling leaves of (L to R): Translocation 10-2 (C83.4), translocation 23-9 (C85.32), W44 (C91.47), TAF2, Agatha (*Sr35*) and Chinese Spring; infected with A. pt. 34-1,2,3,4,5,6,7 [*PTAF2*] and B. pt. 34-1,2,3,6,7,8,9 [*pTAF2*]. The first two lines carry the same gene. The four genetically different sources of resistance can be distinguished on the basis of genetic homology and infection type or stem rust specificity.

Stem Rust Resistance Genes in Triticale (Plate 3-46)

Stem rust became a problem for the Australian triticale industry in the early 1980s (McIntosh *et al.*, 1983; McIntosh and Singh, 1986). In 1982 severe losses were experienced by growers in northern New South Wales and Queensland. McIntosh *et al.* (1983) showed that losses were caused by an increased frequency of a pathotype with virulence for a resistance gene that occurred in several Australian triticales. This gene was present at high frequency in contemporary CIMMYT breeding populations. They showed that the resistance gene in triticale was *Sr27* (see *Sr27*). Triticales with resistance to the *Sr27*-virulent pathotype, designated 34-2,12 (McIntosh *et al.*, 1983), were recommended, but in 1984 severe rusting occurred on some of these and increased rusting occurred on others. This second mutation, which occurred in pt. 34-2,12, overcame a resistance gene known as *SrSatu*. The new pathotype was designated 34-2,12,13.

Genetic studies led to the identification of several genes for resistance to *P. graminis tritici* (McIntosh and Singh, 1986; Singh and McIntosh, 1988). Only *Sr27* carries a wheat designation as a chromosome segment bearing this gene has been transferred to wheat. The second designated wheat gene of rye origin, *Sr31*, has not been identified in triticale. The gene *SrNin* confers a low infection type that is probably too high for *Sr31*. The European cultivar, Lasko, possesses at least two genes (IT ; and IT 2) that are different from those described below. Neither of these genes appear to be *Sr27*. An earlier study in Canada by Morrison *et al.* (1977) led to the postulation of five resistance genes but it is not possible to relate their work to the Australian results.

Sr27

See *Sr27*. Low infection type is ;1.
v: Coorong; Dua; Towan; Tyalla. Tahara *SrNin*. The frequency of *Sr27* has declined in CIMMYT breeding populations since the mid 1980s (McIntosh *et al.*, 1983; Singh and McIntosh, 1988; RA McIntosh, unpublished 1993).

Virulence on triticale with *Sr27* was identified in South Africa in 1988 (Smith and Le Roux, 1992), but virulence on wheats with *Sr27* was reported earlier in Kenya (Harder *et al.*, 1972).

SrSatu

This gene is allelic with *Sr27* hence many CIMMYT-generated lines have one or the other gene (Singh and McIntosh, 1988). The low infection type is ;.
v: Satu, Toort. Bejon (= Tejon-Beagle) *SrBj*. Ningadhu (= Drira) *SrNin*. Currency *SrVen*; Samson *SrVen*; Venus *SrVen*.

SrBj

SrBj occurs in the Australian cultivar Bejon, a selection of Tejon-Beagle produced at CIMMYT. This gene is present in very few lines. The low infection type is $X^=$ to X^-.
v: Bejon *SrSatu*.

PLATE 3-46. Stem Rust Resistance Genes in Triticale ▶

Seedling leaves of triticale cultivars (L to R): 1. Lasko, 2. Tiga Type 1, 3. Tiga Type 2, 4. Bejon, 5. Ningadhu, 6. Tahara, 7. Venus, 8. Satu, 9. Coorong, 10. Juanillo (Type 1), 11. Juanillo 231 (Type 2), 12. Setter and 13. wheat cv. Chinese Spring; infected with A. pt. 34-2 [*P27PSatu*], B. pt. 34-2,12 [*p27PSatu*] and C. pt. 34-2,12,13 [*p27pSatu*]. Lasko and Tiga Type 1 show similar responses with all three cultures, Tiga Type 2, Bejon, Ningadhu, Venus, Satu and Juanillo Type 1 possess *SrSatu* and show more compatible responses with pt. 34-2,12,13. On the basis of IT with this pathotype, it can be postulated that Tiga Type 2 (3), Ningadhu (5) and Tahara (6) carry a common gene that is distinctive from genes in Bejon (4), Venus (7) and Juanillo (10 and 11). Coorong (9) and Tahara (6) carry *Sr27* which is effective against pt. 34-2. Juanillo Type 2 possesses a gene with only moderate resistance to all three cultures. Setter and Chinese Spring wheat give similar susceptible responses with all three cultures. Lasko and Tiga Type 2 give unchanged phenotypes with all three pathotypes indicating the presence of additional genes.

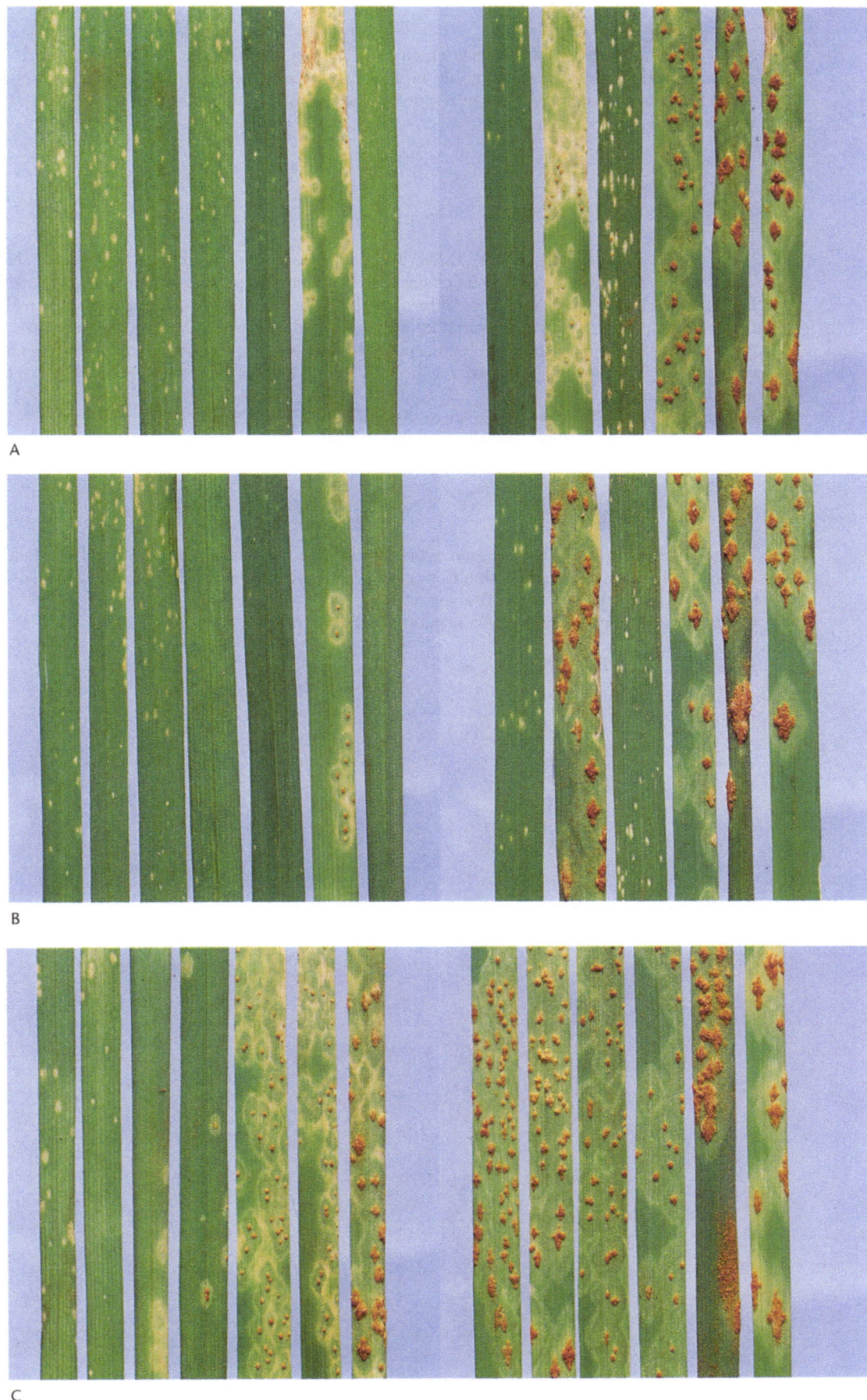

A

B

C

SrNin

SrNin occurs at relatively high frequencies in CIMMYT-generated lines where it is usually combined with *SrSatu* or less commonly with *Sr27*. Infected with pt. 34-2,12,13, lines possessing *SrNin* produce a moderately resistant response. The low infection type is 2^- to 2.

v: Ningadhu *SrSatu*. Tahara *Sr27*.

SrVen

SrVen is relatively frequent among CIMMYT-generated triticale lines that do not possess *SrNin*. The low infection type is usually very characteristic with mixed pustules ranging from IT 1 to 3, but due to lack of necrosis, does not conform to IT X. The relative frequencies of pustules with ITs 1, 2 and 3 on a single leaf are variable. When infected with pt. 34-2,12,13, the responses of genotypes with *SrVen* are moderately susceptible to susceptible with terminal disease ratings of up to 70S when Coorong and Satu have reached 100S. This level of protection is considered to be inadequate.

v: Currency *SrSatu*; Samson *SrSatu*; Venus *SrSatu*.

SrJ

SrJ has been found alone in some sources of Juanillo. The low infection type is 3 and the field response is inadequate for protection from losses. This gene is sometimes difficult to distinguish from *SrVen* on the basis of infection type.

v: Juanillo selections. Some Juanillo selections also have *SrSatu*.

CHAPTER 4

The Genes for Resistance to Stripe Rust in Wheat and Triticale

CATALOGUED STRIPE RUST RESISTANCE GENES

Yr1 (Lupton and Macer, 1962)   (Plate 4-1)

Synonym

L (Zadoks, 1961).

Chromosome Location

2A (Macer, 1966; Xin *et al.*, 1984); 2AL (Bariana and McIntosh, 1993).

Low Infection Type

0;.

Environmental Variability

Low.

Origin

Common wheat. *Yr1* probably originated from Chinese 166.

Pathogenic Variability

Variation has been identified in most geographical areas. Stubbs (1985) noted that virulence is especially high in east Asia where Chinese 166, and thus possibly *Yr1*, originated. Although no pathogenic variant has been found in Australia, pts 109 E141 A– and 111 E143 A– were isolated from New Zealand samples in 1986 and 1988, respectively (Wellings and McIntosh, 1990). Several races of the barley stripe rust pathogen (*P. striiformis* f. sp. *hordei*) give intermediate to high infection types on Chinese 166 (Stubbs and Fuchs, 1992).

Reference Stocks

i: Aroona*5/*Yr1*; Kite*6/*Yr1*; Warigal*6/*Yr1*; other backcross derivatives of Australian wheats produced at PBI, Cobbitty. Hobbit Sib (*Yr1 Yr2 Yr14*) and substitution line Hobbit Sib*4/ *T. macha* 2A (*yr1* presumably *Yr2 Yr14*) can be treated as near-isogenic (RA McIntosh, unpublished 1990).

v: Chinese 166 (Lupton and Macer, 1962).

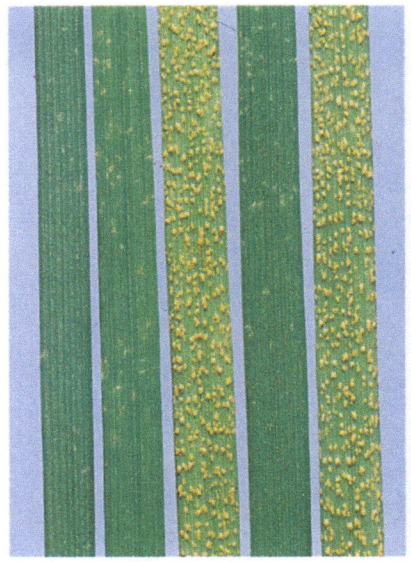

PLATE 4-1. *Yr1*
Seedling leaves of (L to R): Chinese 166, Kite*6/*Yr1*, Kite, Harrier*2/*Yr1* and Harrier; infected with A. pt. 108 E141 A- [*P1*] and B. pt. 109 E141 A- [*p1*]. The low infection type remains unaffected in the susceptible backgrounds, and is completely overcome by the culture with virulence for *Yr1*.

Source Stocks

China: Feng Kang 13 (Xin *et al.*, 1984).

Europe: Dalee (Bayles and Thomas, 1984); Durin (Taylor *et al.*, 1981); Heines 110 (McIntosh, 1988*a*); Maris Templar (Taylor *et al.*, 1981); Corin (de Vallavieille-Pope *et al.*, 1990). Pegasus *Yr6* (Cromey and Munro, 1992). Other combination genotypes are listed in McIntosh (1988*a*). *Yr1* is common in winter wheats (see Taylor *et al.*, 1981).

India: The report that *Yr1* occurred in a group of Indian wheats (Sawhney and Luthra, 1970) is not correct.

Use in Agriculture

Non-durable; the ineffectiveness of cv. Rothwell Perdix in the mid-1960s in the UK (Johnson, 1988) was due to a severe epidemic following increase of a mutant pathotype with virulence for *Yr1*. However, resistance to avirulent pathotypes is very effective and *Yr1* may have a role as a component of multiple gene resistances in countries such as Australia where the frequency of virulence is very low or absent.

Yr2 (Lupton and Macer, 1962)   (Plate 4-2)

Yr2 was originally allocated to a gene present in Heines VII and Soissonais-Desprez and identified with European pathotypes. Singh and Johnson (1988) showed that whereas Kalyansona responded identically to these wheats when tested with British pathotypes, it was susceptible and the European wheats resistant, when tested with certain pathotypes from other geographic areas, for example pt. 6 E16 from Lebanon. Kalyansona shared at least one gene, presumably *Yr2*, with Heines VII and Soissonais-Desprez. In addition to *Yr2*, Heines VII and Soissonais possess a gene (or genes) which was isolated in lines TP981 and TP1295, respectively (Johnson, 1992). An Australian culture avirulent on Heines VII and virulent on Kalyansona produced a low infection type on TP981 and a high response on TP1295, indicating that these genes are not identical (CR Wellings, unpublished 1993). The gene in TP981 was expected to be common in European wheats (Johnson, 1992) and maybe allelic with a gene in Strubes Dickkopf (Johnson and Minchin, 1992). This example illustrates a potential problem when differentials established using the pathogen flora of one geographical area are adopted for testing pathogenic variation in another area. The chromosomal location of the second unnamed gene in Heines VII is unknown.

Synonym

U (Zadoks, 1961).

Chromosome Location

7B (Labrum, 1980).

A

B

C

E

D

PLATE 4-2. *Yr2*

A. Seedling leaves of (L to R): Kalyansona, Heines VII and Soissonais-Desprez; infected with IPSR pt. 108 E9 (L), IPSR pt. 108 E141 (C) and IPSR pt. 6 E16 (R). All three host lines are resistant to the first culture and susceptible to the second, but Kalyansona is clearly differentiated from the other two wheats with the third culture. Note the distinctive low responses of Heines VII and Soissonais-Desprez. Courtesy R Johnson.

B. Infected seedling leaves of (L to R): TP981 (derived from Heines VII), TP1295 (derived from Soissonais-Desprez), TP981, TP1295, Heines VII and Soissonais-Desprez. The first two leaves were inoculated with IPSR pt. 6 E16 and the next four leaves, with IPSR pt. 108 E9. The susceptible responses of the two derivatives with IPSR pt. 108 E9 shows that resistance to IPSR pt. 6 E16 cannot be due to *Yr2*. Courtesy R Johnson.

C. and D. Seedling leaves of Australian wheats (L to R): Bodallin, Katyil, Olympic, Egret and European cv. Vilmorin 23; infected with C. IPO pt. 104 E9 A- and D. pt. 104 E137 A-. *Yr2* appears to be widespread in Australian wheats. The Egret plants used for this comparison did not carry *YrA* (Wellings *et al.*, 1988).

E. and F. Seedling leaves of (L to R): Merlin, Heines VII, TP981, TP1295, Kalyansona and Cleo; infected with E. pt. 104 E9 A+ and F. pt. 104 E137 A+. With the Australian culture selected for avirulence on seedlings of Heines VII, TP981 and TP1295 display distinctive responses indicating that TP981 carries a different gene from TP1295, or that TP981 carries a second gene that is effective against the Australian culture. Similarly, Merlin carries a gene in addition to *Yr2*.

F

Low Infection Type

0; to 2.

Environmental Variability

Reported variability in low infection type may reflect genetic background (Singh *et al.*, 1990) as well as environmental influences (Johnson, 1988).

Origin

Common wheat.

Pathogenic Variability

Common in most geographic areas. The *P. striiformis* f. sp. *tritici* pathotype (104 E137) introduced to Australia was virulent on seedlings of Heines VII, the accepted tester for *Yr2*. Two avirulent isolates (104 E9 A–, 104 E9 A+) were detected on seedlings of Heines VII (Wellings and McIntosh, 1990). Recent research has shown that Heines VII will be a valid tester for *Yr2* only when the pathogen population is uniformly virulent for the second gene isolated in TP981.

Reference Stocks

i: B1-B5 Lemhi 53*5/Soissonais (Griffey and Allan, 1988).
v: **Yr2 only**: Kalyansona (Singh and Johnson, 1988).
   Yr2 + second gene: Heines VII (Zadoks, 1961); Soissonais-Desprez (Lupton and Macer, 1962).
   Second gene only: TP981, TP1295 (Chilosi and Johnson, 1990).

Source Stocks

Merlin. Cleo *Yr3* (Stubbs, 1985). *Yr2* is very common in both winter and spring wheats (Perwaiz and Johnson, 1986; McIntosh, 1988a). The reason for its presence in a range of wheats distributed by CIMMYT is not clear. Singh *et al.* (1990) reported the presence of *Yr2* in several Indian wheats including Sonalika, WL711 and HD2329. Additional genes were also present. Genetic analyses of wheats believed to have *Yr2* have been extremely difficult. Whereas Chilosi and Johnson (1990), Singh *et al.* (1990) and de Vallavieille-Pope *et al.* (1990) concluded that Heines Peko carried *Yr2*, in addition to *Yr6*, Badebo *et al.* (1990) reported cultures virulent on Heines Kolben and Kalyansona and avirulent on Heines Peko. Johnson (1992) interpreted similar data as indicating the presence of *Yr2*, *Yr6* and the TP981 resistance in Heines Peko, and *Yr6* plus a weakly expressed *Yr2* in Heines Kolben. However, Chen and Line (1992b) doubted the presence of *Yr2* in Heines Kolben and suggested that the latter may possess a different gene.

Use in Agriculture

Although ineffective against at least some pathotypes in all geographic areas, *Yr2* is present in a wide range of wheats.

Yr3 (Lupton and Macer, 1962)   (Plate 4-3)

Lupton and Macer (1962) proposed an allelic series at the *Yr3* locus to explain the interaction of parental cultivars and hybrid populations with four pathotypes of *P. striiformis*. The alleles *Yr3a*, *Yr3b* and *Yr3c* were assigned to cultivars Cappelle-Desprez, Hybrid 46 and Minister, respectively. After careful consideration of the data, Wellings (1986) and Johnson (in Knott, 1989) concluded that these designations were doubtful and that pathotypes able to discriminate these alleles were no longer available. Chen and Line (1992b) concluded that Cappelle Desprez, Nord Desprez and Vilmorin 23 each have two genes for resistance, although the relationships between these genes and those reported at the *Yr3* locus were not resolved. Moreover, the proposed allelic designations at this locus were not adopted by some laboratories (Johnson *et al.*, 1972; Dubin *et al.*, 1989; Stubbs and Fuchs, 1992). However, multi-pathotype testing revealed that Nord Desprez and Vilmorin 23 may possess a gene or genes in addition to *Yr3* (Stubbs and Fuchs, 1992; Chen and Line, 1993a).

Synonym

M (Zadoks, 1961); *Yr3a*, *Yr3b* (Lupton and Macer, 1962).

Chromosome Location

5BL (Worland, 1988).

Low Infection Type

;.

PLATE 4-3. Yr3
Paired seedling leaves of Vilmorin 23 and Nord Desprez; infected with (L to R): IPO pt. 82 E0, IPO pt. 6 E150 and IPO pt. 237 E141. Note the different resistant responses to the first two pathotypes. It is assumed that both responses are conferred by *Yr3*, although pt. 6 E150 could be avirulent for a second gene present in each wheat. The responses of Vilmorin 23 and Nord Desprez are usually identical with European pathotypes, although low infection types with certain non-European pathotypes indicate a genetic difference between the two cultivars (Stubbs and Fuchs, 1992).

Environmental Variability

Unknown, probably low. Vilmorin 23 showed no variation in response to differences in light intensity (Stubbs, 1967).

Origin

Common wheat.

Pathogenic Variability

Variation has been detected worldwide, with the suggestion that *Yr3* may be more effective in southeast Asia (Stubbs, 1985). Kumar *et al.* (1993) reported a high frequency of virulence for Vilmorin 23 in India.

Reference Stocks

v: Bon Fermier (Stubbs *et al.*, 1974); Cappelle-Desprez (Lupton and Macer, 1962); Doerfler (Stubbs and Fuchs, 1992); Nord-Desprez (Johnson *et al.*, 1972); Staring (Stubbs and Fuchs, 1992); Vilmorin 23 (Johnson *et al.*, 1972).

Source Stocks

Europe: Felix. Anouska *Yr2*; Cleo *Yr2*; Stella *Yr2*. See Stubbs (1985).
CIMMYT: Inia P-Altar 82 *Yr6* (Dubin *et al.*, 1989).
North America: Druchamp, Stephens (Chen and Line, 1993a).

Use in Agriculture

The resistance of cv. Felix with *Yr3* and a non-designated gene (*YrCV*) in common with that in Carstens V, remained effective for 16 years in Europe. The former became ineffective after the cultivation of Caribo (*YrCV*) led to the development of virulence for *YrCV* and subsequently virulence for *YrCV* and *Yr3* (Stubbs, 1985). However, *Yr3* is not used as an effective source of resistance in breeding programs.

Yr4 (Lupton and Macer, 1962)   (Plate 4-4)

Lupton and Macer (1962) designated alleles at the *Yr4* locus in cultivars Cappelle Desprez (*Yr3a Yr4a*) and Hybrid 46 (*Yr3b Yr4b*). These cultivars showed allelism in intercrosses when tested with three avirulent pathotypes, although tests were not reported with the critical Race 2B which was virulent for Cappelle Desprez (Lupton and Macer, 1962). On the basis of pathotype interactions on these two cultivars in the UK, Johnson (in Knott, 1989) suggested they share a gene in common, with a second gene in Hybrid 46. Stubbs and Fuchs (1992) concluded that Suwon 92/Omar and Hybrid 46 shared *Yr4* as a gene in common, with the latter having additional resistance. Moreover

Chen and Line (1993*a*) suggested that Hybrid 46 did not carry *Yr3* in studies using North American cultures. In view of the unavailability of the original pathogen cultures, and hence the lack of opportunity to repeat the work, we conclude that the resistance genotype of Hybrid 46 should be *Yr4* plus at least one additional gene.

Synonyms

Yr4a, *Yr4b* (Lupton and Macer, 1962).

Chromosome Location

3B (Worland, 1988).

Low Infection Type

;.

Environmental Variability

Low. Variation in chlorosis and pustule necrosis in compatible infection types may reflect environmental factors.

Origin

Common wheat.

Pathogenic Variability

Virulence has been detected in most wheat-growing areas, and is especially frequent in South America (Stubbs, 1985) and Australia (Wellings and McIntosh, 1990). *Yr4* was effective in India until the emergence of new pathotypes in 1989 and 1991 (Kumar *et al.*, 1993).

Reference Stocks

v: Hybrid 46 (Johnson *et al.*, 1972); Opal (Stubbs *et al.*, 1974); Vaillant (Stubbs and Fuchs, 1992).

Source Stocks

Many stocks were listed by McIntosh (1988*a*). These include Mardler *Yr1 Yr2 Yr3 Yr13*, Argent *Yr1 Yr3 Yr6* and Maris Huntsman *Yr2 Yr3 Yr13*.

Use in Agriculture

Yr4 appears to be effective in Asia and North America (Stubbs, 1985). However, de Vallavieille-Pope and Line (1990) noted that two cultures from the USA were virulent on Hybrid 46, casting some doubt on the effectiveness of *Yr4* in North America.

PLATE 4-4. Yr4
Cv. Hybrid 46 inoculated with five cultures of *P. striiformis* f.sp. *tritici*. (L to R): IPO 8 E0 [*p3P4*], IPO 40 E0 [*p3P4*], IPO pt. 82 E0 [*p3P4*], IPO pt. 104 E9 [*p3p4*] and IPO pt. 234 E139 [*p3p4*]. Although infection type variation is evident, the distinction between avirulent and virulent cultures is clear. Variation in low infection types between pathotypes may indicate the presence of additional genes in Hybrid 46.

Yr5 (Macer, 1966) (Plate 4-5)

Chromosome Location

2BL, 21 cM from the centromere (Law, 1976). *Yr5* and *Yr7* are allelic or closely linked. Johnson and Dyck (1984) found susceptible progeny from a cross of *T. spelta* and Thatcher. They proposed that Thatcher carried a dominant inhibitor of *Yr5*. Johnson (1986) confirmed allelism of *Yr5* with *Yr7* in Lee.

Low Infecton Type

0; to ;.

Environmental Variability

Low.

Origin

Spelt wheats (Kema, 1992).

Pathogenic Variability

Isolates of *P. striiformis* with virulence for *Yr5* are extremely rare. Virulent cultures were identified in Australia (Wellings and McIntosh, 1990). Indian workers (Nagarajan *et al.*, 1986a) claimed to have identified a pathotype virulent on seedlings of *T. spelta album*.

Reference Stocks

i: Aroona*6/*Yr5*; Avocet S*6/*Yr5*; Warigal*6/*Yr5*; other Australian backcross derivatives produced at PBI Cobbitty.
v: *T. spelta album*.

Source Stocks

Several spelt accessions are listed in Kema (1992).

Use in Agriculture

Not currently deployed. *Yr5* has been transferred to several Australian spring wheats and was reported to have been used sporadically in Indian and Dutch wheat breeding programs (Kema, 1992).

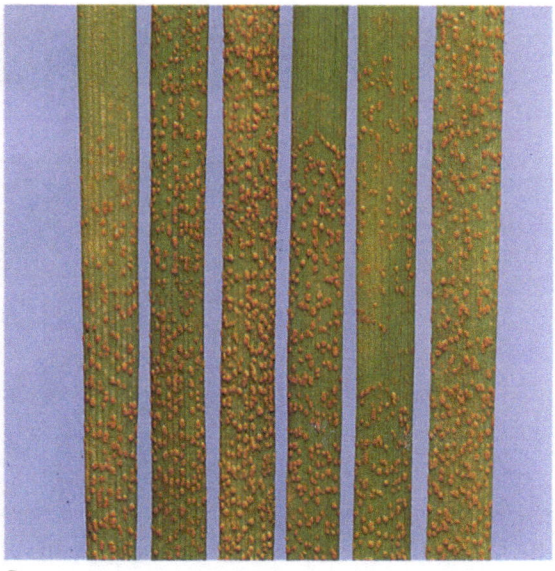

A   B

PLATE 4-5. Yr5
Seedling leaves of (L to R): *T. spelta album*, Avocet S*6/*Yr5*, Avocet S, Kite*6/*Yr5*, Kite and Federation; infected with A. pt. 104 E137 A+ and B. pt. 360 E137 A+. Note that the high level of resistance conferred by *Yr5* was preserved during backcrossing and that the virulent mutant completely overcomes the resistance.

Yr6 (Macer, 1966)   (Plate 4-6)

Synonym

B (Zadoks, 1961).

Chromosome Location

7B (Labrum, 1980); 7BS (El-Bedewy and Röbbelen, 1982).

Low Infection Type

; to ;N1.

Environmental Variability

Moderate; higher responses may be obtained with avirulent pathotypes when tested at low temperatures (Wellings, 1986). In contrast, Dubin *et al.* (1989) found *Yr6* to be less effective at higher greenhouse temperatures, and concluded that optimal expression occurred at intermediate temperatures and was possibly influenced by light intensity.

Origin

Common wheat. *Yr6* was also identified in current durum wheats and advanced lines (Wellings, 1986; Chilosi and Johnson, 1990). It is not clear if *Yr6* is naturally present in tetraploid wheats, or if it was transferred to durums from common wheats.

A

B

C

PLATE 4-6. *Yr6*

A. and B. Seedling leaves of Australian cultivars (L to R): Bindawarra, Jacup, Miling, Millewa, Takari and Federation; infected with A. pt. 104 E137 A+ [*P6*] and B. pt. 108 E141 A+ [*p6*].

C. Responses of non-Australian cultivars (L to R): Pitic 62, Shoshi, Romany, Frontana and Federation; infected with pt. 104 E137 A+ [*P6*].

Note the relatively consistent necrotic response conferred by *Yr6*.

Pathogenic Variability

Virulence is very common (Stubbs, 1985). The *P. striiformis* population of the Indian subcontinent appears to be fixed for virulence (Kumar *et al.*, 1988).

Reference Stocks

v: Oxley (Wellings, 1986). Heines Kolben *Yr2* (Johnson *et al.*, in press); Heines Peko *Yr2* (Johnson *et al.*, in press).
tv: Duilio (Chilosi and Johnson, 1990); Latino (Chilosi and Johnson, 1990); Quadruno (Chilosi and Johnson, 1990).

Source Stocks

Bindawarra (Wellings, 1986); Frontana (Wellings, 1986); Koga II (Labrum, 1980); Maris Dove (McIntosh, 1988*a*); Millewa (Wellings, 1986); Récital (de Vallavieille-Pope *et al.*, 1990). See McIntosh (1988*a*) for list. *Yr6* is relatively frequent in both winter and spring wheats. Many wheats with *Yr6* also carry combinations of several known and unknown resistance genes (Perwaiz and Johnson, 1986; Wellings, 1986; Dubin *et al.*, 1989; Badebo *et al.*, 1990).

Use in Agriculture

Although not used intentionally, *Yr6* is common in spring wheats of South American origin. Many of these wheats, in particular Frontana and Frontiera (Wellings, 1986), were introduced into the CIMMYT program as sources of leaf rust resistance, including *Lr13* and other genes such as *Lr34*. The transfer of *Yr6* into CIMMYT germplasm thus appeared to be fortuitous.

Yr6 is inherited as a recessive or dominant gene depending on the pathogen culture used in the selection process (El-Bedewy and Röbbelen, 1982). Johnson and Minchin (1992) suggested that the dominance reversal of *Yr6* was in some way related to the pathogenicity for *Yr2* in the test culture. The evidence presented indicated that *Yr6* was inherited as a dominant gene when *Yr2* was effective; in contrast, when *Yr2* was ineffective, *Yr6* appeared to be recessive.

Yr7 (Macer, 1966)   (Plate 4-7)

Chromosome Location

2B (Stubbs, 1964; Johnson *et al.*, 1969). 2BL, 21 cM from the centromere (Law, 1976). *Yr7* is very closely linked with *Sr9g* (McIntosh *et al.*, 1981). *Yr7* is allelic or closely linked with *Yr5* (see *Yr5*).

Low Infection Type

;N to 1N.

Environmental Variability

Low.

Origin

Yr7 originates from durum cv. Iumillo, and was transferred to Thatcher wheat, its relatives and derivatives. Cultivar Lee is assumed to be a derivative of Thatcher rather than Hope. Whereas close linkage of *Yr7* and *Sr9g* occurs in Thatcher and its derivatives, the durum wheats Acme and Kubanka carry *Sr9g*, but not *Yr7* (McIntosh *et al.*, 1981).

Pathogenic Variability

Variability occurs in most geographic areas (Stubbs, 1985) except the Indian subcontinent where the population appears fixed for virulence on plants with *Yr7* (Kumar *et al.*, 1988; 1993).

Reference Stocks

i: Avocet S*6/*Yr7*.
s: CS*7/Marquis 2B (McIntosh *et al.*, 1981); the selection of 'Marquis' used by Sheen and Snyder (1964) was apparently Thatcher; CS*4/Thatcher 2B (McIntosh *et al.*, 1981).
v: Lee (Stubbs, 1985); Reichersberg 42 (Singh *et al.*, 1990).

Source Stocks

v: *Yr7* is present in a range of winter and spring wheats and is frequently associated with *Sr9g*. Barani 83 (Perwaiz and Johnson, 1986); PBW12 (Singh *et al.*, 1990); WL2265 (Singh *et al.*, 1990). Seri 82 *Yr2 Yr9* (Badebo *et al.*, 1990). Pavon 76 *Yr6* (Wellings, 1986). Pak 81 *Yr9* (Dubin *et al.*, 1989). See McIntosh (1988*a*).
tv: Iumillo (McIntosh *et al.*, 1981).

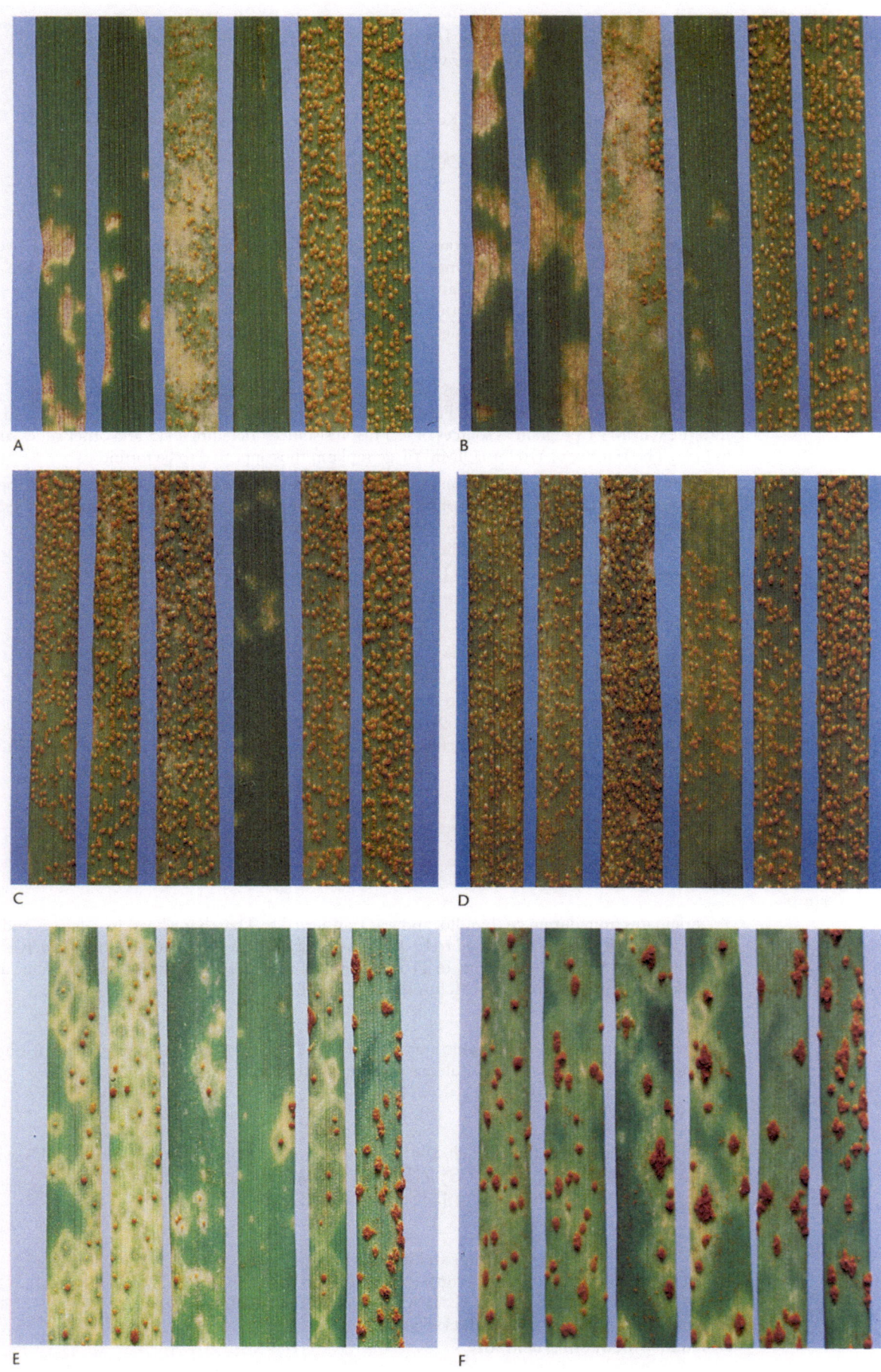

A

B

C

D

E

F

Use in Agriculture

Yr7 was deployed in Europe, Australia and New Zealand but virulent cultures overcame the resistance (Wellings and McIntosh, 1990). Yr7 is often found in combination with other genes, such as Yr6 in cvv. Corella, Dollarbird and Hermosillo 77 (Wellings, 1986).

Yr8 (Riley *et al.*, 1968)   (Plate 4-8)

Chromosome Location

2A [2A/2M] (McIntosh *et al.*, 1982), 2D [2D/2M] (Riley *et al.*, 1968). Several 2A/2M translocations in Hobbit Sib and one 2D/2M translocation in Maris Widgeon are listed in Miller *et al.* (1988).

Low Infection Type

0; to ;.

Environmental Variability

Low.

Origin

T. comosum.

Pathogenic Variability

Virulence in the pathogen population appears to be frequent in southeast Asia including India, moderate in Europe and the Mediterranean, and absent in North and South America (Stubbs, 1985). A low frequency of mutant Yr8-virulent isolates was detected in Australia (Wellings and McIntosh, 1990).

A   B

PLATE 4-8. Yr8
Seedling leaves of (L to R): Compair, Chinese Spring, Avocet S*6/Yr8, Avocet S, Harrier*6/Yr8 and Harrier; infected with A. pt. 104 E137 A+ [P8] and B. pt. 104 E153 A+ [p8]. The expression of resistance conferred by Yr8 is modified by the genetic background, the Avocet S background producing more necrosis than either the Chinese Spring or Harrier backgrounds.

◀ **PLATE 4-7. Yr7**
Seedling leaves of (L to R): Lee, Oroua, Gatcher Selection, Corella, Kubanka (2n = 28) and Federation; infected with A. pt. 104 E137 A- [P6P7], B. pt. 108 E141 A+ [p6P7], C. pt. 106 E139 A- [P6p7] and D. pt. 110 E143 A+ [p6p7]. The first three wheats have only Yr7 and are susceptible with pt. 106 E139 A-, the pathotype that caused epidemics on cv. Oroua in New Zealand. Note that Gatcher Selection gives a higher IT with the avirulent cultures than Lee and Oroua. Corella carries both Yr6 and Yr7 and hence the IT with pt. 104 E137 A- (A.) is lower than with pt. 108 E141 A+ (B.) or 106 E139 A- (C.).

E. and F. The above wheats infected with *P. graminis* f. sp. *tritici* pt. 343-1,2,3,5,6 [P9g] and pt. 98-1,2,3,5,6 [p9g], respectively. The second pathotype is assumed to be a mutant derivative of the first with virulence for Sr9g. Although Kubanka carries Sr9g it does not carry Yr7. Gatcher Selection and Corella carry an additional gene, probably Sr12, which interacts with Sr9g, but also has a small effect on response with the second culture.

Reference Stocks

       i: Aroona*6/Yr8; Avocet S*6/Yr8; other backcross derivatives of Australian wheats.

       s: Chinese Spring 2M(2A) (Riley *et al.*, 1968).

       v: Compair (Riley *et al.*, 1968).

Source Stocks

       CS2A/2M and CS2D/2M (McIntosh *et al.*, 1982); Hobbit Sib 2A/2M; Hobbit Sib 2D/2M and Maris Widgeon 2D/2M (Miller *et al.*, 1988).

Use in Agriculture

       Resistance conferred by *Yr8* was considered non-durable due to the occurrence in Britain of three virulent pathogen cultures ten years after the development of Compair (Johnson *et al.*, 1978). *Yr8* has not been deployed in commercial cultivars. The report that *Yr8* was present in Pakistani wheats (Kirmani *et al.*, 1984) is not correct.

Yr9 (Macer, 1975)   (Plate 4-9)

Chromosome Location

       1B (1BL.1RS) or 1R(1B) (Mettin *et al.*, 1973; Zeller, 1973). Some wheats include both substitution and translocation biotypes (Zeller, 1973).

Low Infection Type

       0;.

Environmental Variability

       Low.

Origin

       S. cereale cv. Petkus. Most, if not all wheats with *Yr9* are derived from wheat x rye derivatives produced in Germany in the 1930s (see Mettin *et al.*, 1973; Zeller, 1973).

Pathogenic Variability

       The deployment of wheats with *Yr9* has usually been followed by an increase of pathotypes with virulence for this gene. Virulence occurs in Africa, China, Europe, South America, (Stubbs, 1985) and New Zealand (Wellings and Burdon, 1992). *Yr9* remains effective in North America (Line and Qayoum, 1991), India (Kumar *et al.*, 1993) and Nepal (Louwers and Sharma, 1992). Severe epidemics in the UK in 1988 and 1989 on several cultivars, principally Slejpner, were due to a rapid increase in virulence for *Yr9* in the pathogen population (Bayles *et al.*, 1990).

Reference Stocks

       i: Avocet S*3/Yr9; Federation*4/Kavkaz; M2435*6/Yr9; Warbler = Oxley*4/Kavkaz.

       v: Aurora (Zeller, 1973); Kavkaz (Zeller, 1973). Clement *Yr2* (Johnson, 1992).

Source Stocks

       Many wheats have *Yr9* which is associated with *Lr26* and *Sr31* and often also with *Pm8*. In addition to the publications of Mettin *et al.* (1973) and Zeller (1973) with emphasis on European wheats, further genotypes carrying *Yr9* can be found in Merker (1982), Bartos *et al.* (1983, 1984), Stubbs (1985), Hu and Roelfs (1986), Perwaiz and Johnson (1986), Bartos and Valkoun (1988), Knott (1989), Badebo *et al.* (1990), Singh and Gupta (1991) and Singh and Rajaram (1991). See also McIntosh (1988*a*) and annual supplements to the Gene Symbols Catalogue for Wheat.

       Stubbs and Yang (1988) discussed evidence suggesting that *Yr9* was present in certain triticales. Preliminary work (Plate 4-25) at PBI Cobbitty supports the possibility that *Yr9* is common in Mexican spring triticales.

Use in Agriculture

       Cultivars with this alien chromatin have had a major impact on global wheat production as indicated by its presence in a wide array of winter and spring wheats. This has arisen not only from the multiple disease resistances that are not always durable, but also from widespread adaptability found in wheats carrying the 1RS chromosome (Zeller and Hsam, 1983). German and Australian workers (see Dhaliwal *et al.*, 1988) have shown that the presence of 1RS causes an undesirable dough stickiness that is aggravated by overmixing. The presence of 1RS in wheat can be confirmed by an array of disease responses as well as cytological and biochemical tests (Javornik *et al.*, 1991; May and Wray, 1991; Gupta and Shepherd, 1992).

PLATE 4-9. Yr9

A. and B. Seedling leaves of (L to R): Veery #5, Skorospelka 35, Kavkaz, Federation*4/Kavkaz and Federation; infected with A. pt. 106 E139 A- [*P9*] and B. pt. 234 E139 A- [*p9*]. Skorospelka 35 carries a resistance gene(s) additional to *Yr9*.

C. Seedling leaf pairs of Clement and Federation*4/Kavkaz; infected with cultures of (L to R): IPO pt. 6 E0 [*P9*], IPO pt. 6 E64 [*p9*] and IPO pt. 234 E139 [*p9*]. This shows that although Clement is resistant to pt. 6 E64, the gene conferring resistance is not *Yr9*. Although Johnson (1992) concluded that Clement possessed *Yr9* and *Yr2*, resistance to pt. 6 E64 which has virulence for both genes must be conferred by an additional gene(s). Johnson (1992) suggested that the gene in TP981 which was derived from Heines VII, may be present in Clement.

Yr10 (Macer, 1975) (Plate 4-10)

Chromosome Location

1BS (Metzger and Silbaugh, 1970). *Yr10* is very closely linked (recombination of 0.02) with *Rg1* for red/brown glume colour (Metzger and Silbaugh, 1970).

Low Infection Type

0;.

Environmental Variability

Low.

Origin

Bread wheat. Kema and Lange (1992) reported the presence of *Yr10* in a spelt wheat designated *T. spelta* 415 which has white glumes.

A   B

PLATE 4-10. Yr10

A. Seedling leaves of (L to R): Moro, Harrier*5/*Yr10*, Harrier, M2435*6/*Yr10*, M2435 and Federation; infected with pt. 104 E137 A+. Virulence is currently not available in Australia, but has been reported in other countries.

B. Seedling leaves of Moro infected with IPO pt. 234 E171, IPO pt. 6 E150 and IPO pt. 82 E0. The clearly different resistance phenotypes on the first two leaves may be conferred by *Yr10*, or alternatively the second response could involve a different gene.

PLATE 4-11. Yr11

Paired adult leaves of Joss Cambier and Heines VII infected with (L to R): IPSR pt. 104 E137 (Type 1) [*P11*] and IPSR pt. 104 E137 (Type 2) [*p11*]. The latter pathotype resulted in extensive stripe rust epidemics on Joss Cambier in the UK in the early 1970s. Some uredia of leaf rust appear on the first two leaves and colonies of powdery mildew are present on the third leaf. Courtesy R Johnson.

Pathogenic Variability

Virulent pathotypes increased in frequency soon after the use of this gene in cv. Moro in the Pacific northwest of the USA (Beaver and Powelson, 1969). Stubbs (1985) reported *Yr10*-virulent cultures from eastern Europe, the eastern Mediterranean region (including East Africa) and North America.

Reference Stocks

i: Avocet S*2/*Yr10*; Eradu*5/*Yr10*; Warigal*5/*Yr10*; other Australian backcross derivatives.
v: P.I.178383 (Metzger and Silbaugh, 1970). *T. spelta* 415 (Kema and Lange, 1992).

Source Stocks

Australia: Angas (Wallwork and Rathjen, 1991).
USA: Crest (Line and Qayoum, 1991); Jacman (Line and Qayoum, 1991); and Moro (Line and Qayoum, 1991).

Use in Agriculture

> *Yr10* has been exploited in the USA in cvv. Moro, Crest and Jacman and more recently in Australia in cv. Angas. Virulent pathotypes emerged soon after the use of *Yr10* in the USA (Line and Qayoum, 1991).

Yr11, Yr12, Yr13 and *Yr14*

The designations *Yr11, Yr12, Yr13* and *Yr14* were applied by McIntosh (1988*a*) to describe presumed adult plant resistance genes in European wheats. This nomenclature was based on specific interactions between cultivars and certain UK pathotypes. Apart from *Yr11* (CR Wellings, unpublished 1993) these genes have not been confirmed by formal genetic and cytogenetic analyses. Additional genes for adult plant resistance may be present in some of the stocks. A further group of specificities were tentatively designated *Yr12, Yr13, Yr14* and *Yr15* by Stubbs (1985), but they are different from the genes *Yr11* to *Yr14* listed below (see Miscellaneous Sources of Resistance).

Pathogenic Variability

> Variants with increased pathogenicity were detected in the UK in 1969 on cultivar Joss Cambier (*Yr11*) (Chamberlain *et al.*, 1971), on Hobbit (*Yr14*) in 1972 (Priestley, 1978), and on Maris Huntsman (*Yr13*) in 1974 (Johnson *et al.*, 1975). Characters such as mycelial growth rates and spore yields from seedling leaves (Johnson and Bowyer, 1974), and differential response of adult plants in replicated field nurseries (Johnson and Taylor, 1977) and polythene tunnels (Priestley *et al.*, 1984*a*) were used to confirm these phenomena. Increased rusting of cv. Brock in New Zealand in 1989 was attributed to pathogen virulence for *Yr7* and *Yr14* (Cromey, 1992).

Use in Agriculture

> Non-durable; these genes have not been intentionally manipulated in breeding programs.

Yr11   (Plate 4-11)

Reference Stocks

> **v**: Joss Cambier (Johnson and Taylor, 1972).

Source Stocks

> **v**: Heines VII *Yr2* (R Johnson, unpublished 1994).

Yr12   (Plate 4-12)

Reference Stocks

> **v**: Mega *Yr3 Yr4* (Priestley, 1978).

Source Stocks

> **v**: Beacon (R Johnson, unpublished 1994); Pride (Priestley, 1978). Nord Desprez *Yr3* (R Johnson, unpublished 1994). Armada *Yr3 Yr4* (Priestley *et al.*, 1984*b*).

Yr13   (Plate 4-13)

Reference Stocks

> **v**: Maris Huntsman *Yr2 Yr3 Yr4* (Priestley *et al.*, 1984*a*).

Source Stocks

> Longbow *Yr1 Yr2 Yr6* (Bayles *et al.*, 1986). Virtue *Yr1 Yr3 Yr4* (Taylor *et al.*, 1981). Maris Nimrod *Yr2 Yr3 Yr4* (Taylor *et al.*, 1981). Brigand *Yr2 Yr3 Yr4 Yr14* (Johnson *et al.*, 1984). Kinsman *Yr3 Yr4 Yr6* (Taylor *et al.*, 1981). Further sources are listed in McIntosh (1988*a*).

A

B

C

Plate 4-12. Yr12

Adult leaves infected with (L to R): IPSR pt. 104 E137 (Type 1) [*p12*], IPSR pt. 104 E137 (Type 2) [*P12*] and IPSR pt. 104 E137 (Type 3) [*P12*]. A. cv. Mega; B. cv. Nord Desprez; and C. cv. Beacon. Pt. 104 E137 (Type 1) is virulent for *Yr12* in Mega (A.). The responses of cv. Nord Desprez and Beacon show less contrast between the virulent and avirulent cultures, although the differences are repeatable in field and greenhouse tests (R Johnson, *pers. comm.* 1994). Courtesy R Johnson.

Plate 4-13. Yr13

Paired adult leaves of (L to R): A. Maris Huntsman and Maris Nimrod; B. Brigand and Kinsman. Members of each leaf pair were infected separately with (L to R): IPSR pt. 41 E136 (Type 2) [*P6P13P14*] and IPSR pt. 108 E141 (Type 3) [*p6p13p14*]. The lower responses to the avirulent culture are due to the interaction of *Yr13* and *Yr14* in Brigand (leaf 5) and *Yr6* and *Yr13* in Kinsman (leaf 7). Although IPSR pt. 108 E141 (Type 3) carries pathogenicity for all three genes, additional adult plant resistance may be present in each cultivar. Courtesy R Johnson.

A

B

A   B

PLATE 4-14. Yr14

A. Three adult leaves of Maris Bilbo (L) and one of Joss Cambier (R). B. Three adult leaves of Hobbit (L) and one of Heines VII (R). The leaves in each figure were infected with (L to R): IPSR pt. 104 E137 (Type 1) [*P11P14*], IPSR pt. 104 E137 (Type 2) [*p11P14*] and the last pair with IPSR pt. 104 E137 (Type 3) [*P11p14*]. IPSR pt. 104 E137 (Type 3) is virulent for *Yr14* present in Maris Bilbo and Hobbit, but is avirulent for *Yr11* which is present in both Joss Cambier and Heines VII. Courtesy R Johnson.

Yr14 (Plate 4-14)

Reference Stocks

v: Hobbit *Yr3 Yr4* (Priestley, 1978).

Source Stocks

Rapier *Yr2 Yr3 Yr4* (Bayles *et al.*, 1986). Brigand *Yr2 Yr3 Yr4 Yr13* (Johnson *et al.*, 1984). Maris Bilbo *Yr3 Yr4* (Taylor *et al.*, 1981).
Further sources are listed in McIntosh (1988*a*).

Yr15 (Gerechter-Amitai *et al.*, 1989*a*) (Plate 4-15)

Chromosome Location

1BS (RA McIntosh and J Silk, unpublished 1990). *Yr15* showed close linkage with the centromere and genetic independence of *Yr10*.

Low Infection Type

0;.

Environmental Variability

Low.

Origin

T. turgidum var. dicoccoides (wild emmer) accession G25 (Gerechter-Amitai and Stubbs, 1970; Grama and Gerechter-Amitai, 1974). *Yr15* is known to be present in other collections of wild emmer (Gerechter-Amitai *et al.*, 1989*b*). Resistance was transferred from G25 to cultivated tetraploid and hexaploid wheats (Grama and Gerechter-Amitai, 1974; van Silfhout *et al.*, 1989*a*).

Pathogenic Variability

Lines with *Yr15* are resistant to most isolates of the pathogen (Gerechter-Amitai and Stubbs, 1970; Gerechter-Amitai *et al.*, 1989*b*). Virulent isolates were reported from Afghanistan (van Silfhout, 1989*b*) and other locations (G Kema, *pers. comm.* 1993).

PLATE 4-15. Yr15
Seedling leaves of (L to R): V763.254, resistant and susceptible F2 segregates from Avocet S*6/V763.254, and Avocet S; infected with pt. 110 E143 A+.

Reference Stocks

i: Aroona*3/Yr15; Avocet S*6/Yr15; Spear*3/Yr15.
v: Israeli lines from the families V761, V763, V766, V879 and V882 with the general pedigree G25/durum cv. Nursit 163//2*T. aestivum (A Grama, pers. comm. 1984).
tv: T. turgidum var. dicoccoides sel. G25.

Source Stocks

tv: Several collections of wild emmer (van Silfhout et al., 1989a, 1989b).

Use in Agriculture

Yr15 is not currently used in commercial cultivars, but some lines possessing the gene must be present in breeders populations, particularly private breeding companies in The Netherlands (G Kema, pers. comm. 1993).

Yr16 (Worland and Law, 1986)   (Plate 4-16)

Adult plant resistance.

Chromosome Location

2DL (Worland and Law, 1986).

Low Infection Type

Worland and Law (1986) reported that flag leaves of Cappelle Desprez showed 2.5% area infected. A line with chromosome 2D of Mara replacing the Cappelle Deprez homologue showed 13.5% area infected. The difference was attributed to the presence of Yr16 in Cappelle Desprez.

Environmental Variability

Not known.

Origin

Common wheat.

Pathogenic Variability

Not known.

Reference Stocks

s: Cappelle Desprez and substitution line Cappelle Desprez/Mara 2D differ in the presence and absence, respectively, of Yr16.

PLATE 4-16. Yr16
Adult leaves of Cappelle Desprez and chromosome substitution lines; infected with IPSR pt. 104 E137 (Type 1) (L to R): Capelle Desprez (CD), CD/Bezostaya 2D, CD/Desprez 80 2D, CD/Mara 2D, CD/Poros 2D and CD/Vilmorin 27 2D. The substitution lines from various chromosome 2D sources in CD resulted in a range of disease responses indicating further genetic variation among the substituted 2D chromosomes. Courtesy R Johnson.

Source Stocks

The distribution of *Yr16* is difficult to ascertain in the absence of contrasting pathotypes or appropriate genetic studies.

Use in Agriculture

Yr16 presumably contributes to the oligogenic resistance and possibly to the durable stripe rust resistance of Cappelle Desprez.

Yr17 (Bariana and McIntosh, 1993)   (Plate 4-17)

Chromosome Location

2AS (Bariana and McIntosh, 1993). In wheat *Lr37* is closely linked in coupling with *Sr38* and *Yr17*, and in repulsion with *Lr17* (Bariana and McIntosh, 1993).

Low Infection Type

; C to ;1.

Environmental Variability

Moderate, seedlings with *Yr17* become more susceptible at lower temperatures and at low light intensities (Bariana, 1991; Bariana and McIntosh, 1994).

Origin

T. ventricosum. Resistance was transferred to the hexaploid wheat line VPM1 (*T. ventricosum/ T. turgidum* var. persicum//3*T. aestivum* cv. Marne) by Dr G Doussinault and colleagues in France (see Doussinault *et al.*, 1988). Although the primary aim was to transfer eyespot resistance to bread wheat, resistance to rust diseases was also achieved.

Pathogenic Variability

Most isolates of *P. striiformis* f. sp. *tritici* are avirulent on seedlings of VPM1. Virulence on cvv. Hyak and Madsen was found in the USA (Line *et al.*, 1992).

Reference Stocks

i: Avocet S*6/VPM1; Kulin*5/VPM1; Spear*5/VPM1. Bindawarra*4/ VPM1 Yr6. Thatcher*8/VPM1, R.L.6081 Yr7 (Bariana and McIntosh, 1993).
v: Hyak (Allan *et al.*, 1990); Madsen (Allan *et al.*, 1991); Rendezvous (Bariana and McIntosh, 1993); Sunbri (Brown *et al.*, 1991); VPM1 (Bariana and McIntosh, 1993).

Source Stocks

Cv. Trident;
Australian backcross derivatives additional to those listed above.

A   B

PLATE 4-17. Yr17

Seedling leaves of (L to R): A. VPM1, Rendezvous, Spear*4/VPM1, Spear, Cook*3/VPM1 and Cook, and B. VPM1, Rendezvous, VPM1*4/Hartog, Hartog, Thatcher*8/VPM1 and Thatcher; infected with pt. 110 E143 A+. The expression of resistance conferred by *Yr17* is significantly influenced by genetic background (see Thatcher, for example) and by temperature and light (Bariana, 1991; Bariana and McIntosh, 1994).

Use in Agriculture

Although interest in VPM1 was primarily directed at the eyespot, (caused by *Pseudocercosporella herpotrichoides*) resistance located in chromosome 7D, additional selection by both European and USA breeders resulted in rust resistance. In Australia, selection was only for rust resistance.

Yr18 (Singh, 1992*a*)   (Plate 4-18)

Adult plant resistance

Chromosome Location

7D, probably 7DL (see *Lr34*). *Yr18* is possibly completely linked with *Lr34* (McIntosh, 1992*a*; Singh, 1992*a*) and also with *Ltn* which controls a distinctive leaf tip necrosis (Dyck, 1991; Singh, 1992*b*).

Low Infection Type

Mature plants of wheats with this gene are distinctly more resistant than closely related counterparts not possessing it. The infection type is 'resistant' with up to 20% flag leaf area affected by symptoms.

Environmental Variability

Singh (1992*a*) mentioned that wheats with *Yr18* displayed inadequate resistance in some locations in Ecuador and Kenya.

Origin

Common wheat. The history of *Yr18* presumably is identical to that of *Lr34*. Many pedigrees trace back to South American wheats, but a Chinese origin is possible (Dyck, 1991). It is of interest that the development of Frontana in South America followed a damaging stripe rust epidemic in the early 1940s.

Pathogenic Variability

Not known.

A

B

C

D

PLATE 4-18. Yr18

A. Flag leaves of field-grown plants of Thatcher and derivatives; infected with stripe rust and leaf rust. (L to R): R.L.6050, R.L.6058, Line 896 and Thatcher. The first three lines were generated in Canada by selection only for resistance to leaf rust: R.L.6050 and R.L.6058 carry *Lr34*, but Line 896 does not. Significant levels of both stripe rust and leaf rust are evident on the last two lines.

B. Flag leaves from field-grown plants at Toluca, Mexico, showing leaf tip necrosis (first three leaves) which is associated with the presence of *Lr34* and *Yr18*. The fourth leaf is from a cultivar not possessing these genes and is unaffected by necrosis. Courtesy RP Singh.

C. and D. Adjacent hill plots of Jupeteco 73R and Jupeteco 73S, at Toluca, Mexico. Most of the visible disease symptoms on both wheats were caused by stripe rust although both stripe rust and leaf rust were present in the nursery. The presence of *Yr18* In Jupeteco 73R reduced the leaf area affected from almost 100% to about 20%. Courtesy RP Singh.

Reference Stocks

i: Thatcher lines with *Lr34* (see *Lr34*) (McIntosh, 1992a; Singh, 1992a). The selections, Jupeteco 73R and Jupeteco 73S (Singh, 1992a) can be considered a near-isogenic pair.

Source Stocks

Wheats with *Lr34* (see *Lr34*).

Use in Agriculture

Yr18 has been widely deployed in spring wheats and at least some winter wheats either by its association with leaf rust resistance or by its phenotypic effect on stripe rust response. Anza = WW15 = Karamu = T4, considered to have durable stripe rust resistance (see Singh, 1992a), was widely used in wheat breeding in Australia and at least some of its derivatives, such as Condor, apparently carry *Yr18* (Singh, 1992a). Selections of another derivative, Avocet, possess no adult plant resistance and have been adopted as susceptible standards for genetic studies in Australia where they repeatedly display responses of 100S in the field. The original cultivar, Avocet, was heterogeneous for the seedling resistance tentatively designated *YrA* (probably complementary genes, Wellings *et al.*, 1988) and also for adult plant response to pathotypes virulent on seedlings with *YrA*.

TEMPORARILY DESIGNATED AND MISCELLANEOUS STRIPE RUST RESISTANCE GENES

There are several partially characterised resistance genes which have not been adequately documented to justify permanent designations. These genes or gene descriptions are arranged under temporary designations, resistance genes in the Johnson *et al.* (1972) differentials, stripe rust resistance genes in triticale and miscellaneous sources of resistance.

Temporary Designations

YrA (Plate 4-19)

The first pathogenic change in *P. striiformis tritici* following its introduction to Australia in 1979 involved increased virulence on Avocet and certain other derivatives of WW15 = Anza (Wellings *et al.*, 1988). Avocet was the worst affected cultivar because it possessed the lowest level of adult plant resistance. Subsequent research indicated that the Avocet cultivar was heterogeneous in respect of a seedling resistance designated *YrA* (Wellings *et al.*, 1988) as well as for adult plant response. Selections with and without *YrA* and with high degrees of adult plant susceptibility to *YrA*-virulent pathotypes (designated A+) were selected as susceptible standards for genetic investigations (RA McIntosh and CR Wellings, unpublished 1985). A monosomic series is being produced for the *YrAYrA* selection (Avocet R).

Chromosome Location

Genetic studies indicated that the *YrA* resistance is conferred by complementary dominant genes (Wellings, 1986), one of which is located in chromosome 3D. The second gene has not been located.

Low Infection Type

;CN1 to 2+. The lower IT on the second leaf, compared to the primary leaf, is a useful feature in distinguishing *YrA* in genetic studies (Wellings *et al.*, 1988).

Environmental Variability

Wellings *et al.* (1988) showed that high light intensity is essential for the expression of resistance in seedling tests.

Origin

The *YrA* specificity occurs alone or in combination with other genes in a range of Australian wheats that are derived from WW15. In certain instances these cultivars are heterogeneous and selections with and without *YrA* were made (Wellings *et al.*, 1988).

G   H

PLATE 4-19. YrA

Seedling leaves of (L to R): WW80, WW15, Oxley, Condor R, CSP44 and Condor S; infected with A. pt. 104 E137 A-, B. pt. 108 E141 A-, C. pt. 104 E137 A+, D. pt. 108 E141 A+. WW80 and WW15 were the parents of the other four wheats. Condor R and Condor S are resistant and susceptible selections of Condor selected with pt. 104 E137 A-; CSP44 was a selection from Condor with superior stem rust resistance. The comparison of A and B shows that WW80 was the source of *Yr6* in Oxley and CSP44, and WW15 contributed *YrA* to Condor R. The presence of these two resistance genes in cultivars with identical pedigree was due to chance fixation before the occurrence of wheat stripe rust in Australia.

E. and F. Seedling leaves of (L to R): Banks R, Condor R, Egret R, Sunstar, Avocet R and Federation; infected with pt. 108 E141 A- and pt. 108 E141 A+, respectively.

G. and H. The expression of resistance conferred by the complementary genes comprising the *YrA* response is significantly influenced by light intensity (Wellings *et al.*, 1988). Seedling leaves of (L to R): Avocet R, WW15, Nainari 60, Inia 66, Sonalika and Federation; infected with pt. 108 E141 A- and incubated at high and low light intensities. The low light intensity was conferred by covering plants with 50% shade cloth.

Pathogenic Variability

Stubbs *et al.* (1974) state that Inia 66 (*YrA*) and Noroesti 66 (*Yr6*) were severely attacked by stripe rust in co-ordinated trials in the USA, Chile, Ecuador, Kenya, Iraq and Tunisia. Other evidence suggests that virulence for *YrA* was absent in Europe in the early 1960s, although perhaps present in Kenya during the same period and now noted to occur at low frequency in some west European pathotypes (Stubbs and Fuchs, 1992).

Reference Stocks

v: Anza = Karamu = WW15 = T4 (Wellings *et al.*, 1988); Avocet (heterogeneous) (Wellings *et al.*, 1988); Banks (heterogeneous) (Wellings *et al.*, 1988); Condor (heterogeneous) (Wellings *et al.*, 1988); Egret (heterogeneous) (Wellings *et al.*, 1988).

Source Stocks

Funo; Inia 66; Lerma Rojo 64; Sonalika; Sunstar. Cajeme 71 *Yr6*; Condor SP44 *Yr6*; Pari 73 *Yr6*; Yecora 70 *Yr6*. See Wellings *et al.* (1988). Perwaiz and Johnson (1986) and R Johnson (*pers. comm.* 1980) indicated that several Australian wheats with *YrA* and/or *Yr6* also carry *Yr2*. These genes apparently occur at relatively high frequencies in Mexican wheats.

Yr Selkirk   (Plate 4-20)

Selkirk, Webster and a number of Mexican wheats, including Opata 85, Crow S, Buckbuck S and Ciano 79, display a distinctive resistant seedling reaction that is characterised by a chlorotic response with small uredinia that are often brown or black (Wellings, 1992). A single gene which appears to be involved (Wellings, 1992) is effective under Australian field conditions. The gene may be derived from McMurachy (Wellings, 1992), a parent of Selkirk. Webster was included as a differential by Gassner and Straib (1932) but was later dropped because of variability. It is included here with Selkirk on the basis of seedling response only (Wellings, 1992). No genetic study on the relationship of Webster and Selkirk has been undertaken or reported.

A   B

PLATE 4-20. *Yr Selkirk*
Seedling leaves of (L to R): Selkirk, McMurachy, Webster, Opata 85, Ciano 79 and Federation; infected with A. pt. 106 E139 A- [culture 821589] and B. pt. 106 E139 A- [911582]. These avirulent and virulent cultures have the same pathotype designation because this specificity is not assessed using the European and international differential sets.

Pathogenic Variability

A culture with virulence on Selkirk seedlings was detected in the Australian pathotype survey (Wellings and Burdon, 1992). Stubbs *et al.* (1974) state that Selkirk has a specific adult plant resistance and that virulence occurred in the Near East, East Africa and Indian subcontinent. Highly avirulent cultures from the Netherlands, collected on weedy *Agropyron repens*, were noted to be virulent on Selkirk and Redman (Stubbs and Fuchs, 1992). The relationship between the wheat and *Agropyron* cultures remains unclear. Redman does not carry the Selkirk factor described here.

Reference Stocks

v: Opata 85 (Wellings, 1992); Selkirk (Wellings, 1992).

Source Stocks

Bluejay S; Buckbuck S; Ciano 79; Crow S; McMurachy; Webster. See Wellings (1992).

Resistance Genes in the Johnson *et al.* (1972) Differentials

Yr Strubes Dickkopf   (Plate 4-21)

Strubes Dickkopf is the sixth differential in the World set and may have derived its resistance from cv. Blé Rouge d'Ecosse which was used as a tester in the early work of Gassner and Straib (Stubbs and Fuchs, 1992). Johnson (1992) concluded that Strubbes Dickkopf, Heines VII and Heines Peko share a gene(s) which is present in line TP981. Virulence has been detected in Europe for many years. Avirulence is more common in eastern and south-eastern Europe, the Middle East, South Asia and Africa (Stubbs *et al.*, 1974). Several French wheats, for example Castan and Festival, were postulated to carry a resistance similar or identical with Strubes Dickkopf (de Vallavieille-Pope *et al.*, 1990).

PLATE 4-21. *Yr Strubes Dickkopf*

Seedling leaves of Strubes Dickkopf; infected with (L to R): IPO pt. 82 E0, IPO pt. 0 E0 , IPO pt. 32 E0, IPO pt. 40 E0, IPO pt. 234 E139 and IPO pt. 40 E8. From the pathotype designations it is evident that the first two responses were considered avirulent and the last four, virulent.

PLATE 4-22. *Yr Suwon 92/Omar*

Seedling leaves of Suwon 92/Omar; infected with (L to R): IPO pt. 32 E0, IPO pt. 40 E8, IPO pt. 40 E0, IPO pt. 6 E0, IPO pt. 82 E0 and pt. 104 E41. The last two cultures would be considered virulent unless there was consistent genetic or host:pathogen interaction data to decide otherwise. All Australian cultures are virulent on seedlings of Suwon 92/Omar.

Yr Suwon 92/Omar   (Plate 4-22)

Suwon 92/Omar (Suwon 92/3*Omar = Selection 63301) is the seventh differential in the World set and was reported to carry a single dominant resistance factor in North American studies (Allan and Purdy, 1970). The Chinese differential Suwon 11 showed an almost identical response to Suwon 92/3*Omar in the studies of Yang and Stubbs (1990). The resistance of Suwon 92/Omar was highly effective in the USA until 1966 when virulence was first detected (Purdy and Allan, 1966). In most other areas Suwon 92/Omar acts as a differential, or is uniformly susceptible. In the survey of Stubbs *et al.* (1974) the frequency of avirulence was highest in southern and eastern Europe. Some evidence suggests that Suwon 92/Omar carries a factor identical to Hybrid 46, that is *Yr4* (Stubbs and Fuchs, 1992). In the USA a closely related line, Paha (= Suwon 92/*4 Omar), is used as the differential (Line and Qayoum, 1991). CR Wellings (unpublished, 1990) noted the identical responses of Paha and Suwon 92/Omar to 10 pathotypes held in the IPO culture collection. Chen and Line (1992b) found that Paha possessed three genes for resistance to CDL-21 and two genes for resistance to other pathotypes.

Yr Nord Desprez   (Plate 4-3)

Nord Desprez is the fourth differential of the European set. Nord Desprez (ND) shares *Yr3* with Cappelle Desprez (CD) *Yr4* (Bayles and Thomas, 1984) and Vilmorin 23 (de Vallavieille-Pope *et al.*, 1990). Stubbs and Fuchs (1992) allocated the genotype *Yr3+* to ND and Vilmorin 23. However, the unknown '+' in each cultivar may not involve the same gene(s). Nord Desprez is susceptible to North American pt. CDL 20 whereas CD is resistant (Chen and Line, 1992b). This indicates that either CD must carry a third gene which is not present in ND or that both cultivars do not carry the same two genes. Chen and Line (1992b) reported that CD, ND and

A

B

PLATE 4-23. *Yr Carstens V*

A. Seedling leaves of (L to R): Cook*3/Carstens V, Cook, Carstens V and Federation; infected with pt. 104 E137 A+. The gene from Carstens V is more effective in the Cook background. Cook frequently shows a degree of seedling resistance relative to controls such as Federation.

B. Seedling leaves of Carstens V; infected with (L to R): IPO pt. 6 E0, IPO pt. 40 E8, IPO pt. 104 E9, IPO pt. 234 E139 and IPO pt. 104 E41. It is difficult to decide if the fourth culture should be considered avirulent or virulent and if each low response is conferred by the same gene.

Vilmorin 23 each carry two genes with each carrying common and different genes. The common gene probably occurred in the North American wheats, Druchamp, Yamhill and Stephens. Johnson *et al.* (1993) provided evidence with UK pathogen cultures that resistance in Yamhill was not related to *Yr3* in CD, ND and Vilmorin 23.

In order to focus on the genes that differentiate between Vilmorin 23, CD and ND it is essential that genetic analyses be firstly undertaken using the critical differential cultures. As a second step, lines segregating for single genes should be screened with other cultures to establish single gene derivatives suitable for collaborative international screening.

According to Stubbs *et al.* (1974) the frequency of avirulence on ND was low in the Middle East, Asia and North Africa. Australian pathotypes are virulent on seedlings of ND, CD and Vilmorin 23.

Yr Carstens V   (Plate 4-23)

Carstens V, the sixth differential in the European supplemental set, appears to have a single gene for resistance to Australian cultures of *P. striiformis* (CR Wellings and RA McIntosh, unpublished 1991). In contrast, Chen and Line (1993*b*) detected up to three genes for resistance to North American cultures. According to Stubbs (1985), who used the provisional designation '12' for Carstens V, the same pattern of response was evident for the resistances of Anouska (plus *Yr2* or *Yr3*), Caribo, Carstens VI, Cyrano, Felix (plus *Yr3*) and Okapi. Virulence on Carstens V is common in western and central Europe and South America, but uncommon in southern Europe and southeast Asia and Australasia (Stubbs, 1985).

Reference Stocks

       i: Avocet S*4/Carstens V; Cook*6/Carstens V.
       v: Carstens V (Johnson *et al.*, 1972).

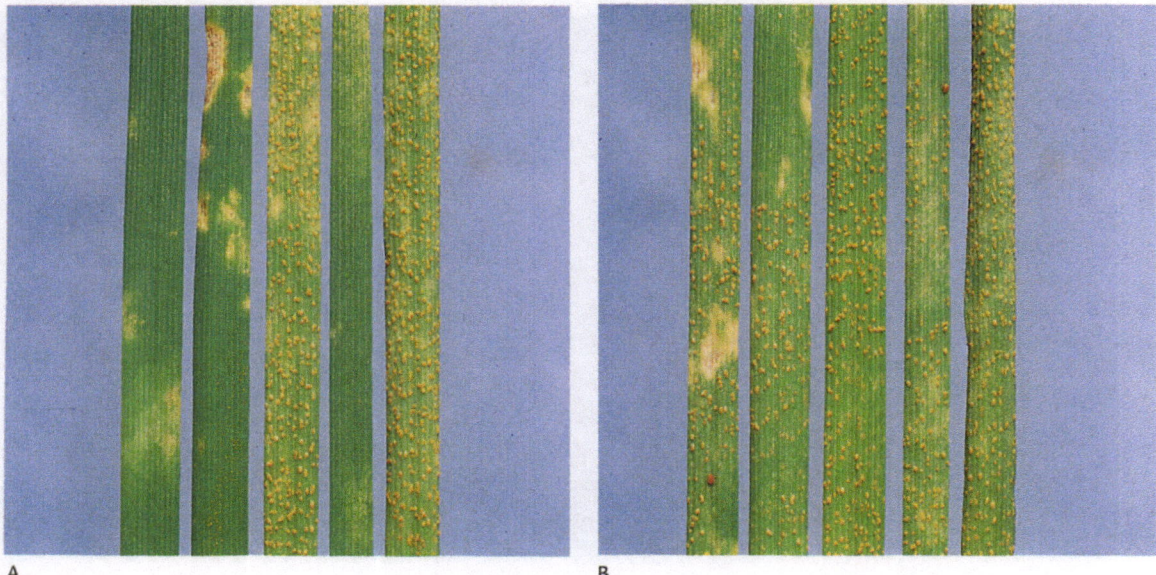

A   B

PLATE 4-24. *Yr Spaldings Prolific*

Seedling leaves of (L to R): Spaldings Prolific, Kite*2/Spaldings Prolific, Kite, Avocet S*2/Spaldings Prolific and Avocet S; infected with
A. pt. 108 E141 A+ (avirulent) and B. pt. 108 E205 A+ (virulent). Spaldings Prolific may have additional resistance to the virulent culture.

Yr Spaldings Prolific   (Plate 4-24)

Spaldings Prolific is the seventh differential in the European set adopted by Johnson *et al.* (1972).
The resistance appears to be inherited as a single dominant gene (CR Wellings, unpublished
1991) and could be related to the resistance in the English cultivar Clovers Red (Stubbs and
Fuchs, 1992). Abnormal plants in F3 families from Avocet*2/ Spaldfings Prolific (CR Wellings,
unpublished 1993) suggest the presence of a reciprocal translocation. R Johnson (*pers. comm.*
1993) indicated that a 5BS.7BS, 5BL.7BL reciprocal translocation is very common among old
UK winter wheats. Pathotypes virulent on Spaldings Prolific have been reported in most
geographic areas (Stubbs *et al.*, 1974).

Reference Stocks
i: Avocet S*2/Spaldings Prolific; Cook*3/Spaldings Prolific; Kite*2/Spaldings Prolific.
v: Spaldings Prolific (Johnson *et al.*, 1972).

Stripe Rust Resistance Genes in Triticale   (Plate 4-25)

Stubbs and Yang (1988) described the dilemmas associated with rust samples collected from
triticale and wheats with the 1BL.1RS translocation carrying *Yr9*. While the pathotypes involved
were identified as 6 E0 (Ecuador), 6 E150 (Rwanda), 140 E12 (Zambia) and 134 E150 (Kenya),
the first three were not identified as virulent on seedlings with *Yr9* until a suitable differential
such as Federation*4/Kavkaz was used. Both Clement and Riebesel 47/51 carried resistance genes
additional to *Yr9*. By contrast, Netherlands race 234 E171, which developed on Granada wheat,
and Chinese race 191 E206 from Lovrin 13, were virulent on Clement. Under field conditions
triticale cultivars Mapache and Rosner were resistant to Dutch isolates avirulent for *Yr9* but
susceptible to isolates possessing virulence for *Yr9* (Stubbs and Yang, 1988).

Further comparisons of seedling responses of triticale cultivars in Australia using an Australian
pathotype and a *Yr9*-virulent pathotype from New Zealand confirmed the likelihood that
Yr9 occurs at high frequency in Mexican triticales. In wheat, the presence of *Yr9* is often
confirmed by demonstrating the concurrent presence of *Lr26* and *Sr31*. Such tests on Mexican
triticales are not valid because neither *Lr26* nor *Sr31* appear present. It appears likely that these
genes can freely recombine in the rye background compared to their behaviour as a linkage block
in wheat where recombination is normally not possible.

A   B

PLATE 4-25. *Stripe rust resistance genes in triticale*

Seedling leaves of triticales (L to R): Bejon = Tejon/Beagle, Satu, Venus, Coorong and NLD Sus., a highly susceptible selection from the field; infected with A. pt. 104 E137 A+ [*P9*] and B. pt. 234 E139A- [*p9*]. The distinctive responses of the first four triticales with the two cultures provides strong evidence for the presence of *Yr9*. Their different responses with the second culture indicates the presence of a second gene in Bejon and possibly a second gene in Venus and Satu. The origin of the second culture is independent of *Yr9*-virulent cultures collected from triticales in Africa and South America (Stubbs and Yang, 1988).

Miscellaneous Sources of Resistance

Lemhi

An otherwise very susceptible wheat, Lemhi possesses resistance to North American pathotype CDL21 (de Vallavieille-Pope and Line, 1990; Chen and Line, 1992*a*). This pathotype (equivalent to pt. 1 E0) was virulent only on Chinese 166 (*Yr1*) of the North American, World and European differentials. Stubbs (1985) noted that several pathotypes of *P. striiformis* f. sp. *hordei* were virulent for *Yr1*, and hence CDL21 may be unrelated to other components of the USA wheat stripe rust pathogen population. The available data fail to reveal if the gene in Lemhi is extremely common in these wheats or if the various genes in the differentials are individually effective against pt. CDL21. However, Chen and Line (1992*b*) reported that crosses of Lemhi with Druchamp (two genes), Stephens (two genes) and Riebesel 47/51 (*Yr9*) did not segregate when tested with pt. CDL21 indicating the possibility of common genes or linkage. No illustration is available.

Hexaploid Wheats

Seedling Genes. Genetic studies in Europe and North America have indicated the presence of some less characterised genes. Whereas many of these genes may not be important on a local basis [e.g. the evolving *P. striiformis tritici* population in Australia was initially virulent on seedlings of Lemhi, Vilmorin 23 (*Yr3*), Hybrid 46 (*Yr4*), Strubes Dickkopf, Suwon 92/Omar, Nord Desprez and Heines VII], a wider knowledge base is necessary for a thorough understanding of global variability.

Stubbs (1988) further described the shortcomings of the current World and European differentials in detecting virulence for *Yr9*. South American cultures identified as races 134 E0 and 6 E0 were avirulent on the *Yr9* differential, Clement, but were virulent on Federation*4/Kavkaz, which carries *Yr9*. This demonstrated firstly the presence of an additional gene(s) in Clement and secondly, the potential lack of critical information if surveys based on those differentials were to be used as early warning systems for cultivar recommendation and deployment.

According to Stubbs (1985) cv. Alba carries a non-catalogued gene for seedling resistance. Chen and Line (1992*b*) presented evidence for the presence of additional genes in many of the World, European and North American testers. Unfortunately, the interrelationships of these genes remain poorly established. The lack of this knowledge, together with the significant influence of temperature, nutritional and light conditions on the data actually obtained indicate that further work is necessary.

Adult Plant Resistances. In addition to the gene in Carstens V, Stubbs (1985) refers to three distinctive specificities for adult plant resistance. The specificities and wheats typifying them were provisionally designated as follows.

13: Arminda, Flamingo (plus *Yr2*, *Yr6*), Ibis (*Yr1 Yr2*), Sylvia and Tadorna (*Yr1 Yr2*).
14: Alba, Apollo, Falco, Lely (*Yr2*, *Yr7*), Manella (*Yr2*), Nautica (*Yr2 Yr9*) and Staring (*Yr3*).
15: Dippes Triumph.

It is presumed that these are different from the genes *Yr11–Yr14* described by UK workers (Johnson, 1992) as well as *Yr16* and *Yr18*.

Tetraploid Wheats   (Plate 4-26)

Studies in Australia (CR Wellings and RA Hare, unpublished 1989) and Europe (Chilosi and Johnson, 1990) showed that a number of durum wheats, especially those from Mexico, carry *Yr6*. It is not clear if *Yr6* is present in older durum populations or whether it was introgressed into durum during the post 'Green Revolution' period when wide crossing practices were more commonly used. Line and Qayoum (1991) postulated that Produra, a durum used as a differential in the USA Pacific Northwest, possessed a gene in common with Fielder which carries *Yr6* (Chen and Line, 1992*b*; CR Wellings, unpublished 1989). Chen and Line (1992*a*) detected two genes in Produra. They suggested the second gene in Fielder may be the same as a second gene in Heines Kolben. That gene was different from the second gene in Produra. However, resistance genes other than *Yr6* are known to be present in HD4530 (see Plate 4-26; CR Wellings, unpublished 1989) and in K733 which was used by R Appels and E Lagudah (CSIRO Division of Plant Industry, Canberra, Australia) as a durum parent for the construction of synthetic hexaploid wheats. One synthetic involving K733 was backcrossed to Meering. A single gene for resistance was located in chromosome 1B (RA McIntosh, unpublished 1993).

A   B   C

PLATE 4-26. *Tetraploid wheats*

Seedling leaves of (L to R): durums HD4530 R, PBW34, hexaploids RAH56430 [Oxley (6x) /820911 durum (4x)], RAH56431 [Oxley/ Steinwedel//Guillemot Sib (4x)] and Oxley; infected with A. pt. 104 E137 A+ and B. pt. 108 E141 A+. The pathotype virulent on *Yr6* (Oxley) is clearly virulent on PBW34 and the hexaploid derivatives whereas HD4530 R possesses a different or additional gene.

C. Seedling leaves of (L to R): Meering*3//K733/*T. tauschii* Selections 8 and 17, Meering, Lee (*Yr7*) and Oxley (*Yr6*); infected with pt. 110 E143 A+. The Meering derivatives carry an unknown resistance gene derived from durum accession K733.

REFERENCES

Abdel Hak TM, El-Sherif NA, Bassiouny AA, Shafik II and El-Dauadi Y 1980. Control of wheat leaf rust by systemic fungicides. Proceedings of the Fifth European and Mediterranean Cereal Rusts Conference, Bari, Italy, 255–266.

Acosta AC 1962. The transfer of stem rust resistance from rye to wheat. *Dissertation Abstracts* **23**, 34–35.

Allan RE and Purdy LH 1970. Reaction of F2 seedlings of several crosses of susceptible and resistant wheat selections to *Puccinia striiformis*. *Phytopathology* **60**, 1368–1372.

Allan RE, Peterson CJ Jr, Rubenthaler GL, Line RF and Roberts DE 1989. Registration of 'Madsen' wheat. *Crop Science* **29**, 1575–1576.

Allan RE, Peterson CJ Jr, Rubenthaler GL, Line RF and Roberts DE 1990. Registration of 'Hyak' wheat. *Crop Science* **30**, 234.

Allard RW and Shands RG 1954. Inheritance of resistance to stem rust and powdery mildew in cytologically stable spring wheats derived from *Triticum timopheevi*. *Phytopathology* **44**, 266–274.

Anderson MK, Williams ND and Maan SS 1971. Monosomic analyses of genes for resistance derived from Marquis and Reliance wheat. *Crop Science* **11**, 556–558.

Anderson RG 1961. The inheritance of leaf rust resistance in seven varieties of common wheat. *Canadian Journal of Plant Science* **41**, 342–359.

Athwal DS and Watson IA 1954. Inheritance and the genetic relationship of resistance possessed by two Kenya wheats to races of *Puccinia graminis tritici*. *Proceedings of the Linnean Society of New South Wales* **79**, 1–14.

Athwal DS and Watson IA 1957. Inheritance studies with certain leaf rust resistant varieties of *Triticum vulgare* Vill. *Proceedings of the Linnean Society of New South Wales* **82**, 272–284.

Atkins RE 1967. Inheritance of reaction to race 15B of wheat stem rust in derivatives of Frontana and *Triticum timopheevii* Zhuk. *Iowa State Journal of Science* **41**, 305–311.

Ausemus ER, Harrington JB, Reitz LP and Worzella WW 1946. A summary of genetic studies in hexaploid and tetraploid wheats. *Journal of the American Society of Agronomy* **38**, 1083–1099.

Badebo A, Stubbs RW, van Ginkel M and Gebeyehu G 1990. Identification of resistance genes to *Puccinia striiformis* in seedlings of Ethiopian and CIMMYT bread wheat varieties and lines. *Netherlands Journal of Plant Pathology* **96**, 199–210.

Baker EP and McIntosh RA 1973. Utilization of marked telocentric chromosomes in more efficient genetic analysis. *In* 'Proceedings of the Fourth International Wheat Genetics Symposium'. (Eds ER Sears and LMS Sears.) pp. 635–636. (Agricultural Experiment Station, University of Missouri: Columbia, USA.)

Baker EP, Sanghi AK, McIntosh RA and Luig NH 1970. Cytogenetical studies in wheat III. Studies of a gene conditioning resistance to stem rust strains with unusual genes for avirulence. *Australian Journal of Biological Sciences* **23**, 369–375.

Ballantyne B, Murray GM and Brennan JP 1994. Assessing the threat to resistant cultivars from public risk diseases and pests. *Euphytica* **72**, 51–59.

Bariana HS 1991. Genetic Studies on Stripe Rust Resistance in Wheat. PhD Thesis, University of Sydney.

Bariana HS and McIntosh RA 1993. Cytogenetic studies in wheat XIV. Location of rust resistance genes in VPM1 and their genetic linkage with other disease resistance genes in chromosome 2A. *Genome* **36**, 476–482.

Bariana HS and McIntosh RA 1994. Characterisation and origin of rust and powdery mildew resistance genes in VPM1 wheat. *Euphytica* **76**, 53–61.

Bartos P and Stuchlikova E 1986. Genes for rust resistance. *Annual Wheat Newsletter* **32**, 65–66.

Bartos P and Valkoun J 1988. Rust resistance genes in Czechoslovak wheats. *Cereal Rusts and Powdery Mildews Bulletin* **16**, 36–40.

Bartos P, Dyck PL and Samborski DJ 1969a . Adult-plant leaf rust resistance in Thatcher and Marquis wheat, a genetic analysis of the host–parasite interaction. *Canadian Journal of Botany* **47**, 267–269.

Bartos P, Samborski DJ and Dyck PL 1969b. Leaf rust resistance of some European varieties of wheat. *Canadian Journal of Botany* **47**, 543–546.

Bartos P, Tersova R and Slovencikova V 1983. Genetics of rust resistance in Czechoslovak wheat cultivars. *Tagungsbericht, Akademie der Landwirtschaftswissenschaften der Deutschen Demokratischen Republik* **216**, 555–560.

Bartos P, Stuchlikova E and Kubova R 1984. Wheat leaf rust epidemics in Czechoslovakia in 1983. *Cereal Rusts Bulletin* **12**, 40–41.

Bayles RA and Thomas JE 1984. Yellow rust of wheat. United Kingdom Cereal Pathogen Virulence Survey. *1983 Annual Report*, 23–31.

Bayles RA, Thomas JE, Parry DW and Herron CM 1986. Yellow rust of wheat. United Kingdom Cereal Pathogen Virulence Survey. *1985 Annual Report*, 13–18.

Bayles RA, Channell MH and Stigwood PL 1989. Yellow rust of wheat. United Kingdom Cereal Pathogen Virulence Survey. *1988 Annual Report*, 11–15.

Bayles RA, Channell MH and Stigwood PL 1990. Yellow rust of wheat. United Kingdom Cereal Pathogen Virulence Survey. *1989 Annual Report*, 11–17.

Beaver RG and Powelson RL 1969. A new race of stripe rust pathogenic on the wheat variety Moro, C.I.13740. *Plant Disease Reporter* **53**, 91–93.

Becker D, Brettschneider R and Lörz H 1994. Fertile transgenic wheat from microprojectile bombardment of scutellar tissue. *The Plant Journal* **5**, 299–307.

Berg LA, Gough FJ and Williams ND 1963. Inheritance of stem rust resistance in two wheat varieties, Marquis and Kota. *Phytopathology* **53**, 904–908.

Bhardwaj SC, Nayar SK, Prashar M, Kumar J, Menon MK and Singh SB 1990. A pathotype of *Puccinia graminis* f. sp. *tritici* on *Sr24* in India. *Cereal Rusts and Powdery Mildews Bulletin* **18**, 35–38.

Brennan JP and Murray GM 1988. Australian wheat diseases — assessing their economic importance. *Agricultural Science New Series* **1**, 26–35.

Brennan PS 1983. Hartog. *Journal of the Australian Institute of Agricultural Science* **49**, 42.

Brennan PS, Martin DJ, The, TT and McIntosh RA 1983. Torres. *Journal of the Australian Institute of Agricultural Science* **49**, 47.

Briggle LW and Sears ER 1966. Linkage of resistance to *Erysiphe graminis* f. sp. *tritici* (*Pm3*) and hairy glume (*Hg*) on chromosome 1A of wheat. *Crop Science* **6**, 559–561.

Broers LHM 1989. Partial resistance to wheat leaf rust in 18 spring wheat cultivars. *Euphytica* **44**, 247–258.

Browder LE 1971. Pathogenic specialization in cereal rust fungi, especially *Puccinia recondita* f. sp. *tritici*, concepts, methods of study, and application. Agricultural Research Service Technical Bulletin No. 1432. (United States Department of Agriculture: Washington, DC.)

Browder LE 1972. Designation of two genes for resistance to *Puccinia recondita* in *Triticum aestivum*. *Crop Science* **12**, 705–706.

Browder LE 1973a. Specificity of the *Puccinia recondita* f. sp. *tritici*, *Triticum aestivum* 'Bulgaria 88' relationship. *Phytopathology* **63**, 524–528.

Browder LE 1973b. Probable genotypes of some *Triticum aestivum* 'Agent' derivatives for reaction to *Puccinia recondita* f. sp. *tritici*. *Crop Science* **13**, 203–206.

Browder LE 1980. A compendium of information about named genes for low reaction to *Puccinia recondita* in wheat. *Crop Science* **20**, 775–779.

Browder LE and Eversmeyer MG 1986. Interactions of temperature and time with some *Puccinia recondita*, *Triticum* corresponding gene pairs. *Phytopathology* **76**, 1286–1288.

Browder LE and Young HC Jr 1975. Further development of an infection type coding system for cereal rusts. *Plant Disease Reporter* **59**, 964–965.

Brown GN 1993. A seedling marker for gene *Sr2* in wheat. Proceedings of the Tenth Australian Plant Breeding Conference Volume 2. (Eds BC Imrie and JB Hacker.) pp. 139–140. (Conference Organising Committee: Gold Coast, Australia.)

Brown GN, Ellison FW and Mares DJ 1991. *Triticum aestivum* spp. *vulgare* (bread wheat) cv. Sunbri. *Australian Journal of Experimental Agriculture* **31**, 734.

Bushnell WR and Roelfs AP (Eds) 1984. 'The Cereal Rusts. Vol. I. Origins, Specificity, Structure and Physiology.' (Academic Press: Orlando, FL, USA.)

Caldwell RM 1968. Breeding of general and/or specific plant disease resistance. *In* 'Proceedings of the Third International Wheat Genetics Symposium'. (Eds KW Findlay and KW Shepherd.) pp. 263–272. (Australian Academy of Science: Canberra.)

Caldwell RM, Roberts JJ and Eyal Z 1970. General resistance ('slow rusting') in *Puccinia recondita* f. sp. *tritici* in winter and spring wheats. *Phytopathology* **60**, 1287 (Abstract).

Campbell AB and Czarnecki EM 1981. Benito hard red spring wheat. *Canadian Journal of Plant Science* **61**, 145–146.

Carefoot GL and Sprott ER 1967. 'Famine on the Wind, Man's Battle Against Plant Disease.' (Rand McNally and Co.: USA.)

Cauderon Y, Saigne B and Dauge M 1973. The resistance to wheat rusts of *Agropyron intermedium* and its use in wheat improvement. *In* 'Proceedings of the Fourth International Wheat Genetics Symposium'. (Eds ER Sears and LMS Sears.) pp. 401–407. (Agricultural Experiment Station, University of Missouri: Columbia, USA.)

Chamberlain NH, Doodson JK and Johnson R 1971. The occurrence of two new physiologic races of *Puccinia striiformis* Westend. in Britain. *Plant Pathology* **20**, 92–95.

Chen X and Line RF 1992a. Inheritance of stripe rust resistance in wheat cultivars used to differentiate races of *Puccinia striiformis* in North America. *Phytopathology* **82**, 633–637.

Chen X and Line RF 1992b. Identification of stripe rust resistance genes in wheat genotypes used to differentiate North American races of *Puccinia striiformis*. *Phytopathology* **82**, 1428–1434.

Chen X and Line RF 1993a. Inheritance of stripe rust resistance in wheat cultivars postulated to have resistance genes at *Yr3* and *Yr4* loci. *Phytopathology* **83**, 382–388.

Chen X and Line RF 1993b. Inheritance of stripe rust (yellow rust) resistance in the wheat cultivar Carstens V. *Euphytica* **71**, 107–113.

Chester KS 1946. 'The Nature and Prevention of the Cereal Rusts as Exemplified in the Leaf Rust of Wheat.' (Chronica Botanica Co.: Waltham, Massachusetts, USA.)

Chilosi G and Johnson R 1990. Resistance to races of *Puccinia striiformis* in seedlings of Italian wheats and possible presence of the *Yr6* gene in some durum cultivars. *Journal of Genetics and Breeding* **44**, 13–20.

Choudhuri HC 1958. The inheritance of stem and leaf rust resistance in common wheat. *Indian Journal of Genetics* 18, 90–115.

Christensen CM 1984. 'EC Stakman, Statesman of Science.' (American Phytopathological Society: St Paul, Minnesota, USA.)

Cox TS 1991. The contribution of introduced germplasm to the development of US wheat cultivars. *In* 'Use of Plant Introduction in Cultivar Development, Part 1.' Special Publication 17. pp. 25–47. (Crop Science Society of America: Madison, Wisconsin, USA.)

Cox TS, Sears RG and Gill BS 1992. Registration of KS90WGRC10 leaf rust-resistant hard red winter wheat germplasm. *Crop Science* **32**, 506.

Cox TS, Raupp WJ and Gill BS 1994. Leaf rust-resistance genes *Lr41*, *Lr42* and *Lr43* transferred from *Triticum tauschii* to common wheat. *Crop Science* **34**, 339–343.

Craigie JH 1927. Experiments on sex in rust fungi. *Nature* **120**, 116.

Cromey MG 1992. A new pathotype of *Puccinia striiformis* in New Zealand with increased pathogenicity to wheat cultivars with the adult plant resistance gene *Yr14*. *Australasian Plant Pathology* **21**, 172–174.

Cromey MG and Munro CA 1992. Stripe rust resistance in New Zealand wheat cultivars. *Vorträge für Pflanzenzüchtung* **24**, 276–278.

Czarnecki EM and Lukow OM 1992. Linkage of stem rust resistance gene *Sr33* and the gliadin (*Gli-D1*) locus on chromosome 1DS. *Genome* **35**, 565–568.

Day PR 1974. 'Genetics of Host-Parasite Interactions.' (WH Freeman: San Francisco, USA.)

Day PR (Ed.) 1977. 'The Genetic Basis of Epidemics in Agriculture.' Annals of the New York Academy of Sciences Volume 287. (The New York Academy of Sciences: New York, USA.)

Dhaliwal AS, Mares DJ, Marshall DR and Skerritt JH 1988. Protein composition and pentosan content in relation to dough stickiness of 1B/1R translocation wheats. *Cereal Chemistry* **65**, 143–149.

Doussinault G, Dosba F and Tanguy AM 1981. Analyse monosomique de la résistance a la rouille jaune du géniteur blé tendre VPM1. Académie D'Agriculture de France. *Extrait du procés-verbal de la Séance du 14 Janvier 1981*, 133–138.

Doussinault G, Dosba F and Jahier J 1988. Use of a hybrid between *Triticum aestivum* L. and *Aegilops ventricosa* Tausch in wheat breeding. *In* 'Proceedings of the Seventh International Wheat Genetics Symposium'. (Eds TE Miller and RMD Koebner.) pp. 253–258. (Institute of Plant Science Research: Cambridge, UK.)

Drijepondt SC, Pretorius ZA and Rijkenberg FHJ 1991. Expression of two wheat leaf rust resistance gene combinations involving *Lr34*. *Plant Disease* **75**, 526–528.

Driscoll CJ and Anderson LM 1967. Cytogenetic studies of Transec — a wheat–rye translocation line. *Canadian Journal of Genetics and Cytology* **9**, 375–380.

Driscoll CJ and Bielig LM 1968. Mapping of the Transec wheat–rye translocation. *Canadian Journal of Genetics and Cytology* **10**, 421–425.

Driscoll CJ and Jensen NF 1964. Characteristics of leaf rust resistance transferred from rye to wheat. *Crop Science* **4**, 372–374.

Driscoll CJ and Jensen NF 1965. Release of a wheat–rye translocation stock involving leaf rust and powdery mildew resistances. *Crop Science* **5**, 279–280.

Dubin HJ, Johnson R and Stubbs RW 1989. Postulated genes for resistance to stripe rust in selected CIMMYT and related wheats. *Plant Disease* **73**, 472–475.

Dvorak J 1977. Transfer of leaf rust resistance from *Aegilops speltoides* to *Triticum aestivum*. *Canadian Journal of Genetics and Cytology* **19**, 133–141.

Dvorak J and Knott DR 1977. Homoeologous chromatin exchange in a radiation-induced gene transfer. *Canadian Journal of Genetics and Cytology* **19**, 125–131.

Dvorak J and Knott DR 1980. Chromosome location of two leaf rust resistance genes transferred from *Triticum speltoides* to *T. aestivum*. *Canadian Journal of Genetics and Cytology* **22**, 381–389.

Dvorak J and Knott DR 1990. Location of a *Triticum speltoides* chromosome segment conferring resistance to leaf rust in *Triticum aestivum*. *Genome* **33**, 892–897.

Dvorak J, Resta P and Kota RS 1990. Molecular evidence on the origin of wheat chromosome 4A and 4B. *Genome* **33**, 30–39.

Dyck PL 1977. Genetics of leaf rust reaction in three introductions of common wheat. *Canadian Journal of Genetics and Cytology* **19**, 711–716.

Dyck PL 1979. Identification of the gene for adult plant leaf rust resistance in Thatcher. *Canadian Journal of Plant Science* **59**, 499–501.

Dyck PL 1987. The association of a gene for leaf rust resistance with the chromosome 7D suppressor of stem rust resistance in common wheat. *Genome* **29**, 467–469.

Dyck PL 1991. Genetics of adult plant leaf rust resistance in 'Chinese Spring' and 'Sturdy' wheats. *Crop Science* **31**, 309–311.

Dyck PL 1992. Transfer of a gene for stem rust resistance from *Triticum araraticum* to hexaploid wheat. *Genome* **35**, 788–792.

Dyck PL and Green GJ 1970. Genetics of stem rust resistance in wheat cultivar Red Bobs. *Canadian Journal of Plant Science* **50**, 229–232.

Dyck PL and Jedel PE 1989. Genetics of resistance to leaf rust in two accessions of common wheat. *Canadian Journal of Plant Science* **69**, 531–534.

Dyck PL and Johnson R 1983. Temperature sensitivity of genes for resistance in wheat to *Puccinia recondita*. *Canadian Journal of Plant Pathology* **5**, 229–234.

Dyck PL and Kerber ER 1970. Inheritance in hexaploid wheat of adult-plant leaf rust resistance derived from *Aegilops squarrosa*. *Canadian Journal of Genetics and Cytology* **12**, 175–180.

Dyck PL and Kerber ER 1971. Chromosome location of three genes for leaf rust resistance in common wheat. *Canadian Journal of Genetics and Cytology* **13**, 480–483.

Dyck PL and Kerber ER 1977a. Inheritance of leaf rust resistance in wheat cultivars Rafaela and EAP 26127 and chromosome location of gene *Lr17*. *Canadian Journal of Genetics and Cytology* **19**, 355–358.

Dyck PL and Kerber ER 1977b. Chromosome location of gene *Sr29* for reaction to stem rust. *Canadian Journal of Genetics and Cytology* **19**, 371–373.

Dyck PL and Kerber ER 1981. Aneuploid analysis of a gene for leaf rust resistance derived from the common wheat cultivar Terenzio. *Canadian Journal of Genetics and Cytology* **23**, 405–409.

Dyck PL and Samborski DJ 1968a. Host-parasite interactions involving two genes for leaf rust resistance in wheat. *In* 'Proceedings of the Third International Wheat Genetics Symposium'. (Eds KW Findlay and KW Shepherd.) pp. 245–250. (Australian Academy of Science: Canberra, Australia.)

Dyck PL and Samborski DJ 1968b. Genetics of resistance to leaf rust in the common wheat varieties Webster, Loros, Brevit, Carina, Malakof and Centenario. *Canadian Journal of Genetics and Cytology* **10**, 7–17.

Dyck PL and Samborski DJ 1970. The genetics of two alleles for leaf rust resistance at the *Lr14* locus in wheat. *Canadian Journal of Genetics and Cytology* **12**, 689–694.

Dyck PL and Samborski DJ 1974. Inheritance of virulence in *Puccinia recondita* of alleles at the *Lr2* locus for resistance in wheat. *Canadian Journal of Genetics and Cytology* **16**, 323–332.

Dyck PL and Samborski DJ 1982. The inheritance of resistance to *Puccinia recondita* in a group of common wheat cultivars. *Canadian Journal of Genetics and Cytology* **24**, 273–283.

Dyck PL, Samborski DJ and Anderson RG 1966. Inheritance of adult plant leaf rust resistance derived from the common wheat varieties Exchange and Frontana. *Canadian Journal of Genetics and Cytology* **8**, 665–671.

Dyck PL, Samborski DJ and Martens JW 1985. Inheritance of resistance to leaf rust and stem rust in the wheat cultivar Glenlea. *Canadian Journal of Plant Pathology* **7**, 351–354.

Dyck PL, Kerber ER and Lukow OM 1987. Chromosome location and linkage of a new gene (*Lr33*) for reaction to *Puccinia recondita* in common wheat. *Genome* **29**, 463–466.

Eizenga GC 1987. Locating the *Agropyron* segment in wheat — *Agropyron* transfer no. 12. *Genome* **29**, 365–366.

El-Bedewy R and Röbbelen G 1982. Chromosomal location and change of dominance of a gene for resistance against yellow rust, *Puccinia striiformis* West., in wheat, *Triticum aestivum* L. *Zeitschrift für Pflanzenzüchtung* **89**, 145–157.

Endo TR 1988. Induction of chromosomal structural changes by a chromosome of *Aegilops cyclindrica* L. in common wheat. *Journal of Heredity* **79**, 366–370.

Fang CT 1944. Physiologic specialization of *Puccinia glumarum* Erikss. and Henn. in China. *Phytopathology* **34**, 1020–1024.

Fitzgerald PM, Caldwell RM and Nelson OE 1957. Inheritance of resistance to certain races of leaf rust in wheat. *Agronomy Journal* **49**, 539–543.

Fletcher RJ 1983. Takari. *Journal of the Australian Institute of Agricultural Science* **49**, 46.

Flor HH 1942. Inheritance of pathogenicity in *Melampsora lini*. *Phytopathology* **32**, 653–659.

Flor HH 1956. The complementary gene systems in flax and flax rust. *Advances in Genetics* **8**, 29–54.

Forsyth FR 1956. Interaction of temperature and light on the seedling reaction of McMurachy wheat to race 15B of stem rust. *Canadian Journal of Botany* **34**, 745–749.

Friebe B, Heun M and Bushuk W 1989. Cytological characterization, powdery mildew resistance and storage protein composition of tetraploid and hexaploid 1BL/1RS wheat–rye translocation lines. *Theoretical and Applied Genetics* **78**, 425–432.

Friebe B, Zeller FJ, Mukai Y, Forster BP, Bartos P and McIntosh RA 1992. Characterization of rust resistant wheat–*Agropyron intermedium* derivatives by C-banding, *in situ* hybridization and isozyme analysis. *Theoretical and Applied Genetics* **83**, 775–782.

Friebe B, Jiang J, Gill BS and Dyck PL 1993. Radiation-induced nonhomoeologous wheat–*Agropyron intermedium* chromosomal translocations conferring resistance to leaf rust. *Theoretical and Applied Genetics* **86**, 141–149.

Gassner G and Straib W 1932. Die bestimmung der biologischen rassen des weizengelbrostes (*Puccinia glumarum* f. sp. *tritici* (Schmidt.) Erikss. und Henn.). *Arbeiten der Biologischen Reichsanstalt für Land und Forstwirtschaft, Berlin*, **20**, 141–163.

Gerechter-Amitai ZK and Stubbs RW 1970. A valuable source of yellow rust resistance in Israeli populations of wild emmer, *Triticum dicoccoides* Koern. *Euphytica* **19**, 12–21.

Gerechter-Amitai ZK, Wahl I, Vardi A and Zohary D 1971. Transfer of stem rust seedling resistance from wild diploid einkorn to tetraploid *durum* wheat by means of a triploid hybrid bridge. *Euphytica* **20**, 281–285.

Gerechter-Amitai ZK, van Silfhout CH, Grama A and Kleitman F 1989a. *Yr15* — a new gene for resistance to *Puccinia striiformis* in *Triticum dicoccoides* sel. G-25. *Euphytica* **43**, 187–190.

Gerechter-Amitai ZK, Grama A, van Silfhout CH, and Kleitman, F 1989b. Resistance to yellow rust in *Triticum dicoccoides*. II. Crosses with resistant *Triticum dicoccoides* sel. G25. *Netherlands Journal of Plant Pathology* **95**, 79–83.

German SE and Kolmer JA 1992. Effect of *Lr34* in the enhancement of resistance to leaf rust of wheat. *Theoretical and Applied Genetics* **84**, 97–105.

Gill BS, Friebe B and Endo TR 1991. Standard karyotype and nomenclature system for description of chromosome bands and structural observations in wheat (*Triticum aestivum*). *Genome* **34**, 830–839.

Gough FJ and Merkle OG 1971. Inheritance of stem and leaf rust resistance in Agent and Argus cultivars of *Triticum aestivum*. *Phytopathology* **61**, 1501–1505.

Gousseau HDM, Deverall BJ and McIntosh RA 1985. Temperature-sensitivity of the expression of resistance to *Puccinia graminis* conferred by the *Sr15*, *Sr8b* and *Sr14* genes in wheat. *Physiological Plant Pathology* **27**, 335–343.

Grama A and Gerechter-Amitai ZK 1974. Inheritance of resistance to stripe rust (*Puccinia striiformis*) in crosses between wild emmer (*Triticum dicoccoides*) and cultivated tetraploid and hexaploid wheats II. *Triticum aestivum*. *Euphytica* **23**, 393–398.

Green GJ 1971. Physiologic races of wheat stem rust in Canada from 1919 to 1969. *Canadian Journal of Botany* **49**, 1575–1588.

Green GJ and Campbell AB 1979. Wheat cultivars resistant to *Puccinia graminis tritici* in western Canada, their development, performance and economic value. *Canadian Journal of Plant Pathology* **1**, 3–11.

Green GJ and Dyck PL 1979. A gene for resistance to *Puccinia graminis* f. sp. *tritici* that is present in wheat cultivar H–44 but not in cultivar Hope. *Phytopathology* **69**, 672–675.

Green GJ and Knott DR 1962. Adult plant reaction to stem rust of lines of Marquis wheat with substituted genes for resistance. *Canadian Journal of Plant Science* **42**, 163–168.

Green GJ, Knott DR, Watson IA and Pugsley AT 1960. Seedling reactions to stem rust of lines of Marquis wheat with substituted genes for rust resistance. *Canadian Journal of Plant Science* **40**, 524–538.

Griffey CA and Allan RE 1988. Inheritance of stripe rust resistance among near-isogenic lines of spring wheat. *Crop Science* **28**, 48–54.

Gupta RB and Shepherd KW 1992. Identification of rye chromosome 1R translocations and substitutions in hexaploid wheats using storage proteins as genetic markers. *Plant Breeding* **109**, 130–140.

Gyarfas J 1978. Transference of Disease Resistance from *Triticum timopheevii* to *Triticum aestivum*. MSc Thesis, University of Sydney.

Gyarfas J 1983. Suneca. *Journal of the Australian Institute of Agricultural Science* **49**, 43–44.

Haggag MEA and Dyck PL 1973. The inheritance of leaf rust resistance in four common wheat varieties possessing genes at or near the *Lr3* locus. *Canadian Journal of Genetics and Cytology* **15**, 127–134.

Haggag MEA, Samborski DJ and Dyck PL 1973. Genetics of pathogenicity in three races of leaf rust on four wheat varieties. *Canadian Journal of Genetics and Cytology* **15**, 73–82.

Harder DE and Dunsmore KM 1990. Incidence and virulence of *Puccinia graminis* f.sp. *tritici* on wheat and barley in Canada in 1989. *Canadian Journal of Plant Pathology* **12**, 424–427.

Harder DE, Mathenge GR and Mwaura LK 1972. Physiologic specialization and epidemiology of wheat stem rust in east Africa. *Phytopathology* **62**, 166–171.

Hare RA and McIntosh RA 1979. Genetic and cytogenetic studies of durable adult-plant resistances in 'Hope' and related cultivars to wheat rusts. *Zeitschrift für Pflanzenzüchtung* **83**, 350–367.

Hart GE, McMillin DE and Sears ER 1976. Determination of the chromosomal location of a glutamate oxaloacetate transaminase structural gene using *Triticum–Agropyron* translocations. *Genetics* **83**, 49–61.

Hart, Helen 1931. Morphologic and Physiologic Studies on Stem Rust Resistance in Cereals. Technical Bulletin 266. (United States Department of Agriculture: Washington, DC.)

Hawthorn WM 1984. Genetic Analyses of Leaf Rust Resistance in Wheat. PhD Thesis, University of Sydney.

Hayes HK, Parker JH and Kurtzweil 1920. Genetics of rust resistance in crosses of varieties of *Triticum vulgare* with varieties of *T. durum* and *T. dicoccum*. *Journal of Agricultural Research* **19**, 523–542.

Heerman RM, Smith GW, Briggle LW and Schwinghamer EA 1956. Inheritance of reaction to stem rust in certain durum and emmer wheats. *Proceedings of the Third International Rust Conference, Mexico City*, 82–83.

Heun M and Fischbeck G 1987*a*. Identification of wheat powdery mildew resistance genes by analysing host–pathogen interactions. *Plant Breeding* **98**, 124–129.

Heun M and Fischbeck G 1987*b*. Genes for powdery mildew resistance in cultivars of spring wheat. *Plant Breeding* **99**, 282–288.

Heyne EG and Johnston CO 1954. Inheritance of leaf rust reaction and other characters in crosses among Timstein, Pawnee and Redchief wheats. *Agronomy Journal* **46**, 81–85.

Heyne EG and Livers RW 1953. Monosomic analysis of leaf rust reaction, awnedness, winter injury and seed colour in Pawnee wheat. *Agronomy Journal* **45**, 54–58.

Holmes RJ and Dennis JI 1985. Accessory hosts of wheat stripe rust in Victoria, Australia. *Transactions of the British Mycological Society* **85**, 159–160.

Hovmøller MS 1989. Race specific powdery mildew resistance in 31 Northwest European wheat cultivars. *Plant Breeding* **103**, 228–234.

Howes NK 1986. Linkage between the *Lr10* gene conditioning resistance to leaf rust, two endosperm proteins and hairy glumes in hexaploid wheat. *Canadian Journal of Genetics and Cytology* **28**, 595–600.

Hsam SLK and Zeller FJ 1982. Relationships of *Agropyron intermedium* chromosomes determined by chromosome pairing and alcohol dehydrogenase isozymes in common wheat background. *Theoretical and Applied Genetics* **63**, 213–217.

Hu CC and Roelfs AP 1985. The wheat rusts in the People's Republic of China. *Cereal Rusts Bulletin* **13**, 11–28.

Hu CC and Roelfs AP 1986. Postulation of genes for stem rust resistance in 13 Chinese wheat cultivars. *Cereal Rusts Bulletin* **14**, 68–74.

Hu CC and Roelfs AP 1989. Races and virulence of *Puccinia recondita* f. sp. *tritici* in China in 1986. *Plant Disease* **73**, 499–501.

Huerta-Espino J 1992. Analysis of Wheat Leaf and Stem Rust Virulence on a Worldwide Basis. PhD Thesis, University of Minnesota, USA.

Huerta-Espino J and Singh RP 1994. First report of virulence for wheat leaf rust gene *Lr19* in Mexico. *Plant Disease* **78**, 640.

Hussain M, Hassan SF and Kirmani MAS 1980. Virulences in *Puccinia recondita* Rob. ex Desm. f. sp. *tritici* in Pakistan during 1978 and 1979. Proceedings of the Fifth European and Mediterranean Cereal Rusts Conference, Bari, Italy, 179–184.

Javornik B, Sinkovic T, Vapa L, Koebner RMD and Rogers WJ 1991. A comparison of methods for identifying and surveying the presence of 1BL.1RS translocations in bread wheat. *Euphytica* **54**, 45–53.

Johnson R 1981. Durable resistance, definition of, genetic control, and attainment in plant breeding. *Phytopathology* **71**, 567–568.

Johnson R 1986. Genetics of the interaction between plants and their parasites. *In* 'Natural Antimicrobial Systems. Part 1. Antimicrobial Systems in Plants and Animals'. (Eds GW Gould, ME Rhodes-Roberts, AK Charnley, RM Copper and RG Board.) pp. 131–145. (Bath University Press: Bath, UK.)

Johnson R 1988. Durable resistance to yellow (stripe) rust in wheat and its implications in plant breeding. *In* 'Breeding Strategies for Resistance to the Rusts of Wheat'. (Eds NW Simmonds and S Rajaram.) pp. 63–75. (CIMMYT: Mexico.)

Johnson R 1992. Reflections of a plant pathologist on breeding for disease resistance, with emphasis on yellow rust and eyespot of wheat. British Society for Plant Pathology Presidential Address 1991. *Plant Pathology* **41**, 239–254.

Johnson R and Bowyer DE 1974. A rapid method of measuring production of yellow rust spores on single seedlings to assess differential interactions of wheat cultivars with *Puccinia striiformis*. *Annals of Applied Biology* **77**, 251–258.

Johnson R and Dyck PL 1984. Resistance to yellow rust in *Triticum spelta* var. *album* and bread wheat cultivars Thatcher and Lee. Proceedings of the Sixth European and Mediterranean Cereal Rusts Conference. (Ed. P Auriau.) pp. 71–74. (Grignon: France.)

Johnson R and Minchin PN 1992. Genetics of resistance to yellow (stripe) rust of wheat in some differential cultivars. *Vorträge für Pflanzenzüchtung* **24**, 227–229.

Johnson R and Taylor AJ 1972. Isolates of *Puccinia striiformis* collected in England from the wheat varieties Maris Beacon and Joss Cambier. *Nature* **238**, 105–106.

Johnson R and Taylor AJ 1977. Yellow rust of wheat. Plant Breeding Institute, Cambridge. *Annual Report 1976*, 119–122.

Johnson R, Wolfe MS and Scott PR 1969. Plant Breeding Institute, Cambridge. *Annual Report 1968*, 113–123.

Johnson R, Stubbs RW, Fuchs E and Chamberlain NH 1972. Nomenclature for physiologic races of *Puccinia striiformis* infecting wheat. *Transactions of the British Mycological Society* **58**, 475–480.

Johnson R, Taylor AJ and Smith GMB 1975. Samples of *Puccinia striiformis* from yellow rust infections on Maris Huntsman in 1974. *Cereal Rusts Bulletin* **3**, 4–6.

Johnson R, Priestley RH and Taylor EC 1978. Occurrence of virulence in *Puccinia striiformis* for Compair wheat in England. *Cereal Rusts Bulletin* **6**, 11–13.

Johnson R, Taylor AJ and Smith GMB 1984. Plant Breeding Institute, Cambridge. *Annual Report 1983*, 82–85.

Johnson R, Calonnec A, Bimb HP and Singh H. In Press. Identification and distribution of genes for resistance to *Puccinia striiformis* in differential and related cultivars. In 'Proceedings of the Eighth Wheat Genetics Symposium'. (Beijing, China).

Johnston CO and Browder LE 1966. Seventh revision of the international register of physiologic races of *Puccinia recondita* f. sp. *tritici*. *Plant Disease Reporter* **50**, 756–760.

Johnston CO and Heyne EG 1964. Wichita wheat back-cross lines for differential hosts in identifying physiologic races of *Puccinia recondita*. *Phytopathology* **54**, 385–388.

Johnston CO and Levine MN 1955. Fifth revision of the international register of physiologic races of *Puccinia rubigo-vera* (DC) Wint. f. sp. *tritici* (Eriks) Carleton=(*P. triticina Erikss*). *Plant Disease Reporter Supplement* **233**, 104–120.

Jones DR and Deverall BJ 1977. The effect of the *Lr20* resistance gene in wheat on the development of leaf rust, *Puccinia recondita*. *Physiological Plant Pathology* **10**, 275–285.

Jones ERL, Clifford BC and Willoughby DB 1992 Brown rust of wheat. United Kingdom Cereal Pathogen Virulence Survey. *1991 Annual Report*, 19–23.

Jones SS, Dvorak J and Qualset CO 1990. Linkage relations of *Gli-D1*, *Rg2* and *Lr21* on the short arm of chromosome 1D in wheat. *Genome* **33**, 937–940.

Jones SS Dvorak J, Knott DR and Qualset CO 1991. Use of double-ditelosomic and normal chromosome 1D recombinant lines to map *Sr33* on chromosome arm 1DS in wheat. *Genome* **34**, 505–508.

Jorgensen JH and Jensen CJ 1973. Gene *Pm6* for resistance to powdery mildew in wheat. *Euphytica* **22**, 4–23.

Joshi LM 1976. Recent contributions towards epidemiology of wheat rusts in India. *Indian Phytopathology* **29**, 1–16.

Kao KN and Knott DR 1969. The inheritance of pathogenicity in races 111 and 29 of wheat stem rust. *Canadian Journal of Genetics and Cytology* **11**, 266–274.

Keed BR and White NH 1971. Quantitative effects of leaf and stem rusts on yield and quality of wheat. *Australian Journal of Experimental Agriculture and Animal Husbandry* **11**, 550–555.

Kema GHJ 1992. Resistance in spelt wheat to yellow rust I. Formal analysis and variation for gliadin patterns. *Euphytica* **63**, 207–217.

Kema GHJ and Lange W 1992. Resistance in spelt wheat to yellow rust II. Monosomic analysis of the Iranian accession 415. *Euphytica* **63**, 219–224.

Kenaschuk EO, Anderson RG and Knott DR 1959. The inheritance of stem rust resistance to race 15B of stem rust in ten varieties of durum wheat. *Canadian Journal of Plant Science* **39**, 316–328.

Kerber ER 1987. Resistance to leaf rust in hexaploid wheat, *Lr32*, a third gene derived from *Triticum tauschii*. *Crop Science* **27**, 204–206.

Kerber ER 1988. Telocentric mapping in wheat of the gene *Lr32* for resistance to leaf rust. *Crop Science* **28**, 178–179.

Kerber ER and Dyck PL 1973. Inheritance of stem rust resistance transferred from diploid wheat (*Triticum monococum*) to tetraploid and hexaploid wheat and chromosome location of the gene involved. *Canadian Journal of Genetics and Cytology* **15**, 397–409.

Kerber ER and Dyck PL 1979. Resistance to stem rust and leaf rust of wheat in *Aegilops squarrosa* and transfer of a gene for stem rust resistance to hexaploid wheat. In 'Proceedings of the Fifth International Wheat Genetics Symposium'. (Ed. S Ramanujam.) pp. 358–364. (Indian Society of Genetics and Plant Breeding: New Delhi, India.)

Kerber ER and Dyck PL 1990. Transfer to hexaploid wheat of linked genes for adult-plant leaf rust and seedling stem rust resistance from an amphiploid of *Aegilops speltoides* x *Triticum monococcum*. *Genome* **33**, 530–537.

Kibirige-Sebunya I and Knott DR 1983. Transfer of stem rust resistance to wheat from an *Agropyron* chromosome having a gametocidal effect. *Canadian Journal of Genetics and Cytology* **25**, 215–221.

Kirmani MAS, Risvi SSA and Stubbs RW 1984. Postulated genotypes for stripe rust resistance in wheat cultivars of Pakistan. In 'Proceedings of the Sixth European and Mediterranean Cereal Rusts Conference'. (Ed. P Auriau.) pp. 81–85. (Grignon, France.)

Kislev ME 1982. Stem rust of wheat 3300 years old found in Israel. *Science* **216**, 993–994.

Knott DR 1957a. The inheritance of rust resistance. II. The inheritance of stem rust resistance in six additional varieties of common wheat. *Canadian Journal of Plant Science* **37**, 177–192.

Knott DR 1957b. The inheritance of rust resistance. III. The inheritance of stem rust resistance in nine Kenya varieties of common wheat. *Canadian Journal of Plant Science* **37**, 366–384.

Knott DR 1959. The inheritance of rust resistance. IV. Monosomic analysis of rust resistance and some other characters in six varieties of wheat including Gabo and Kenya Farmer. *Canadian Journal of Plant Science* **39**, 215–228.

Knott DR 1961. The inheritance of rust resistance. VI. The transfer of stem rust resistance from *Agropyron elongatum* to common wheat. *Canadian Journal of Plant Science* **41**, 109–123.

Knott DR 1962a. Inheritance of rust resistance. VIII. Additional studies on Kenya varieties of wheat. *Crop Science* **2**, 130–132.

Knott DR 1962b. The inheritance of rust resistance. IX. The inheritance of resistance to races 15B and 56 of stem rust in the wheat variety Khapstein. *Canadian Journal of Plant Science* **42**, 415–419.

Knott DR 1965. A comparison of the reaction to stem rust of wheat lines backcrossed five and nine times to Marquis that carry the same resistance genes. *Canadian Journal of Plant Science* **45**, 106–107.

Knott DR 1966. The inheritance of stem rust resistance in wheat. Proceedings of the Second International Wheat Genetics Symposium. (Ed. J MacKey.) Lund, Sweden 1963. *Hereditas Supplement* **2**, 156–166.

Knott DR 1968. The inheritance of resistance to stem rust races 56 and 15B–1L (Can.) in the wheat varieties Hope and H–44. *Canadian Journal of Genetics and Cytology* **10**, 311–320.

Knott DR 1971. Genes for stem rust resistance in wheat varieties Hope and H–44. *Canadian Journal of Genetics and Cytology* **13**, 186–188.

Knott DR 1977. Studies on general resistance to stem rust in wheat. *In* 'Induced Mutations against Plant Diseases'. pp. 81–88. (International Atomic Energy Agency: Vienna, Austria.)

Knott DR 1980. Mutation of a gene for yellow pigment linked to *Lr19* in wheat. *Canadian Journal of Genetics and Cytology* **22**, 651–654.

Knott DR 1983. The inheritance of resistance to stem rust races 15B-1 and 56 in 'French Peace' wheat. *Canadian Journal of Genetics and Cytology* **25**, 283–285.

Knott DR 1984. The inheritance of resistance to race 56 of stem rust in 'Marquillo' wheat. *Canadian Journal of Genetics and Cytology* **26**, 174–176.

Knott DR 1989. 'The Wheat Rusts — Breeding for Resistance.' Monographs on Theoretical and Applied Genetics 12. (Springer Verlag: Berlin.)

Knott DR 1990. Near-isogenic lines of wheat carrying genes for stem rust resistance. *Crop Science* **30**, 901–905.

Knott DR and Anderson RG 1956. The inheritance of rust resistance. I. The inheritance of stem rust resistance in ten varieties of common wheat. *Canadian Journal of Agricultural Science* **36**, 174–195.

Knott DR and McIntosh RA 1978. The inheritance of stem rust resistance in the common wheat cultivar Webster. *Crop Science* **17**, 365–369.

Knott DR and Shen I 1961. The inheritance of rust resistance. VII. The inheritance of resistance to races 15B and 56 of stem rust in eleven common wheat varieties of diverse origin. *Canadian Journal of Plant Science* **41**, 587–601.

Koebner RMD and Shepherd KW 1986. Controlled introgression to wheat of genes from rye chromosome arm 1RS by induction of allosyndesis I. Isolation of recombinants. *Theoretical and Applied Genetics* **73**, 197–208.

Kolmer JA 1991. Physiologic specialization of *Puccinia recondita* f. sp. *tritici* in Canada in 1990. *Canadian Journal of Plant Pathology* **13**, 371–373.

Kolmer JA 1992a. Virulence heterozygosity and gametic phase disequilibria in two populations of *Puccinia recondita* (wheat leaf rust fungus). *Heredity* **68**, 505–513.

Kolmer JA 1992b. Enhanced leaf rust resistance in wheat conditioned by resistance gene pairs with *Lr13*. *Euphytica* **61**, 123–130.

Kolmer JA 1993. Physiologic specialisation of *Puccinia recondita* f. sp. *tritici* in Canada in 1991. *Canadian Journal of Plant Pathology* **15**, 34–36.

Kolmer JA, Dyck PL and Roelfs AP 1991. An appraisal of stem and leaf rust resistance in North American hard red spring wheats and the probability of multiple mutations to virulence in populations of cereal rust fungi. *Phytopathology* **81**, 237–239.

Kumar J, Nayar SK, Bahadur P, Nagarajan S, Bhardwaj SC, Prashar M and Singh SB 1988. Virulence survey of yellow rust of wheat (*Puccinia striiformis* f.sp. *tritici*) and barley (*P. striiformis* f.sp. *hordei*) during 1985–87. *Cereal Rusts and Powdery Mildews Bulletin* **16**, 30–35.

Kumar J, Nayar SK, Prashar M, Bhardwaj SC and Bhatnagar R 1993. Virulence survey of *Puccinia striiformis* in India during 1990–1992. *Cereal Rusts and Powdery Mildews Bulletin* **21**, 17–24.

Labrum KE 1980. The location of *Yr2* and *Yr6* genes conferring resistance to yellow rust. Proceedings of the Fifth European and Mediterranean Cereal Rusts Conference, Bari, Italy, 41–45.

Large EC 1940. 'The Advance of the Fungi.' (Jonathon Cape: London.)

Larkin PJ, Spindler LH and Banks PM 1990. Cell culture of alien chromosome addition lines to induce somatic recombination and gene introgression. *In* 'Progress in Plant Cellular and Molecular Biology'. (Eds HJJ Nijkamp, LHW Van Der Plas and J Van Aartrijk.) pp. 163–168. (Kluwer Academic Publishers: Dordrecht.)

Law CN 1976. Genetic control of yellow rust resistance in *T. spelta album*. Plant Breeding Institute, Cambridge. Annual Report 1975, 108–109.

Law CN and Johnson R 1967. A genetic study of leaf rust resistance in wheat. *Canadian Journal of Genetics and Cytology* **9**, 805–822.

Law CN and Wolfe MS 1966. Location of genetic factors for mildew resistance and ear emergence time on chromosome 7B of wheat. *Canadian Journal of Genetics and Cytology* **8**, 462–470.

Le Roux J 1985. First report of a *Puccinia graminis* f.sp. *tritici* race with virulence for *Sr24* in South Africa. *Plant Disease* **69**, 1007.

Le Roux J and Rijkenberg FHJ 1987a. Occurrence and pathogenicity of *Puccinia graminis* f.sp. *tritici* in South Africa during the period 1981–1985. *Phytophylactica* **19**, 456–472.

Le Roux J and Rijkenberg FHJ 1987b. Pathotypes of *Puccinia graminis* f.sp. *tritici* with increased virulence for *Sr24*. *Plant Disease* **71**, 1115–1119.

Limpert E, Clifford B, Dreiseitl A, Johnson R, Müller K, Roelfs A and Wellings C 1994. Systems of designation of pathotypes of plant pathogens. *Journal of Phytopathology* **140**, 359–362.

Line RF and Qayoum A 1991. Virulence, aggressiveness, evolution and distribution of races of *Puccinia striiformis* (the cause of stripe rust of wheat) in North America, 1968–1987. Technical Bulletin No. 1788. (United States Department of Agriculture: Washington.)

Line RF, Chen XM and Qayoum A 1992. Races of *Puccinia striiformis* in North America, identification of resistance genes and durability of resistance. *Vorträge für Pflanzenzüchtung* **24**, 280–282.

Livers RW 1978. Registration of Sage wheat. *Crop Science* **18**, 917.

Loegering WQ 1966. The relationship between host and pathogen in stem rust of wheat. Proceedings of the Second International Wheat Genetics Symposium. (Ed. J Mackey.) Lund, Sweden, 1963. *Hereditas Supplement* **2**, 167–177.

Loegering WQ 1968. A second gene for resistance to *Puccinia graminis* f.sp. *tritici* in the Red Egyptian 2D wheat substitution line. *Phytopathology* **58**, 584–586.

Loegering WQ 1972. Specificity in plant diseases In 'Biology of Rust Resistance in Forest Trees'. Miscellaneous Publication No. 1221. pp. 29–37. (United States Department of Agriculture: Washington.)

Loegering WQ 1975. An allele for low reaction to *Puccinia graminis tritici* in Chinese Spring wheat. *Phytopathology* **65**, 925.

Loegering WQ and Harmon DL 1969. Wheat lines near-isogenic for reaction to *Puccinia graminis tritici*. *Phytopathology* **59**, 456–459.

Loegering WQ and Powers HR Jr 1962. Inheritance of pathogenicity in a cross of physiological races 111 and 36 of *Puccinia graminis* f.sp. *tritici*. *Phytopathology* **52**, 547–554.

Loegering WQ and Sears ER 1966. Relationships among stem-rust genes on wheat chromosomes 2B, 4B and 6B. *Crop Science* **6**, 157–160.

Loegering WQ and Sears ER 1973. The gene for low reaction to *Puccinia graminis tritici* in the Thatcher–3B substitution line. *Crop Science* **13**, 282.

Long DL and Kolmer JA 1989. A North American system of nomenclature for *Puccinia recondita* f. sp. *tritici*. *Phytopathology* **79**, 525–529.

Long DL, Roelfs AP and Roberts JJ 1992. Virulence of *Puccinia recondita* f. sp. *tritici* in the United States during 1988–1990. *Plant Disease* **76**, 495–499.

Long DL, Roelfs AP, Leonard KJ and Roberts JJ 1993. Virulence and diversity of *Puccinia recondita* f. sp. *tritici* in the United States in 1991. *Plant Disease* **77**, 786–791.

Louwers J and Sharma S 1992. Postulation of resistance genes to yellow rust (*Puccinia striiformis* f. sp. *tritici*) in advanced wheat lines from Nepal. *Vorträge für Pflanzenzüchtung* **24**, 279.

Luig NH 1960. Differential transmission of gametes in wheat. *Nature* **185**, 636–637.

Luig NH 1964. Heterogeneity in segregation data from wheat crosses. *Nature* **204**, 260–261.

Luig NH 1968. Mechanism of differential transmission of gametes in wheat. In 'Proceedings of the Third International Wheat Genetics Symposium'. (Eds KW Finlay and KW Shepherd.) pp. 322–323. (Australian Academy of Science: Canberra.)

Luig NH 1983. A survey of virulence genes in wheat stem rust, *Puccinia graminis* f. sp. *tritici*. Advances in Plant Breeding, Supplement 11 to *Journal of Plant Breeding*. (Paul Parey: Berlin.)

Luig NH and McIntosh RA 1968. Location and linkage of genes on wheat chromosome 2D. *Canadian Journal of Genetics and Cytology* **10**, 99–105.

Luig NH and Rajaram S 1972. The effect of temperature and genetic background on host gene expression and interaction to *Puccinia graminis tritici*. *Phytopathology* **62**, 1171–1174.

Luig NH and Watson IA 1965. Studies on the genetic nature of resistance to *Puccinia graminis* var. *tritici* in six varieties of common wheat. *Proceedings of the Linnean Society New South Wales* **90**, 299–327.

Luig NH and Watson IA 1967. Vernstein — a *Triticum aestivum* derivative with Vernal-emmer-type stem rust resistance. *Crop Science* **7**, 31–33.

Luig NH and Watson IA 1970. The effect of complex genetic resistance in wheat on the variability of *Puccinia graminis* f.sp. *tritici*. *Proceedings of the Linnean Society of New South Wales* **95**, 22–45.

Luig NH and Watson IA 1972. The role of wild and cultivated grasses in the hybridization of formae speciales of *Puccinia graminis*. *Australian Journal of Biological Sciences* **25**, 335–342.

Luig NH, McIntosh RA and Watson IA 1973. Genes for resistance to *P. graminis* in the standard wheat stem rust differentials. *In* 'Proceedings of the Fourth International Wheat Genetics Symposium'. (Eds ER Sears and LMS Sears.) pp. 423–424. (Agricultural Experiment Station, University of Missouri: Columbia, USA.)

Luig NH, Burdon JJ and Hawthorn WM 1985. An exotic strain of *Puccinia recondita tritici* in New Zealand. *Canadian Journal Plant Pathology* **7**, 173–176.

Lupton FGH and Macer RCF 1962. Inheritance of resistance to yellow rust (*Puccinia glumarum* Erikss. & Henn.) in seven varieties of wheat. *Transactions of the British Mycological Society* **45**, 21–45.

Lutz J, Limpert E, Bartos P and Zeller FJ 1992. Identification of powdery mildew resistance genes in common wheat (*Triticum aestivum* L.) I. Czechoslovakian cultivars. *Plant Breeding* **108**, 33–39.

Macer RCF 1966. The formal and monosomic genetic analysis of stripe rust (*Puccinia striiformis*) resistance in wheat. *In* 'Proceedings of the Second International Wheat Genetics Symposium'. (Ed. J MacKey.) Lund, Sweden 1963. *Hereditas Supplement* **2**, 127–142.

Macer RCF 1975. Plant pathology in a changing world. *Transactions of the British Mycological Society* **65**, 351–374.

Macindoe SL and Walkden Brown C 1968. Wheat Breeding and Varieties in Australia. Science Bulletin No. 76. (New South Wales Department of Agriculture: Sydney.)

Mains EB and Jackson HS 1926. Physiologic specialization in the leaf rust of wheat; *Puccinia triticina* Erikss. *Phytopathology* **16**, 89–120.

Mains EB, Leighty CE and Johnston CO 1926. Inheritance of resistance to leaf rust *Puccinia triticina* Erikss., in crosses of common wheat, *Triticum vulgare* Vill. *Journal of Agricultural Research* (Washington, DC) **32**, 931–972.

Marais GF 1990. Preferential transmission in bread wheat of a chromosome segment derived from *Thinopyrum distichum* (Thunb.) Löve. *Plant Breeding* **104**, 152–159.

Marais GF 1992*a*. Gamma irradiation induced deletions in an alien chromosome segment of the wheat 'Indis' and their use in gene mapping. *Genome* **35**, 225–229.

Marais GF 1992*b*. The modification of a common wheat—*Thinopyrum distichum* translocated chromosome with a locus homoeoallelic to *Lr19*. *Theoretical and Applied Genetics* **85**, 73–78.

Marais GF, Roux HS, Pretorius ZA and de V Pienaar R 1988. Resistance to leaf rust of wheat derived from *Thinopyrum distichum* (Thunb.) Löve. *In* 'Proceedings of the Seventh International Wheat Genetics Symposium'. (Eds TE Miller and RMD Koebner.) pp. 369–373. (Institute of Plant Science Research: Cambridge, UK.)

Marshall D 1992 Virulence of *Puccinia recondita* in Texas from 1988 to 1990. *Plant Disease* **76**, 296–299.

May C and Wray F 1991. A rapid technique for the detection of wheat–rye translocation chromosomes. *Genome* **34**, 486–488.

McFadden ES 1930. A successful transfer of emmer characters to *vulgare* wheat. *Journal of the American Society of Agronomy* **22**, 1020–1034.

McIntosh RA 1972. Cytogenetical studies in wheat VI. Chromosome location and linkage studies involving *Sr13* and *Sr8* for reaction to *Puccinia graminis* f. sp. *tritici*. *Australian Journal of Biological Sciences* **25**, 765–773.

McIntosh RA 1977. Nature of induced mutations affecting disease reaction in wheat. *In* 'Induced Mutations Against Plant Disease'. pp. 551–565. (International Atomic Energy Agency: Vienna.)

McIntosh RA 1978. Cytogenetical studies in wheat X. Monosomic analysis and linkage studies involving genes for resistance to *Puccinia graminis* f. sp. *tritici* in cultivar Kota. *Heredity* **41**, 71–82.

McIntosh RA 1980. Chromosome location and linkage studies involving the wheat stem rust resistance gene *Sr14*. *Cereal Research Communications* **8**, 315–320.

McIntosh RA 1981. A gene for stem rust resistance in non–homoeologous chromosomes of hexaploid wheat progenitors. *In* 'Proceedings of the Thirteenth International Botanical Congress'. Abstract p. 274. (University of Sydney: Sydney.)

McIntosh RA 1983*a*. Genetic and cytogenetic studies involving *Lr18* for resistance to *Puccinia recondita*. *In* 'Proceedings of the Sixth International Wheat Genetics Symposium'. (Ed. S Sakamoto.) pp. 777–783. (Faculty of Agriculture, Kyoto University: Japan.)

McIntosh RA 1983*b*. Induced mutations of rust resistance genes in wheat. *In* 'Induced Mutations for Disease Resistance in Crop Plants II'. pp. 115–118. (International Atomic Energy Agency: Vienna.)

McIntosh RA 1987. Gene location and gene mapping in hexaploid wheat. *In* 'Wheat and Wheat Improvement'. (Ed. EG Heyne.) pp. 269–287. (American Society of Agronomy: Madison, Wisconsin,USA.)

McIntosh RA 1988*a*. Catalogue of gene symbols for wheat. Proceedings of the Seventh International Wheat Genetics Symposium Vol 2. (Eds TE Miller and RMD Koebner.) pp. 1225–1323. (Institute of Plant Science Research: Cambridge, UK.)

McIntosh RA 1988*b*. The role of specific genes in breeding for durable stem rust resistance in wheat and triticale. *In* 'Breeding Strategies for Resistance to the Rusts of Wheat'. (Eds NW Simmonds and S Rajaram.) pp. 1–9. (CIMMYT: Mexico.)

McIntosh RA 1992*a*. Close genetic linkage of genes conferring adult-plant resistance to leaf rust and stripe rust in wheat. *Plant Pathology* **41**, 523–527.

McIntosh RA 1992*b*. Pre–emptive breeding to control wheat rusts. *In* 'Breeding for Resistance'. (Eds R Johnson and GJ Jellis.) pp. 103–113. (Kluwer Academic Publications: Dordrecht.)

McIntosh RA and Baker EP 1966. Chromosome location of mature plant leaf rust resistance in Chinese Spring wheat. *Australian Journal of Biological Sciences* **19**, 943–944.

McIntosh RA and Baker EP 1968. A linkage map for chromosome 2D. *In* 'Proceedings of the Third International Wheat Genetics Symposium'. (Eds KW Finlay and KW Shepherd.) pp. 305–309. (Australian Academy of Science: Canberra.)

McIntosh RA and Baker EP 1969. Telocentric mapping of a second gene for grass–clump dwarfism. *Wheat Information Service* **29**, 6–7.

McIntosh RA and Baker EP 1970*a*. Cytogenetical studies in wheat IV. Chromosome location and linkage studies involving the *Pm2* locus for powdery mildew resistance. *Euphytica* **19**, 71–77.

McIntosh RA and Baker EP 1970*b*. Cytogenetical studies in wheat V. Monosomic analysis of Vernstein stem rust resistance. *Canadian Journal of Genetics and Cytology* **12**, 60–65.

McIntosh RA and Dyck PL 1975. Cytogenetical studies in wheat VII. Gene *Lr23* for reaction to *Puccinia recondita* in Gabo and related cultivars. *Australian Journal of Biological Sciences* **28**, 201–211.

McIntosh RA and Gyarfas J 1971. *Triticum timopheevi* as a source of resistance to wheat stem rust. *Zeitschrift für Pflazenzüchtung* **66**, 240–248.

McIntosh RA and Luig NH 1973*a*. Recombination between genes for reaction to *P. graminis* at or near the *Sr9* locus. *In* 'Proceedings of the Fourth International Wheat Genetics Symposium'. (Eds ER Sears and LMS Sears.) pp. 425–432. (Agricultural Experiment Station, University of Missouri: Columbia, Missouri, USA.)

McIntosh RA and Luig NH 1973*b*. Linkage of genes for reaction to *Puccinia graminis* f. sp. *tritici* and *P. recondita* in Selkirk wheat and related cultivars. *Australian Journal of Biological Sciences* **26**, 1145–1152.

McIntosh RA and Singh SJ 1986. Rusts — Real and potential problems for triticale. *In* 'Proceedings of the International Triticale Symposium'. (Compiler NL Darvey.) Occasional Publication No. 24. pp. 199–207. (Australian Institute of Agricultural Science: Sydney.)

McIntosh RA, Baker EP and Driscoll CJ 1965. Cytogenetic studies in wheat. I Monosomic analysis of leaf rust resistance in cultivars Uruguay and Transfer. *Australian Journal of Biological Sciences* **18**, 971–977.

McIntosh RA, Luig NH and Baker EP 1967. Genetic and cytogenetic studies of stem rust, leaf rust and powdery midew resistances in Hope and related wheat cultivars. *Australian Journal of Biological Sciences* **20**, 1181–1192.

McIntosh RA, Dyck PL and Green GJ 1974. Inheritance of reaction to stem rust and leaf rust in the wheat cultivar Etoile de Choisy. *Canadian Journal of Genetics and Cytology* **16**, 571–577.

McIntosh RA, Dyck PL and Green GJ 1976. Inheritance of leaf rust and stem rust resistances in wheat cultivars Agent and Agatha. *Australian Journal of Agricultural Research* **28**, 37–45.

McIntosh RA, Partridge M and Hare RA 1980. Telocentric mapping of *Sr12* in wheat chromosome 3B. *Cereal Research Communications* **8**, 321–324.

McIntosh RA, Luig NH, Johnson R and Hare RA 1981. Cytogenetical studies in wheat XI. *Sr9g* for reaction to *Puccinia graminis tritici*. *Zeitschrift für Pflanzenzüchtung* **87**, 274–289.

McIntosh RA, Miller TE and Chapman V 1982. Cytogenetical studies in wheat XII. *Lr28* for resistance to *Puccinia recondita* and *Sr34* for resistance to *P. graminis tritici*. *Zeitschrift für Pflanzenzüchtung* **89**, 295–306.

McIntosh RA, Luig NH, Milne DL and Cusick JE 1983. Vulnerability of triticales to wheat stem rust. *Canadian Journal of Plant Pathology* **5**, 61–69.

McIntosh RA, Dyck PL, The TT, Cusick JE and Milne DL 1984. Cytogenetical studies in wheat XIII. *Sr35* – a third gene from *Triticum monococcum* for resistance to *Puccinia graminis tritici*. *Zeitschrift für Pflazenzüchtung* **92**, 1–14.

McIntosh RA, Hart GE and Gale MD 1989. Catalogue of gene symbols for wheat (1989 supplement). *Wheat Information Service* **69**, 49–63.

McIntosh RA, Hart GE and Gale MD 1990. Catalogue of gene symbols for wheat (1990 supplement). *Wheat Information Service* **70**, 29–50.

McNeal FH, Konzak CF, Smith EP, Tate WS and Russell TS 1971. 'A Uniform System for Recording and Processing Cereal Research Data.' Agricultural Research Service Bulletin 34–121. (United States Department of Agriculture: Washington.)

McVey DV 1989. Verification of infection–type data for identification of genes for resistance to leaf rust in some hard red spring wheats. *Crop Science* **29**, 304–307.

McVey DV 1990. Reaction of 578 spring spelt wheat accessions to 35 races of wheat stem rust. *Crop Science* **30**, 1001–1005.

McVey DV and Roelfs AP 1975. Postulation of genes for stem rust resistance in the entries of the Fourth International Winter Wheat Performance Nursery. *Crop Science* **15**, 335–337.

McVey DV and Roelfs AP 1978. Stem rust resistance of the cultivar 'Waldron'. *In* 'Proceedings of the Fifth International Wheat Genetics Symposium'. (Ed. S. Ramanujam.) pp. 1061–1065. (Indian Society of Genetics and Plant Breeding: New Delhi, India.)

Mendel G 1865. 'Experiments in Plant Hybridization'. Translated and reprinted in, Peters, JA (Ed.) 1959. 'Classic Papers in Genetics'. (Prentice-Hall: Englewood Cliffs, N.J., USA.)

Merker A 1982. 'Veery' — a CIMMYT spring wheat with the 1B/1R chromosome translocation. *Cereal Research Communications* **10**, 105–106.

Mettin D, Blüthner WD and Schlegel G 1973. Additional evidence on spontaneous 1B/1R wheat–rye substitutions and translocations. *In* 'Proceedings of the Fourth International Wheat Genetics Symposium'. (Eds ER Sears and LMS Sears.) pp. 179–184. (Agricultural Experiment Station, University of Missouri: Columbia, Missouri, USA.)

Metzger RJ and Silbaugh BA 1970. Inheritance of resistance to stripe rust and its association with brown glume colour in *Triticum aestivum* L. 'PI 178383'. *Crop Science* **10**, 567–568.

Miller TE and Koebner RMD (Eds) 1988. 'Proceedings of the Seventh International Wheat Genetics Symposium.' Vol. II. p. 1211. (Institute of Plant Science Research: Cambridge, UK.)

Miller TE, Reader SM and Singh D 1988. Spontaneous non–Robertsonian translocations between wheat chromosomes and an alien chromosome. *In* 'Proceedings of the Seventh International Wheat Genetics Symposium'. (Eds TE Miller and RMD Koebner.) pp. 387–390. (Institute of Plant Science Research: Cambridge, UK.)

Modawi RS, Browder LE and Heyne EG 1985. Use of infection type data to identify genes for low reaction to *Puccinia recondita* in several winter wheat cultivars. *Crop Science* **25**, 9–13.

Morrison JR, Larter EN and Green GJ 1977. The genetics of resistance to *Puccinia graminis tritici* in hexaploid triticale. *Canadian Journal of Genetics and Cytology* **19**, 683–693.

Moseman JG 1970. Genetics of disease and insect resistance. *In* 'Barley Genetics II, Proceedings of the Second International Barley Genetics Symposium'. (Ed. RA Nilan.) pp. 450–456. (Washington State University Press: Pullman, USA.)

Mukai J and Gill BS 1991. Detection of barley chromatin added to wheat by genomic *in situ* hybridisation. *Genome* **34**, 448–452.

Nagarajan S, Nayar SK and Bahadur P 1986a. Race 13 (67 S8) virulent on *Triticum spelta* var. *album* in India. *Plant Disease* **70**, 173.

Nagarajan S, Nayar SK, Bahadur P and Kumar J 1986b. 'Wheat Pathology and Wheat Improvement.' (Azad Hind Stores: Chandrigarh, India.)

Nayar SK, Prashar M, Kumar J, Bhardwaj SC and Bhatnagar R 1991. Pathotypes of *Puccinia recondita* f. sp. *tritici* virulent for *Lr26* (1BL.1RS translocation) in India. *Cereal Research Communications* **19**, 327–331.

Nehra NS, Chibbar RN, Leung N, Caswell K, Mallard C, Steinhauer L, Baga M and Kartha KK 1994. Self–fertile transgenic wheat plants regenerated from isolated scutellar tissues following microprojectile bombardment with two distinct gene constructs. *The Plant Journal* **5**, 285–297.

Nyquist NE 1957. Monosomic analysis of stem rust resistance of a common wheat strain derived from *Triticum timopheevi*. *Agronomy Journal* **49**, 222–223.

Nyquist NE 1962. Differential fertilization in the inheritance of stem rust resistance in hybrids involving a common wheat strain derived from *Triticum timopheevi*. *Genetics* **47**, 1109–1124.

O'Brien L, Brown JS, Young RM and Pascoe T 1980. Occurrence and distribution of wheat stripe rust in Victoria and susceptibility of commercial wheat cultivars. *Australasian Plant Pathology* **9**, 14.

Odintsova IG and Peusha KhO 1982. (Inheritance of resistance to brown rust in bread wheat varieties.) Trudy po Prikladnoi Botanike, Genetikei Selektsii **71**, 41–47. Cited in *Plant Breeding Abstracts* **55**, 7658, p. 841.

Okamoto M 1957. Asynaptic effect of chromosome V. *Wheat Information Service* **5**, 6.

Park RF and McIntosh RA 1994. Studies of single gene adult plant resistances to *Puccinia recondita* f. sp. *tritici* in wheat. *New Zealand Journal of Crop and Horticultural Science* **22**, 151–158.

Park RF and Wellings CR 1992. Pathogenic specialisation of wheat rusts in Australia and New Zealand in 1988 and 1989. *Australasian Plant Pathology* **21**, 61–69.

Park RF, Burdon JJ and McIntosh RA 1993. Origin of an important group of pathotypes of *Puccinia recondita* f. sp. *tritici* in Australasia. *In* 'Proceedings of the Tenth Australian Plant Breeding Conference'. (Eds BC Imrie and JB Hacker.) pp. 135–136. (Conference Organising Committee, Gold Coast: Australia.)

Perwaiz MS and Johnson R 1986. Genes for resistance to yellow rust in seedlings of wheat cultivars from Pakistan tested with British isolates of *Puccinia striiformis*. *Plant Breeding* **97**, 289–296.

Peterson RF, Campbell AB and Hannah AE 1948. A diagrammatic scale for estimating rust intensity of leaves and stems of cereals. *Canadian Journal of Research Section C* **26**, 496–500.

Plessers AG 1954. The genetics of stem and leaf rust reactions and other characters in Lee wheat with Chinese monosomic testers. *Dissertation Abstracts* **15**, 323.

Pretorius ZA and Le Roux J 1988. Occurrence and pathogenicity of *Puccinia recondita* f. sp. *tritici* on wheat in South Africa during 1986 and 1987. *Phytophylactica* **20**, 349–352.

Pretorius ZA, Wilcoxson RD, Long DL and Schaffer, JF 1984. Detecting wheat leaf rust resistance gene *Lr13* in seedlings. *Plant Disease* **68**, 585–586.

Pretorius ZA, Rijkenberg FHJ and Wilcoxson RD 1987. Occurrence and pathogenicity of *Puccinia recondita* f. sp. *tritici* on wheat in South Africa from 1983 through 1985. *Plant Disease* **71**, 1133–1137.

Pretorius ZA, Le Roux J and Drijenpondt SC 1990. Occurrence and pathogenicity of *Puccinia recondita* f. sp. *tritici* on wheat in South Africa during 1988. *Phytophylactica* **22**, 225–228.

Pridham JT 1939. A successful cross between *Triticum vulgare* and *Triticum timopheevi*. *Journal of the Australian Institute of Agricultural Science* **5**, 160–161.

Priestley RH 1978. Detection of increased virulence in populations of wheat yellow rust. *In* 'Plant Disease Epidemiology'. (Eds PR Scott and A Bainbridge.) pp. 63–70. (Blackwell Scientific Publications: Oxford.)

Priestley RH and Bayles RA 1982. Evidence that varietal diversification can reduce the spread of cereal diseases. *Journal of the National Institute of Agricultural Botany* **16**, 31–38.

Priestley RH and Bayles RA 1988. The contribution and value of resistant cultivars to disease control in cereals. *In* 'Control of Plant Diseases, Costs and Benefits'. (Eds BC Clifford and E Lester.) pp. 53–65. (Blackwell Scientific Publications: Oxford.)

Priestley RH, Bayles RA and Thomas JE 1984a. Identification of specific resistances against *Puccinia striiformis* (yellow rust) in winter wheat varieties I. Establishment of a set of type varieties for adult plant tests. *Journal of the National Institute of Agricultural Botany* **16**, 469–476.

Priestley RH, Bayles RA and Ryall J 1984b. Identification of specific resistances against *Puccinia striiformis* (yellow rust) in winter wheat varieties II. Use of cluster analysis. *Journal of the National Institute of Agricultural Botany* **16**, 477–485.

Pugsley AT 1956. The gene *SrKa1* in relation to the resistance of wheat to *Puccinia graminis tritici*. *The Empire Journal of Experimental Agriculture* **24**, 178–184.

Pugsley AT and Carter MV 1953. Resistance of twelve varieties of *Triticum vulgare* to *Erysiphe graminis tritici*. *Australian Journal of Biological Sciences* **6**, 335–346.

Purdy LH and Allan RE 1966. A stripe rust race pathogenic to 'Suwon 92' wheat. *Plant Disease Reporter* **50**, 205–207.

Purdy LH, Peterson CJ Jr and Allan RE 1966. Illustrating pedigrees of small grain cereals. Agricultural Research Service, Crops Research Bulletin CR–22–66. (United States Department of Agriculture: Washington DC.)

Quinones MA, Larter EN and Samborski DJ 1972. The inheritance of resistance to *Puccinia recondita* in hexaploid triticale. *Canadian Journal of Genetics and Cytology* **14**, 495–505.

Rajaram S, Luig NH and Watson IA 1971a. The inheritance of leaf rust resistance in four varieties of common wheat. *Euphytica* **20**, 574–585.

Rajaram S, Luig NH and Watson IA 1971b. Genetic analysis of stem rust resistance in three cultivars of wheat. *Euphytica* **20**, 441–452.

Rajaram S, Singh RP and Torres E 1988. Current CIMMYT approaches in breeding wheat for rust resistance. *In* 'Breeding Strategies for Resistance to the Rusts of Wheat'. (Eds NW Simmonds and S Rajaram.) pp. 101–117. (CIMMYT: Mexico.)

Rees RG and Platz GJ 1975. Control of wheat leaf rust with 4–n–butyl–1,2,4–triazole. *Australian Journal of Experimental Agriculture and Animal Husbandry* **15**, 276–280.

Riley R and Chapman V 1958. Genetic control of the cytologically diploid behaviour of hexaploid wheat. *Nature* (London) **182**, 713–715.

Riley R, Chapman V and Johnson R 1968. The incorporation of alien disease resistance in wheat by genetic interference with the regulation of meiotic chromosome synapsis. *Genetical Research, Cambridge* **12**, 199–219.

Roelfs AP 1978. Estimated losses caused by rust in small grain cereals in the United States — 1918–76. Miscellaneous Publication 1363. (United States Department of Agriculture: Washington DC.)

Roelfs AP 1982. Effects of barberry eradication on stem rust in the United States. *Plant Disease* **66**, 177–181.

Roelfs AP 1984. Race specificity and methods of study. *In* 'The Cereal Rusts'. Vol. I. (Eds WR Bushnell and AP Roelfs.) pp. 131–164. (Academic Press: Orlando, Florida, USA.)

Roelfs AP 1985. Wheat and rye stem rust. *In* 'The Cereal Rusts'. Vol. II. (Eds AP Roelfs and WR Bushnell.) pp. 3–37. (Academic Press: Orlando, Florida, USA.)

Roelfs AP 1988. Resistance to leaf and stem rusts in wheat. *In* 'Breeding Strategies for Resistance to the Rusts of Wheat'. (Eds NW Simmonds and S Rajaram.) pp. 10–22. (CIMMYT: Mexico.)

Roelfs AP and Bushnell WR (Eds) 1985. 'The Cereal Rusts. Vol. II. Diseases, Distribution, Epidemiology and Control.' (Academic Press: Orlando.)

Roelfs AP and Martens JW 1988. An international system of nomenclature for *Puccinia graminis* f. sp. *tritici*. *Phytopathology* **78**, 526–533.

Roelfs AP and McVey DV 1975. Races of *Puccinia graminis* f. sp. *tritici* in the USA during 1974. *Plant Disease Reporter* **59**, 681–685.

Roelfs AP and McVey DV 1979. Low infection types produced by *Puccinia graminis* f.sp. *tritici* and wheat lines with designated genes for resistance. *Phytopathology* **69**, 722–730.

Roelfs AP, Long DL and Casper DH 1983. Races of *Puccinia graminis* f. sp. *tritici* in the United States and Mexico in 1981. *Plant Disease* **67**, 82–84.

Roelfs AP, Casper DH, Long DL and Roberts JJ 1987. Races of *Puccinia graminis* in the United States and Mexico during 1986. *Plant Disease* **71**, 903–907.

Roelfs AP, Casper DH, Long DL and Roberts JJ 1990. Races of *Puccinia graminis* in the United States in 1988. *Plant Disease* **74**, 555–557.

Roelfs AP, Casper DH, Long DL and Roberts JJ 1991. Races of *Puccinia graminis* in the United States in 1989. *Plant Disease* **75**, 1127–1130.

Roelfs AP, Singh RP and Saari EE 1992. 'Rust Diseases of Wheat, Concepts and Methods of Disease Management.' (CIMMYT: Mexico.)

Rondon MR, Gough FJ and Williams ND 1966. Inheritance of stem rust resistance in *Triticum aestivum* ssp. *vulgare* 'Reliance' and P.I.94701 of *Triticum durum*. *Crop Science* **6**, 177–179.

Rowell JB 1981*a*. Relation of post penetration events in Idaed 59 wheat seedlings to low receptivity to infection by *Puccinia graminis* f. sp. *tritici*. *Phytopathology* **71**, 732–736.

Rowell JB 1981*b*. The relationship between slow rusting and a specific resistance gene for wheat stem rust. *Phytopathology* **71**, 1184–1186.

Rowell JB 1982. Control of wheat stem rust by low receptivity to infection conditioned by a single dominant gene. *Phytopathology* **72**, 297–299.

Rowell JB and McVey DV 1979. A method for field evaluation of wheats for low receptivity to infection by *Puccinia graminis* f.sp. *tritici*. *Phytopathology* **69**, 405–409.

Rowland GG 1972. A Cytogenetic Study in Hexaploid Wheat of Characters Derived from *Aegilops squarrosa*. PhD Thesis, University of Manitoba, Winnipeg.

Rowland GG and Kerber ER 1974. Telocentric mapping in hexaploid wheat of genes for leaf rust resistance and other characters derived from *Aegilops squarrosa*. *Canadian Journal of Genetics and Cytology* **16**, 137–144.

Saari EE and Prescott JM 1985. World distribution in relation to economic losses. *In* 'The Cereal Rusts'. Vol II. (Eds AP Roelfs and WR Bushnell.) pp. 259–298. (Academic Press: Orlando, Florida, USA.)

Samborski DJ 1963. A mutation in *Puccinia recondita* Rob. ex Desm. f. sp. *tritici* to virulence on Transfer, Chinese Spring x *Aegilops umbellulata* Zhuk. *Canadian Journal of Botany* **41**, 475–479.

Samborski DJ 1984. Occurrence and virulence of *Puccinia recondita* in Canada in 1983. *Canadian Journal of Plant Pathology* **6**, 238–242.

Samborski DJ and Dyck PL 1976. Inheritance of virulence in *Puccinia recondita* on six backcross lines of wheat with single genes for resistance to leaf rust. *Canadian Journal of Botany* **54**, 1666–1671.

Samborski DJ and Dyck PL 1982. Enhancement of resistance to *Puccinia recondita* by interactions of resistance genes in wheat. *Canadian Journal of Plant Pathology* **4**, 152–156.

Sanghi AK and Baker EP 1972. Genetic bases for resistance in two common wheat cultivars to stem rust strains of unusual avirulence. *Proceedings of the Linnean Society New South Wales* **97**, 56–71.

Sawhney RN 1992. The role of *Lr34* in imparting durable resistance to wheat leaf rust through gene interaction. *Euphytica* **61**, 9–12.

Sawhney RN and Luthra JK 1970. New resistance genes of wheat to Indian races of stripe rust (*Puccinia striiformis*). SABRAO Newsletter, Mishima 2, 155–156. Cited in *Plant Breeding Abstracts* **41**, 7312, p. 935.

Sawhney RN, Sharma JB and Sharma DN 1992. Genetic diversity for adult plant resistance to leaf rust (*Puccinia recondita*) in near–isogenic lines and in Indian wheats. *Plant Breeding* **109**, 248–254.

Schafer JF and Long DL 1988. Relations of races and virulences of *Puccinia recondita* f. sp. *tritici* to wheat cultivars and areas. *Plant Disease* **72**, 25–27.

Schmidt JW, Johnson VA, Mattern PJ, Dreier AF, McVey DV and Hatchett JH 1985. Registration of 'Siouxland' wheat. *Crop Science* **25**, 1130–1131.

Sears ER 1953. Nullisomic analysis in common wheat. *American Naturalist* **87**, 245–252.

Sears ER 1954. The aneuploids of common wheat. *Missouri Agricultural Experimental Station Bulletin* **572**.

Sears ER 1956. The transfer of leaf-rust resistance from *Aegilops umbellulata* to wheat. *Brookhaven Symposia in Biology* **9**, 1–22.

Sears ER 1961. Identification of the wheat chromosome carrying leaf rust resistance from *Aegilops umbellulata*. *Wheat Information Service* **12**, 12–13.

Sears ER 1966. Chromosome mapping with the aid of telocentrics. Proceedings of the Second International Wheat Genetics Symposium. (Ed. J Mackey.) Lund 1963. *Hereditas Supplement* **2**, 370–381.

Sears ER 1972*a*. Reduced proximal crossing-over in telocentric chromosomes of wheat. *Genetica Iberia* **24**, 233–239.

Sears ER 1972*b*. Chromosome engineering in wheat. *In* 'Stadler Genetics Symposia'. Vol. 4. (Eds G Kimber and GR Rédei.) pp. 23–38. (University of Missouri: Columbia, Missouri, USA.)

Sears ER 1973. *Agropyron*–wheat transfers induced by homoeologous pairing. *In* 'Proceedings of the Fourth International Wheat Genetics Symposium'. (Eds ER Sears and LMS Sears.) pp. 191–199. (Agricultural Research Station, University of Missouri: Columbia, Missouri, USA.)

Sears ER 1977. Analysis of wheat — *Agropyron* recombinant chromosomes. *In* 'Proceedings of The Eighth European Association for Research in Plant Breeding Congress'. (Eds E Sanchez Monge and F Garcia-Olmedo.) pp. 63–72. (Madrid, Spain.)

Sears ER and Briggle LW 1969. Mapping the gene *Pm1* for resistance to *Erysiphe graminis* f. sp. *tritici* on chromosome 7A of wheat. *Crop Science* **9**, 96–97.

Sears ER and Loegering WQ 1961. A pollen-killing gene in wheat. *Genetics* **46**, 897.

Sears ER and Loegering WQ 1968. Mapping of stem-rust genes *Sr9* and *Sr16* of wheat. *Crop Science* **8**, 371–373.

Sears ER, Loegering WQ and Rodenhiser HA 1957. Identification of chromosomes carrying genes for stem rust resistance in four varieties of wheat. *Agronomy Journal* **49**, 208–212.

Shaner G, Roberts JJ and Finney RE 1972. A culture of *Puccina recondita* virulent to the wheat cultivar Transfer. *Plant Disease Reporter* **56**, 827–830.

Shang HS, Dyck PL and Samborski DJ 1986. Inheritance of resistance to *Puccinia recondita* in a group of resistant accessions of common wheat. *Canadian Journal of Plant Pathology* **8**, 123–131.

Sharma D and Knott DR 1966. The transfer of leaf-rust resistance from *Agropyron* to *Triticum* by irradiation. *Canadian Journal of Genetics and Cytology* **8**, 137–143.

Sharma HC and Gill BS 1983. Current status of wide hybridization in wheat. *Euphytica* **32**, 17–31.

Sheen SJ and Snyder LA 1964. Studies on the inheritance of resistance to six stem rust cultures using chromosome substitution lines of a Marquis wheat selection. *Canadian Journal of Genetics and Cytology* **6**, 74–82.

van Silfhout CH, Grama A, Gerechter–Amitai ZK and Kleitman F 1989a. Resistance to yellow rust in *Triticum dicoccoides* I. Crosses with susceptible *Triticum durum*. *Netherlands Journal of Plant Pathology* **95**, 73–78.

van Silfhout CH, Kema GHJ and Gerechter–Amitai ZK 1989b. Major genes for resistance to yellow rust in wild emmer wheat. *In* 'Identification and Characterisation of Resistance to Yellow Rust and Powdery Mildew in Wild Emmer Wheat and Their Transfer to Bread Wheat'. pp. 5–15. Ph.D. Thesis. Research Institute for Plant Protection, Wageningen, The Netherlands.

Simmonds NW and Rajaram S (Eds) 1988. 'Breeding Strategies for Resistance to the Rusts of Wheat.' (CIMMYT: Mexico.)

Singh H and Johnson R 1988. Genetics of resistance to yellow rust in Heines VII, Soissonais and Kalyansona. *In* 'Proceedings of the Seventh International Wheat Genetics Symposium'. (Eds TE Miller and RMD Koebner.) pp. 885–890. (Institute of Plant Science Research: Cambridge, UK.)

Singh H, Johnson R and Seth D 1990. Genes for race–specific resistance to yellow rust (*Puccinia striiformis*) in Indian wheat cultivars. *Plant Pathology* **39**, 424–433.

Singh RP 1991. Pathogenicity variations of *Puccinia recondita* f. sp. *tritici* and *P. graminis* f. sp. *tritici* in wheat-growing areas of Mexico during 1988 and 1989. *Plant Disease* **75**, 790–794.

Singh RP 1992a. Genetic association of leaf rust resistance gene *Lr34* with adult plant resistance to stripe rust in bread wheat. *Phytopathology* **82**, 835–838.

Singh RP 1992b. Genetic association between gene *Lr34* for leaf rust resistance and leaf tip necrosis in bread wheats. *Crop Science* **32**, 874–878.

Singh RP 1993. Resistance to leaf rust in 26 Mexican wheat cultivars. *Crop Science* **33**, 633–637.

Singh RP and Gupta AK 1991. Genes for leaf rust resistance in Indian and Pakistani wheats tested with Mexican pathotypes of *Puccinia recondita* f. sp. *tritici*. *Euphytica* **57**, 27–36.

Singh RP and Gupta AK 1992. Expression of wheat leaf rust resistance gene *Lr34* in seedlings and adult plants. *Plant Disease* **76**, 489–491.

Singh RP and McIntosh RA 1984a. Complementary genes for resistance to *Puccinia recondita tritici* in *Triticum aestivum* I. Genetic and linkage studies. *Canadian Journal of Genetics and Cytology* **26**, 723–735.

Singh RP and McIntosh RA 1984b. Complementary genes for resistance to *Puccinia recondita tritici* in *Triticum aestivum* II. Cytogenetic studies. *Canadian Journal of Genetics and Cytology* **26**, 736–742.

Singh RP and McIntosh RA 1985. Genetic basis of leaf rust resistance in wheat cultivar Mediterranean. *Cereal Rusts Bulletin* **13**, 31–36.

Singh RP and McIntosh RA 1986a. Cytogenetical studies in wheat XIV. *Sr8b* for resistance to *Puccinia graminis tritici*. *Canadian Journal of Genetics and Cytology* **28**, 189–197.

Singh RP and McIntosh RA 1986b. Genetics of resistance to *Puccinia graminis tritici* and *Puccinia recondita tritici* in Kenya Plume wheat. *Euphytica* **35**, 245–256.

Singh RP and McIntosh RA 1987. Genetics of resistance to *Puccinia graminis tritici* in 'Chris' and 'W3746' wheats. *Theoretical and Applied Genetics* **73**, 846–855.

Singh RP and Rajaram S 1991. Resistance to *Puccinia recondita* f.sp. *tritici* in 50 Mexican bread wheat cultivars. *Crop Science* **31**, 1472–1479.

Singh RP and Rajaram S 1992. Genetics of adult-plant resistance of leaf rust in 'Frontana' and three CIMMYT wheats. *Genome* **35**, 24–31.

Singh RP and Saari EE 1990. Biotic stresses in triticale. *In* 'Proceedings of the Second International Triticale Symposium'. pp. 171–181. (CIMMYT: Mexico.)

Singh RP, Bechere E and Abdalla O 1992. Genetic analysis of resistance to stem rust in ten durum wheats. *Phytopathology* **82**, 919–922.

Singh SJ and McIntosh RA 1988. Allelism of two genes for stem rust resistance in triticale. *Euphytica* **38**, 185–189.

Singh SJ and McIntosh RA 1990. Linkage and expression of genes for resistance to leaf rust and stem rust in triticale. *Genome* **33**, 115–118.

Smith EL, Schlehuber AM, Young HC Jr and Edwards LH 1968. Registration of Agent wheat. *Crop Science* **8**, 511–512.

Smith GS 1957. Inheritance of stem rust reaction in tetraploid wheat hybrids I. Allelic genes in Mindum Durum and Vernal Emmer. *Agronomy Journal* **49**, 134–137.

Smith J and Le Roux J 1992. First report of wheat stem rust virulence for *Sr27* in South Africa. *Vorträge für Pflanzenzüchtung* **24**, 109–110.

· Soliman AS, Heyne EG and Johnston CO 1963. Resistance to leaf rust in wheat derived from *Aegilops umbellulata* translocation lines. *Crop Science* **3**, 254–256.

Soliman AS, Heyne EG and Johnston CO 1964. Genetic analysis of leaf rust resistance in eight differential varieties of wheat. *Crop Science* **4**, 246–248.

Sorrells ME and Jensen NF 1987. Registration of 'Geneva' winter wheat. *Crop Science* **27**, 1314–1315.

Stakman EC and Levine MN 1922. The determination of biologic forms of *Puccinia graminis* on *Triticum* spp. *Minnesota Agricultural Research Station Bulletin* **8**.

Stakman EC and Piemeisel FJ 1917. Biologic forms of *Puccinia graminis* on cereals and grasses. *Journal of Agricultural Research* (Washington DC) **10**, 429–495.

Stakman EC, Piemeisel FJ and Levine MN 1918. Plasticity of biologic forms of *Puccinia graminis*. *Journal of Agricultural Research* (Washington DC) **15**, 221–250.

Stakman EC, Stewart DM and Loegering WQ 1962. Identification of physiologic races of *Puccinia graminis* var. *tritici*. Agricultural Research Service E617. (United States Department of Agriculture: Washington DC.)

Statler GD 1984. Probable genes for leaf rust resistance in several hard red spring wheats. *Crop Science* **24**, 883–886.

Statler GD and Christianson T 1993. Temperature studies with wheat leaf rust. *Canadian Journal of Plant Pathology* **15**, 97–101.

Stewart DM, Gilmore EC Jr and Ausemus ER 1968. Resistance to *Puccinia graminis* derived from *Secale cereale* incorporated into *Triticum aestivum*. *Phytopathology* **58**, 508–511.

Stubbs RW 1966. Recent aspects of the physiological specialisation of yellow rust in the Netherlands. *In* 'Proceedings of the Third European Yellow Rust Conference'. (Eds RCF Macer and MS Wolfe.) pp. 47–54. (Plant Breeding Institute: Cambridge.)

Stubbs RW 1967. Influence of light intensity on the reactions of wheat and barley seedlings to *Puccinia striiformis*. *Phytopathology* **57**, 615–617.

Stubbs RW 1985. Stripe rust. *In* 'The Cereal Rusts'. Vol II. (Eds AP Roelfs and WR Bushnell.) pp. 61–101. (Academic Press: Orlando, Florida, USA.)

Stubbs RW 1988. Pathogenicity analysis of yellow (stripe) rust of wheat and its significance in a global context. *In* 'Breeding Strategies for Resistance to the Rusts of Wheat'. (Eds NW Simmonds and S Rajaram.) pp. 23–38. (CIMMYT: Mexico.)

Stubbs RW and Yang H-A 1988. Pathogenicity of *Puccinia striiformis* for wheat cultivars with resistance derived from rye. *In* 'Proceedings of the Seventh European and Mediterranean Cereal Rusts Conference'. (Ed. B Zwatz.) pp. 110–112. (Federal Institute of Plant Protection: Vienna, Austria.)

Stubbs RW, Fuchs E, Vecht H and Basset EJW 1974. The international survey of factors of virulence of *Puccinia striiformis* Westend. in 1969, 1970 and 1971. Technische Bericht Nr.21, Nederlands Graan–Centrum, Wageningen, The Netherlands.

Stubbs RW, Prescott JM, Saari EE and Dubin HJ 1986. 'Cereal Disease Methodology Manual.' (CIMMYT: Mexico.)

Sunderwirth SD and Roelfs AP 1980. Greenhouse evaluation of the adult plant resistance of *Sr2* to wheat stem rust. *Phytopathology* **70**, 634–637.

Syme JR 1983. Flinders. *Journal of Australian Institute of Agricultural Science* **49**, 42.

Syme JR, Law DP, Martin DJ and Rees RG 1983. Bass. *Journal of the Australian Institute of Agricultural Science* **49**, 46–47.

Taylor AJ, Smith GMB and Johnson R 1981. Race-specific genetic factors for resistance to *Puccinia striiformis* in wheat cultivars from the Plant Breeding Institute. *Cereal Rusts Bulletin* **9**, 33–45.

The TT 1973*a*. Chromosome location of genes conditioning stem rust resistance transferred from diploid to hexaploid wheat. *Nature New Biology* **241**, 256.

The TT 1973*b*. Transference of Resistance to Stem Rust from *Triticum monococcum* L. to Hexaploid Wheat. PhD Thesis, The University of Sydney.

The TT and Baker EP 1970. Homoeologous relationships between two *Agropyron intermedium* chromosomes and wheat. *Wheat Information Service* **31**, 29–31.

The TT and McIntosh RA 1975. Cytogenetical studies in wheat VIII. Telocentric mapping and linkage studies involving *Sr22* and other genes in chromosome 7AL. *Australian Journal of Biological Sciences* **28**, 531–538.

The TT, McIntosh RA and Bennett FGA 1979. Cytogenetical studies in wheat IX. Monosomic analyses, telocentric mapping and linkage relationships of genes *Sr21*, *Pm4*, and *Mle*. *Australian Journal of Biological Sciences* **32**, 115–125.

The TT, Latter BDH, McIntosh RA, Ellison FW, Brennan PS, Fisher J, Hollamby GJ, Rathjen AJ and Wilson RE 1988. Grain yields of near–isogenic lines with added genes for stem rust resistance. *In* 'Proceedings of the Seventh International Wheat Genetics Symposium'. (Eds TE Miller and RMD Koebner.) pp. 901–906. (Institute of Plant Science Research: Cambridge.)

The TT, Gupta RB, Dyck PL, Appels R, Höhmann U and McIntosh RA 1992. Characterisation of stem rust resistant derivatives of wheat cultivar Amigo. *Euphytica* **58**, 245–252.

Townley–Smith TF, Czarnecki EM, Campbell AB, Dyck PL and Samborski DJ 1993. AC Minto hard red spring wheat. *Canadian Journal of Plant Science* **73**, 1091–1094.

Valkoun J, Kucerova D and Bartos P 1986. Transfer of stem rust resistance from *Triticum monococcum* L. to *T. aestivum* L. Sbornik UVTIZ, Genetika a Slechteni **22**, 9–16. Cited in *Plant Breeding Abstracts* **56**, 4701, p. 504.

deVallavieille–Pope C and Line RF 1990. Virulence of North American and European races of *Puccinia striiformis* in North American, world, and European differential wheat cultivars. *Plant Disease* **74**, 739–743.

de Vallavieille–Pope C, Picard–Formery H, Radulovic S and Johnson R 1990. Specific resistance factors to yellow rust in seedlings of some French wheat varieties and races of *Puccinia striiformis* Westend. in France. *Agronomie* **2**, 103–113.

Vasil V, Castillo AM, Fromm ME and Vasil IK 1992. Herbicide resistant fertile transgenic wheat plants obtained by microprojectile bombardment of regenerable embryogenic callus. *Bio/Technology* **10**, 667–674.

Villareal RL and Rajaram S 1988. 'Semi Dwarf Bread Wheats, Names, Parentages, Pedigrees and Origins.' (CIMMYT: Mexico.)

Walker JC 1976. 'Plant Pathology.' (Tata McGraw-Hill Publishing: New Delhi.)

Waterhouse WL 1933. On the production of fertile hybrids from crosses between vulgare and Khapli emmer wheats. *Proceedings of the Linnean Society of New South Wales* **58**, 99–104.

Waterhouse WL 1952. Australian rust studies IX. Physiologic race determinations and surveys of cereal rusts. *Proceedings of the Linnean Society New South Wales* **77**, 209–258.

Watson IA 1957. Further studies on the production of new races from mixtures of races of *Puccinia graminis* var. *tritici* on wheat seedlings. *Phytopathology* **47**, 510–512.

Watson IA 1962. 'Wheat Leaf Rust in Australia. Strain variations for the period 1951–61. Commonwealth Cereal Rust Survey Report No. 1.' (The University of Sydney: Sydney.)

Watson IA and Baker EP 1943. Linkage of resistance to *Erysiphe graminis tritici* and *Puccinia triticinia* in certain varieties of *Triticum vulgare* Vill. *Proceedings of the Linnean Society New South Wales* **68**, 150–152.

Watson IA and Butler FC 1984. 'Wheat Rust Control in Australia, National Conferences and Other Initiatives and Developments.' (University of Sydney: Sydney.)

Watson IA and Luig NH 1958. Somatic hybridization in *Puccinia graminis tritici*. *Proceedings of the Linnean Society of New South Wales* **83**, 190–195.

Watson IA and Luig NH 1961. Leaf rust on wheat in Australia: a systematic scheme for the classification of strains. *Proceedings of the Linnean Society of New South Wales* **86**, 241–250.

Watson IA and Luig NH 1963. The classification of *Puccinia graminis* var. *tritici* in relation to breeding resistant varieties. *Proceedings of the Linnean Society of New South Wales* **88**, 235–258.

Watson IA and Luig NH 1966. Sr15 — a new gene for use in the classification of *Puccina graminis* var. *tritici*. *Euphytica* **15**, 239–250.

Watson IA and Luig NH 1968. Progressive increase in virulence in *Puccinia graminis* f. sp. *tritici*. *Phytopathology* **58**, 70–73.

Watson IA and de Sousa CNA 1983. Long distance transport of spores of *Puccinia graminis tritici* in the southern hemisphere. *Proceedings of the Linnean Society of New South Wales* **106**, 311–321.

Watson IA and Stewart DM 1956. A comparison of the rust reaction of wheat varieties Gabo, Timstein, and Lee. *Agronomy Journal* **48**, 514–516.

Weeks JT, Anderson OD and Belchl AE 1993. Rapid production of multiple independent lines of fertile transgenic wheat (*Triticum aestivum*). *Plant Physiology* **102**, 1077–1084.

Wellings CR 1986. Host : Pathogen Studies of Wheat Stripe Rust in Australia. PhD Thesis, University of Sydney.

Wellings CR 1992. Resistance to stripe (yellow) rust in selected spring wheats. *Vorträge für Pflanzentüchtung* **24**, 273–275.

Wellings CR and Burdon JJ 1992. Variability in *Puccinia striiformis* f.sp. *tritici* in Australasia. *Vorträge für Pflanzenzüchtung* **24**, 114.

Wellings CR and Luig NH 1984. Wheat rusts, the old and the new. *Australian Institute of Agricultural Science, Occasional Publication* **15**, 5–15.

Wellings CR and McIntosh RA 1990. *Puccinia striiformis* f.sp. *tritici* in Australasia: pathogenic changes during the first 10 years. *Plant Pathology* **39**, 316–325.

Wellings CR, Wong PTW and Murray GM 1985. Use of multiple regression to examine the effect of leaf rust and yellow spot on yield of wheat in northern New South Wales. *Australasian Plant Pathology* **14**, 62–64.

Wellings CR, McIntosh RA and Walker J 1987. *Puccinia striiformis* f. sp. *tritici* in Australia — possible means of entry and implications for plant quarantine. *Plant Pathology* **36**, 239–241.

Wellings CR, McIntosh RA and Hussain M 1988. A new source of resistance to *Puccinia striiformis* f. sp. *tritici* in spring wheats (*Triticum aestivum*). *Plant Breeding* **100**, 88–96.

Wells DG, Bonnemann JJ, Gardiner WS, Finney KF, Giese HA and Stymiest CE 1983. Nell wheat. *Crop Science* **23**, 804–805.

Wiese MV 1977. 'Compendium of Wheat Diseases.' (American Phytopathology Society: St Paul, Minnesota, USA.)

Wiggin HC 1955. Monosomic analysis of stem rust reaction and awn expression in Kentana 52 wheat. *Journal of Heredity* **46**, 239–242.

Williams E Jr and Johnston CO 1965. Effect of certain temperatures on identification of physiologic races of *Puccinia recondita* f. sp. *tritici*. *Phytopathology* **55**, 1317–1319.

Williams ND and Kaveh H 1976. Relationships of genes for reaction to stem rust from 'Marquis' and 'Reliance' wheat to other *Sr* genes. *Crop Science* **16**, 561–564.

Williams ND and Maan SS 1973. Telosomic mapping of genes for resistance to stem rust of wheat. *In* 'Proceedings of the Fourth International Wheat Genetics Symposium'. (Eds ER Sears and LMS Sears.) pp. 765–770. (University of Missouri: USA.)

Wilson J and Shaner G 1989. Inheritance of the leaf rust resistance of four triticale cultivars. *Phytopathology* **79**, 731–736.

Wolfe MS, Brändle U, Koller B, Limpert E, McDermott JM, Müller K and Schaffner D 1992. Barley mildew in Europe, population biology and host resistance. *In* 'Breeding for Disease Resistance'. (Eds R Johnson and GJ Jellis.) pp. 125–139. (Kluwer Academic Press: Dordrecht.)

Worland AJ 1988. Studies of the resistance of wheat to yellow rust. 1987 Annual Report, pp. 8–9. (Institute of Plant Science Research: Cambridge.)

Worland AJ and Law CN 1986. Genetic analysis of chromosome 2D of wheat I. The location of genes affecting height, day-length insensitivity, hybrid dwarfism and yellow-rust resistance. *Zeitschrift für Pflanzenzüchtung* **96**, 331–345.

Worland AJ, Law CN, Hollins TW, Koebner RMD and Giura A 1988. Location of a gene for resistance to eyespot (*Pseudocercosporella herpotrichoides*) on chromosome 7D of wheat. *Plant Breeding* **101**, 43–51.

Xin ZY, Johnson R, Law CN and Worland AJ 1984. A genetic analysis of genes for yellow rust resistance in the winter wheat variety Feng–Kang 13. *Acta Agronomica Sinica* **10**, 217–222.

Yang H–A and Stubbs RW 1990. Gene postulation for wheat stripe rust resistance on Chinese differential hosts. *Acta Phytophylacica Sinica* **17**, 67–72.

Zadoks JC 1961. Yellow rust on wheat. Studies in epidemiology and physiologic specialization. *Tijdschrift Over Plantenziekten* **67**, 69–256.

Zeller FJ 1973. 1B/1R wheat–rye chromosome substitutions and translocations. *In* 'Proceedings of the Fourth International Wheat Genetics Symposium'. (Eds ER Sears and LMS Sears.) pp. 209–221. (Agricultural Experiment Station, University of Missouri: Columbia, Missouri, USA.)

Zeller FJ and Hsam SLK 1983. Broadening the genetic variability of cultivated wheat by utilizing rye chromatin. *In* 'Proceedings of the Sixth International Wheat Genetics Symposium'. (Ed. S Sakamoto.) pp. 161–173. (Faculty of Agriculture, Kyoto University: Japan.)

Zeller FJ and Oppitz K 1977. Monosomic analysis for localizing the gene *SrEC* for resistance to stem rust in the wheat cv. 'Etoile de Choisy'. *Zeitschrift für Pflanzenzüchtung* **78**, 79–82.

Zeven AC and Zeven–Hissink NCh 1976. 'Genealogies of 14,000 Wheat Varieties.' (Netherlands Cereal Centre–NGC: Wageningen, The Netherlands.)

Zwer PK, Park RF and McIntosh RA 1992. Wheat stem rust in Australia — 1969–1985. *Australian Journal of Agricultural Research* **43**, 399–431.

APPENDIX I

Species Names Used in the Text, Genomic Formulae and Synonyms

SPECIES	GENOMIC FORMULA	SYNONYM
1. DIPLOIDS (2n = 14)		
Triticum monococcum L.	AA	*T. boeoticum* Boiss.
Triticum comosum (Sibth. & Sm.) Richter	MM	*Aegilops comosa* Sibth. & Sm.
Triticum speltoides (Tausch) Gren ex Richter	SS	*Ae. speltoides* Tausch
Triticum tauschii (Coss.) Schal.	DD	*Ae. squarrosa* L.
Triticum umbellulatum (Zhuk.) Bowden	UU	*Ae. umbellulata* Zhuk.
Secale cereale L.	RR	
Thinopyrum distichum (Thunb.) Löve	JJ	*Agropyron distichum* (Thunb.) Beauv.
2. TETRAPLOIDS (2n = 28)		
Triticum turgidum L.	AABB	*T. dicoccoides* Körn *T. dicoccum* Schronk *T. durum* Desf.
Triticum timopheevii (Zhuk.)	AAGG	*T. timopheevi* Zhuk. *T. araraticum* Jakubz.
Triticum ventricosum Ces.	DDUnUn	*Ae. ventricosa* Tausch
3. HEXAPLOIDS (2n = 42)		
Triticum aestivum L.	AABBDD	*T. macha* Dek. & Men., *T. spelta* L., *T. vulgare* Host
Thinopyrum intermedium (em Thell. Host) Barkworth & Dewey	$E_1E_1E_2E_2XX$	*Agropyrum intermedium* (Host) P. Beauv.
4. DECAPLOID (2n = 70)		
Thinopyrum ponticum		*Agropyron elongatum* (Host) Beauv. *Lophopyrum ponticum* (Podp.) Löve

APPENDIX II *Pathotypes and Accessions (Cultures) used in Photography*

LEAF RUST — *Puccinia recondita* f. sp. *tritici*

PATHOTYPE	ACCESSION NUMBER
Plant Breeding Institute, Cobbitty, Australia	
10-1,2,3,	63846
10-1,2,3,4	72469
26-1,3	67028
53-1,(6),(7),10,11	81043
53-(6),(7),9,10,11	87115
64-11	90053
76-0	63666
76-1,2,3,6	70109
104-1,2,3,6	81-L-2
104-2,3,6,(7)	76694
104-2,3,6,7	791021
104-2,3,6,9	84412
104-2,3,6,(7),8	80-L-3
104-2,3,(6),(7),11	84045
104-2,3,(6),(7),11	91026a
122-1,2,(3)	63-L-1
122-1,2,3	66-L-3
122-1,3,4,6,7	89172
135-1,2,3,4,5	64-L-3
162-1,2	62408
162-1,2,3,4	61835
162-1,2,3,4	79-L-4b

STEM RUST — *Puccinia graminis* f. sp. *tritici*

PATHOTYPE	ACCESSION NUMBER
Plant Breeding Institute, Cobbitty, Australia	
17-1,2,3,7	64404
21-0	54129
21-(1),2	77114
21-1,2	66535
21-1,2,3,7	61165
21-2,4,5	67034
34-1,2,3,4,5,6,7	74-L-1
34-1,2,3,4,5,6,7	82-L-2c
34-1,2,3,5,6,7	791140
34-1,2,3,6,7,8,9	76-L-7
34-2	69924
34-2,4,5,7,11	64231
34-2,12	82246
34-2,12,13	84552
34-(4),(7),10	57096
98-1,2,3,5,6	781219
Undesignated-1,2,3,5,6	82-L-1d
116-2,3,7	61352
116-4,5	64726
126-1,5,6,7,11	7316
126-5,6,7,11	334
194-2	69313
222-1,2,3,5,6	70-L-2
343-1,2,3,5,6	73879
Undesignated	80-E-2
University of Missouri, Columbia, USA	
Race 17	17-53B
Race 59	59-51A
Undesignated	111 x 36 =97
USDA Cereal Rust Laboratory, St Paul, USA	
BDCN	—
LCBB = race 111	111-SS2
LBBL	—

STRIPE RUST — *Puccinia striiformis* f. sp. *tritici*

PATHOTYPE	ACCESSION NUMBER	
Plant Breeding Institute, Cobbitty, Australia		
104 E9 A+	881534	
104 E137 A−	821559	
104 E137 A+	821552	
104 E153 A+	851595	
106 E139 A−	821589	
106 E139 A−	911582	Virulent on Selkirk
108 E141 A−	832002	
108 E141 A+	831917	
108 E205 A+	841943	
109 E141 A−	861773	
110 E143 A+	861725	
234 E139 A−	911586	
360 E137 A+	841924	
Research Institute for Plant Protection (IPO), Wageningen, The Netherlands		
0 E0	75121	
6 E0	86012	
6 E64	89054	
6 E150	86054	
8 E0	84044	
32 E0	60120	
40 E0	72019	
40 E8	74028	
82 E0	89088	
104 E9 A−	75713	
104 E9	80104	
104 E41	77549	
234 E139	80585	
234 E171	85564	
237 E141	89559	
Institute for Plant Science Research, Norwich, UK		
6 E16	WYR 85/23	
108 E9	WYR 88/10	
108 E141	WYR 75/25	
41 E136 (Type 2)	WYR 72/40	
104 E137 (Type 1)	WYR 69/10	
104 E137 (Type 2)	WYR 71/2	
104 E137 (Type 3)	WYR 72/23	
108 E141 (Type 3)	WYR 81/12	

a Intermediate response on lines with Lr26.   b Intermediate response on lines with Lr9.
c Mutant of 74-L-1 with virulence for Sr35.   d Mutant of 781219 with virulence for Sr9e.

INDEX OF GENES FOR RUST RESISTANCE